Global Perspectives on Air Pollution Prevention and Control System Design

G. Venkatesan
Anna University, India

Jaganthan Thirumal
Anna University, India

A volume in the Advances in Environmental Engineering and Green Technologies (AEEGT) Book Series

IGI Global
DISSEMINATOR OF KNOWLEDGE

Published in the United States of America by
IGI Global
Engineering Science Reference (an imprint of IGI Global)
701 E. Chocolate Avenue
Hershey PA, USA 17033
Tel: 717-533-8845
Fax: 717-533-8661
E-mail: cust@igi-global.com
Web site: http://www.igi-global.com

Library of Congress Cataloging-in-Publication Data

Names: Venkatesan, G., 1972- editor. | Thirumal, Jaganathan, 1978- editor.
Title: Global perspectives on air pollution prevention and control system design / G. Venkatesan and Jaganathan Thirumal, editors.
Description: Hershey, PA : Engineering Science Reference, an imprint of IGI Global, [2019] | Includes bibliographical references and index.
Identifiers: LCCN 2018022064| ISBN 9781522572893 (hardcover) | ISBN 9781522572909 (ebook)
Subjects: LCSH: Air--Pollution. | Air quality management.
Classification: LCC TD883 .G5645 2019 | DDC 628.5/3--dc23 LC record available at https://lccn.loc.gov/2018022064

This book is published in the IGI Global book series Advances in Environmental Engineering and Green Technologies (AEEGT) (ISSN: 2326-9162; eISSN: 2326-9170)

British Cataloguing in Publication Data
A Cataloguing in Publication record for this book is available from the British Library.

Advances in Environmental Engineering and Green Technologies (AEEGT) Book Series

ISSN:2326-9162
EISSN:2326-9170

Editor-in-Chief: Sang-Bing Tsai, University of Electronic Science and Technology of China Zhongshan Institute, China & Ming-Lang Tseng, Lunghwa University of Science and Technology, Taiwan & Yuchi Wang, University of Electronic Science and Technology of China Zhongshan Institute, China

MISSION

Growing awareness and an increased focus on environmental issues such as climate change, energy use, and loss of non-renewable resources have brought about a greater need for research that provides potential solutions to these problems. Research in environmental science and engineering continues to play a vital role in uncovering new opportunities for a "green" future.

The **Advances in Environmental Engineering and Green Technologies (AEEGT)** book series is a mouthpiece for research in all aspects of environmental science, earth science, and green initiatives. This series supports the ongoing research in this field through publishing books that discuss topics within environmental engineering or that deal with the interdisciplinary field of green technologies.

COVERAGE

- Industrial Waste Management and Minimization
- Sustainable Communities
- Electric Vehicles
- Biofilters and Biofiltration
- Green Transportation
- Green Technology
- Air Quality
- Contaminated Site Remediation
- Cleantech
- Pollution Management

IGI Global is currently accepting manuscripts for publication within this series. To submit a proposal for a volume in this series, please contact our Acquisition Editors at Acquisitions@igi-global.com or visit: http://www.igi-global.com/publish/.

The Advances in Environmental Engineering and Green Technologies (AEEGT) Book Series (ISSN 2326-9162) is published by IGI Global, 701 E. Chocolate Avenue, Hershey, PA 17033-1240, USA, www.igi-global.com. This series is composed of titles available for purchase individually; each title is edited to be contextually exclusive from any other title within the series. For pricing and ordering information please visit http://www.igi-global.com/book-series/advances-environmental-engineering-green-technologies/73679. Postmaster: Send all address changes to above address.

Titles in this Series

For a list of additional titles in this series, please visit:
https://www.igi-global.com/book-series/advances-environmental-engineering-green-technologies/73679

Advanced Multi-Criteria Decision Making for Addressing Complex Sustainability Issues
Prasenjit Chatterjee (MCKV Institute of Engineering, India) Morteza Yazdani (Universidad Loyola Andalucía, Spain) Shankar Chakraborty (Jadavpur University, India) Dilbagh Panchal (Dr. B. R. Ambedkar National Institute of Technology (NIT) Jalandhar, India) and Siddhartha Bhattacharyya (RCC Institute of Information Technology Kolkata, India)
Engineering Science Reference • ©2019 • 360pp • H/C (ISBN: 9781522585794) • US $195.00

Amelioration Technology for Soil Sustainability
Ashok K. Rathoure (Biohm Consultare Pvt Ltd, India)
Engineering Science Reference • ©2019 • 280pp • H/C (ISBN: 9781522579403) • US $185.00

Advanced Agro-Engineering Technologies for Rural Business Development
Valeriy Kharchenko (Federal Scientific Agroengineering Center VIM, Russia) and Pandian Vasant (Universiti Teknologi PETRONAS, Malaysia)
Engineering Science Reference • ©2019 • 484pp • H/C (ISBN: 9781522575733) • US $195.00

Spatial Planning in the Big Data Revolution
Angioletta Voghera (Politecnico di Torino, Italy) and Luigi La Riccia (Politecnico di Torino, Italy)
Engineering Science Reference • ©2019 • 359pp • H/C (ISBN: 9781522579274) • US $195.00

Global Initiatives for Waste Reduction and Cutting Food Loss
Aparna B. Gunjal (Asian Agri Food Consultancy Services Ltd, India) Meghmala S. Waghmode (Annasaheb Magar Mahavidyalaya, India) Neha N. Patil (Annasaheb Magar Mahavidyalaya, India) and Pankaj Bhatt (Dolphin (P.G) College of Biomedical and Natural Sciences Dehradun, India)
Engineering Science Reference • ©2019 • 328pp • H/C (ISBN: 9781522577065) • US $195.00

Green Public Procurement Strategies for Environmental Sustainability
Rajesh Kumar Shakya (The World Bank, USA)
Engineering Science Reference • ©2019 • 228pp • H/C (ISBN: 9781522570837) • US $185.00

For an entire list of titles in this series, please visit:
https://www.igi-global.com/book-series/advances-environmental-engineering-green-technologies/73679

Table of Contents

Detailed Table of Contents

Climate records kept worldwide clearly show that ongoing changes are happening in our eco-systems. Such climate changes include temperature, precipitation, or sea level, all of which are expected to keep changing well into the future, thereby affecting human health, the environment, and the economy. The natural causes by themselves are not able to describe these changes, so to understand these, scientists are using a combination of state-of-the-science measurements and models. Human activities are a major contributor due to the release of different air contaminants through various activities. Air pollution is one case-in-point, a human-made factor that contributes to climate change by affecting the amount of incoming sunlight that is either reflected or absorbed by the atmosphere. An overview of modeling techniques used to relate air quality and climate change is presented. The discussion includes the role of air pollution levels affecting the climate. Emerging topics such as black carbon (BC), fine particulate matters (PMs), role of cook stove, and risk assessment are also covered.

The objective of this chapter is to present the central concepts, parameters, and methods for the monitoring of climate changes, with a focus on air pollution, and the possible global and regional impacts of climate changes as well. There are plant species used as bioindicators that have a high sensitivity or ability to accumulate environmental pollutants. Another method that this chapter will present is the use of receiver models that employ both mathematical and statistical approaches to quantify the individual contribution of a given number of emission sources in the composition of a sample. The data presented in this chapter will provide reliable bases and methodologies for environmental control, supporting the adoption of more restrictive policies.

Environmental pollution and remediation on a global scale have drawn attention to the vital need for new environmentally friendly, clean chemical technologies and processes. Ultraviolet LED (UVLEDs) are mainly employed for the photo catalytic degradation of organic pollutants present in air and water. LEDs are available as both chip and bulb types. As LEDs are energy efficient and mercury free, Shengottai municipality proposed to change all exiting streetlights to LED lights in order to reduce the electrical energy as well as the air pollution.

Biodiversity is severely affected by the various anthropogenic activities. Humans depend on natural resources for survival. The resources include food, clean drinking water, timber for construction, fibers for clothing, natural gas, and coal for industries. The biggest threat to the atmosphere is pollutants caused by different human activities. The earth atmosphere is a layer of gases that surrounds the planet and allows us to breathe and live. The atmosphere is made up of Nitrogen, Oxygen, and others types of gases. This mixture of gases is known as air. Air pollution results from a variety of pollutants like NO2, SO2, CO, CFCs, and Ozone. Carbon dioxide, which is a greenhouse gas, warms the earth. This chapter will explore the effects of atmospheric pollutants on biodiversity due to various types of adulterants in the atmosphere and its effects on the survival of earth. Additionally, it will sketch various tools that can reduce these harmful activities.

Air pollution is a major environmental problem. It is divided into indoor air pollution and outdoor air pollution. The pollutants released in the outdoor atmosphere are under the control of meteorological conditions prevailing at the time of emission and after. Thus, the subject, air pollution meteorology, has emerged. It explains the significance of meteorological aspects and their influences on air pollutants emitted in the outdoor atmosphere. Prevailing winds and atmospheric stability are the two major factors that determine the dispersion of the pollutants. In this chapter, a brief outline is presented on composition and structure of atmosphere, and its processes such as radiation through the atmosphere, winds, global circulation, and atmospheric stability. Mixing height and ventilation coefficients are explained. The prediction of air quality using Gaussian dispersion model with the input of data of emission and meteorology is also explained.

Environmental pollution states the toxification by different forms of pollutants in the surroundings. This causes the natural system imbalance and affects the life on earth in different ways. The emissions of pollutants into air result in climate changes. The rising levels of CO2 and other air-polluting gases increase the greenhouse effect which results in temperature elevations and affects global weather patterns. Bill Gates also warned "Climate change is a terrible problem, and it absolutely needs to be solved; it deserves to be a huge priority". Though climate change and air pollution are closely linked, these are treated as separate problems. Hence, it is aimed to understand the aspects of association between climate change and air pollution in this chapter with the objectives: (i) To know the basics of climate change and air pollution, (ii) To review the active background studies on climate change as well as on air pollution, (iii) To identify the issues, controversies, problems which include the effects of global warming and greenhouse gases followed by the types of air pollutants and its harmful effects, (iv) To find the solutions and recommendations, the discussions are made on climate laws and policies, (v) To retrieve, analyze and store the climate change related information, the description about models, tools and databases are given, (vi) To have a flawless insights on the association of climate change and air

Foreword

I am delighted to recommend the book titled *Global Perspectives on Air Pollution Prevention and Control System Design*. The air pollution related subjects of Meteorological parameters, their influence on air pollution dispersion, mathematical modeling, air pollution control methods, effects of pollutants on property, life and economy have been dealt very well. Annually more than 2 million peoples were losing their precious life because of air pollution and millions suffer from various respiratory related health problems. The fundamentals and chemistry of air pollution is well explained.

This book explains the causal relationship between emissions, meteorology, atmospheric concentrations, deposition, and other aspects of air pollution. Exhaustive discussions are found in primary and secondary pollutants, meteorological influence on pollutants dispersion and control strategies to be adopted for air pollution mitigation. The air pollution models, its application in pollution abatement is of great worth. This book is a worthwhile contribution to society.

G. Swaminathan
National Institute of Technology, Trichy, India

Preface

In a ridiculous manner, the population grown over the past 60 years, and these days people are looking for a better way of life, provoking the industrialization and automation process. Therefore, this habitat is badly damaged by direct or indirect methods of relieving air pollutants. Because of this, earth's biomass nitrogen, sulfur dioxide, carbon dioxide, carbon monoxide and a high molecular hydrocarbons and the complex thunderstorm of the darker intravenous-tone source (rotations) of the black gunk are mixed.

The academics and diplomats who want to examine the reviews of air pollution samples. It may be premature money in regulatory effort that is progressive in the pollution model. Air pollution and its control are partially strengthened, decorated with samples used to control remaining air pollutants.

The World Health Organization estimates that, in developing countries yearly more than 2 million peoples are dying because of air pollution and millions are suffering from various respiratory illnesses. Further, emerging industrial rebellion and domestic activities contributes to the deterioration of air quality by accessing a huge amount of emission sources along with meteorological processes, chemical transformations. Ambient concentration of air pollutants depends on emission rate and meteorology as well as the morphology of geography. Oxides of iron, aluminum, manganese, magnesium, zinc and other metals have adverse effects due to deposition of dust on plants during mining operations and metallurgical processes. They create physiological, biochemical and developmental disorders in plants and also contribute towards reproductive failure in plants. In order to protect the human health, air quality management should be crucially taken into deliberation.

Ambient concentrations are either measured or modeled. To characterize air quality at specific locations and times, air pollution measurements give perceptible intelligence about ambient concentrations and deposition. Only those measurements decline to accord the clear conveyance on assimilation of the causes of an air quality problem. Hence to describe the complex relationship between emissions, meteorology, atmospheric concentrations, deposition, and other factors, air pollution modelling is predominantly taken into account.

Modelling provides a complete deterministic description of the air quality problem, including an analysis of physical and chemical changes, emission sources, meteorological processes, and provides few guidance on the implementation of mitigation measures. It is the only method that quantifies the consequences of past and future scenarios and the determination of the effectiveness of abatement strategies. Modelling plays an important role in air pollution control science and technology, because of their capability to assess the relative importance of the relevant processes. This makes air pollution models, indispensable in regulatory, research, and forensic applications.

This text gives an abrupt interpretation of air pollution models and their application in systematizing with analytical and sensible aspects. Mathematical formula is formulated in air pollution modeling tools in order to simulate all the atmospheric processes over various time and space scales with rapidly developed computer technology. The more refined time scale of modelling is accomplished by the power of computer technology in detail relating to present climate change and other global issues including transboundary movements of air pollution are also discussed in detail in this folio.

G. Venkatesan
Anna University, India

Jaganthan Thirumal
Anna University, India

Chapter 1
Air Pollution and Climate Change:
Relationship Between Air Quality and Climate Change

Ashok Kumar
The University of Toledo, USA

Hamid Omidvarborna
The University of Toledo, USA

Kaushik K. Shandilya
The University of Toledo, USA

ABSTRACT

Climate records kept worldwide clearly show that ongoing changes are happening in our eco-systems. Such climate changes include temperature, precipitation, or sea level, all of which are expected to keep changing well into the future, thereby affecting human health, the environment, and the economy. The natural causes by themselves are not able to describe these changes, so to understand these, scientists are using a combination of state-of-the-science measurements and models. Human activities are a major contributor due to the release of different air contaminants through various activities. Air pollution is one case-in-point, a human-made factor that contributes to climate change by affecting the amount of incoming sunlight that is either reflected or absorbed by the atmosphere. An overview of modeling techniques used to relate air quality and climate change is presented. The discussion includes the role of air pollution levels affecting the climate. Emerging topics such as black carbon (BC), fine particulate matters (PMs), role of cook stove, and risk assessment are also covered.

DOI: 10.4018/978-1-5225-7289-3.ch001

INTRODUCTION

It is important to understand climate in order to study climate change, which includes the harm caused by air pollutants. These can occur through natural processes or through human activities, which in turn affect human health and the ecosystem, resulting in millions of annual premature deaths, and hospitalizations (Schreiner et al., 2006; Shandilya and Khare, 2012). Further, climate change highly impacts on agriculture and fisheries (farmers and ranchers). Increases in temperature and CO_2 level in the atmosphere can change in the frequency and severity of droughts and floods, which can threat food safety. In addition, warmer water temperatures are likely to cause the habitat ranges of many fish and shellfish species to shift, which could disrupt ecosystems (USEPA, 2018a).

Climate change is able to alter how primary pollutants are dispersed and to intensify secondary pollutant formation. Therefore, it is helpful to review the impact of climate change. In particular, it is necessary to note that there is a relation between weather and climate. Weather is defined as the state of the atmosphere at any given time and place, while climate is the long-term average (e.g., decades) of the weather in any place (Rosenzweig et al., 2001). Figure 1 shows the relationship between the weather variability and climate change (Solomon et al., 2007).

As shown in Figure 1, the climate system is comprised of five natural components, (1) the atmosphere, (2) the hydrosphere, (3) the cryosphere, (4) the land surface, and (5) the biosphere.

The atmosphere includes gases that surrounds Earth, including the naturally occurring greenhouse gases or GHGs that warm the planet's surface. The hydrosphere includes all of Earth's liquid and gaseous water (water vapor), while the cryosphere includes all frozen water (ice). The land surface only consists of all vegetated and non-vegetated surfaces. The biosphere is the global ecological system integrating all living beings and their relationships.

Several external forces influence the five climate system components, with radiation from the Sun being most important. The Sun is the primary energy source for Earth's climate. Some incoming sunlight is reflected into space (via bright surfaces like ice and clouds), while the rest is absorbed by the surface and the atmosphere. A considerable amount of this absorbed solar energy is re-emitted as heat through long-wave or infrared radiation, but some heat escapes when the atmosphere absorbs and re-radiates heat. Disturbing this balance (incoming and outgoing energy) results in climate change (Wittneben and Kiyar, 2009). Additionally, climate scientists consider the impact of human activities on the climate system as another example of external forcing.

The instability of the atmosphere within the climate system is evident in how it is composed of nitrogen, oxygen, and argon with 78.1, 20.9, and 0.93% vol. mixing

Figure 1. Schematic view of the major components of the climate system, main processes and key interactions
Source: Solomon et al. (2007)

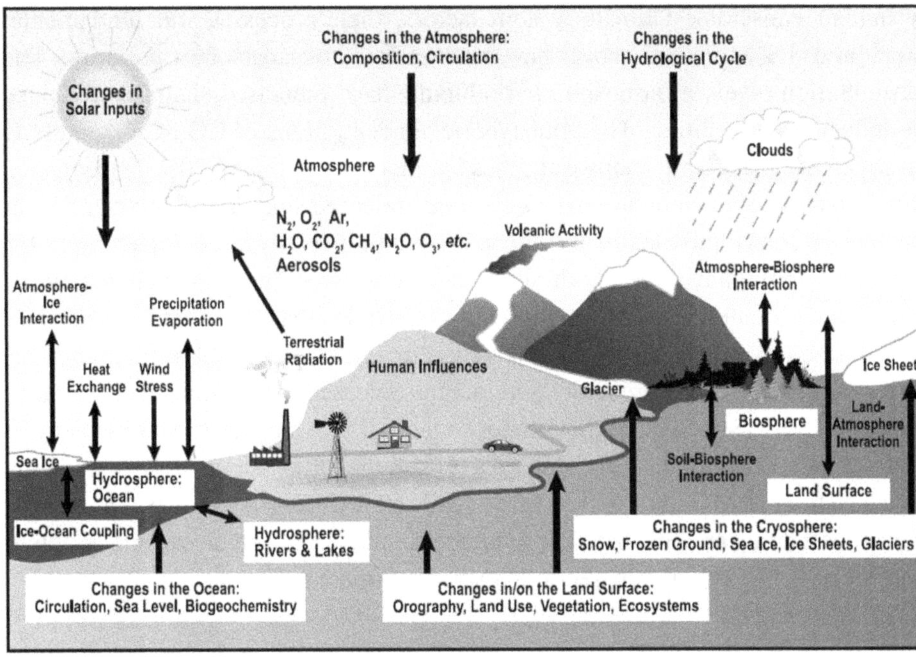

ratio, respectively. In the atmosphere, these three elements have limited interaction with the solar radiation and do not interact with long-wave or infrared radiation. Also, GHGs (e.g. water vapor, CO_2, CH_4, O_3, and N_2O (<0.1% by volume)), can absorb and emit these radiations keeping Earth's surface and lower atmosphere warm. Likewise, solid and liquid particles (aerosols) interact with radiation in a complex and spatial manner (Shandilya et al., 2009).

The oceans are also crucial as they are able to store energy and CO_2. The ice sheets (of Greenland and Antarctica), continental glaciers, snowfields, sea ice and permafrost derive their importance to the climate system through high reflectivity for solar radiation, low thermal conductivity, large thermal inertia, and the driving of deep ocean water circulation. Forests play a major role on the atmosphere's composition as they store large amounts of CO_2 through the photosynthetic process. Vegetation and soils control energy balance (received from the Sun and returned to the atmosphere) while serving to evaporate water (either in the soil or in the leaves of plants). By considering all those factors, it can be concluded that life could not have evolved without this greenhouse effect; however, human activities (through emitting GHGs and changing the nature of the land surfaces) have altered Earth's climate balance toward warming, with the biggest contributor being increases in CO_2 (Wittneben and Kiyar, 2009).

Regardless of natural effects, climate change and air pollution are still united by human emissions. Climate system factors, their processes and fundamental interactions (as explained above) have been affected because of air pollution. The concentration levels of atmospheric pollutants have increased significantly since the Industrial Revolution. The atmospheric concentrations of CO_2, CH_4, and N_2O have increased since pre-industrial times by about 40, 150, and 20%, respectively. More than half of the increase in CO_2 has been released since 1970. Increases in all these gases contribute to climate change, with an especially huge impact by CO_2.

Climate pollutant types (both short- and long-lived) vary in their reactivity. Short-lived climate pollutants are defined as particles that contribute to warming; the duration typically ranges from a few days to a full decade. Table 1 summarizes the residence time of different air contaminants released in the atmosphere because of human activities as well as their major sources. The residence time is how long it takes for an air contaminant to be removed from the atmosphere. For example, the residence time of N_2O is about 120 years, while for BC it is a few weeks. Short-lived climate pollutants are known to cause greater impacts and can be extremely potent in terms of their GWP compared to long-lived climate pollutants.

Table 1. Residence time for major contaminants in air (NOAA website)

Gases and Particles	Residence Time in Atmosphere (Avg.)	Major Sources
	Long-Lived Climate Pollutants	
CO_2	100+ years	Fossil fuel combustion, deforestation, cement production
N_2O	120 years	Fertilizer application, fossil fuel and biomass combustion, industrial processes
	Short-Lived Climate Pollutants	
CH_4	10 years	Fossil fuel production, agriculture, landfills
CFCs	15 years	Refrigerants
Tropospheric O_3	Days to weeks	Indirectly related to the emissions from industrial facilities and electric utilities, motor vehicle exhaust, gasoline vapors, and chemical solvents
BC	1 to 4 weeks	Diesel engines
OC	1 week	Incomplete combustion of biomass fuels, such as wood
Nitrate	1 week	Several natural and anthropogenic sources
Sulfate	1 week	Mines, power plants, and refineries

Human activities are responsible for almost all of the increase in GHGs in the atmosphere over the last 150 years (Kahn Ribeiro et al., 2007). The largest source of greenhouse gas emissions from human activities in the US is from burning fossil fuels for electricity, heat, and transportation (USEPA, 2018b.

METHODOLOGY

A rigorous literature review was carried out for examinations and synthesizing literatures on environment, ecology, GHGs, climate change, climate modeling, global warming, and risk assessment in the electronic bibliographic databases, including Scopus, Web of Science, and governmental databases. The search was limited to most recent published materials and research reports those themes that are related to our research aims.

MEASURING CLIMATE CHANGE

To carry out the experiments, it is important to make climate measurements, taking into account the atmosphere, land surface, snow and ice, oceans, and living things, all of which constitute average weather. No single instrument measures climate change as thousands of measuring devices exist across the globe. Specified regional climate is determined through analysis of climate variables within that region for a 30-year time period. Different parameters such as temperature, wind speed, relative humidity, amount/frequency of rainfall, atmospheric pressure, and sunshine hours are included. Air quality components such as pollutant transport and emissions play a vital role in field measurements and laboratory studies. The consequences of changing the natural atmospheric greenhouse are difficult to predict, but the key factors in measuring climate change are broadly categorized in Table 2 (based on the NIST website):

AIR POLLUTION AND CLIMATE CHANGE RELATIONSHIPS

Air pollution includes GHGs (Table 1) and among them, CO_2 is a common part of the exhaust emitted from cars and trucks. GHGs cause global warming by trapping heat from the Sun in the Earth's atmosphere. Emissions of air pollutants can result in changes to the climate. However, air pollution and climate change impact each other and can provide mutual benefits. On one hand, climate change mitigation is able to reduce air pollution; contrariwise, clean air measure potentially shortens

Table 2. Parameters that play a major role on weather condition and climate change

Key Factors	Descriptions
Temperature	• A primary parameter, which can be measured/reconstructed for the Earth's surface and sea surface temperature. • Extreme temperature conditions are becoming more common as a result of climate change (Frich et al., 2002). • Occurrence of more intense storms (above normal level) as a result of high temperature due to global warming (Van Aalst, 2006).
Precipitation	• An indicator of relative climate variation. • It may include humidity or water balance, and water quality. • Total annual precipitation has increased over land areas worldwide since 1901 (0.08 inches per decade on average) (Blunden and Arndt, 2016). • Higher percentage of heavy precipitation, especially in the United States, has been observed in the form of intense single-day events since 1980s (Kunkel et al., 1999). • Increasing/decreasing in the frequency and magnitude of river flooding become larger and more frequent across the world in many countries (Slater, 2016; Shandilya and Sitaram, 2018).
Biomass and vegetation patterns	• Provide evidence of how ecosystems change to adapt to climate change. • Average drought conditions across the globe have varied. A recently developed index in the United States shows that 20 to 70% of land areas were abnormally dry during 2000-2015 (National Drought Mitigation Center website).
Sea level	• Processes are slow and would take hundreds of years. • Reflect changes in shoreline and regularly relate to the degree of ice coverage in high latitudes and elevations. • Global warming would cause a rise in the sea level which would affect coastal communities through more frequent flooding and increased ground water salinity.
Solar activity	• It could influence climate, mainly through changes in the intensity of solar radiation.
Volcanic eruptions	• It can alter climate due to the aerosols emitted into the atmosphere and alter climate patterns (Shandilya et al., 2009).
Chemical composition of air/water	• It can be measured by tracking levels of GHGs (CO_2 and CH_4) and measuring ratios of oxygen isotopes. • A direct and strong correlation exists between the percent of CO_2 in the atmosphere and the Earth's mean temperature has been detected.

GHG emissions to reduce global warming (Kumar and Omidvarborna, 2018a). Therefore, this relationship is examined here.

Complex interactions exist between air pollution and climate change in the atmosphere. For example, increasing the concentration levels of GHGs alter the energy balance, which can change in surface temperature and chemical composition of the atmosphere. Moreover, this energy balance can be impacted by some of the air pollutants (e.g., BC, PMs, sulfate and O_3) directly.

Successfully implementing air pollution control strategies requires understanding the chemistry-meteorology-climate interactions of gas-phase and particulate pollutants (Kumar and Omidvarborna, 2018b). The model projections indicate climate changes for both the near- (2020-2050) and long-term (2070-2099) globally and regionally.

They indicate that using recent climate conditions may no longer be advisable to estimate future air quality (Steiner et al., 2006). For the 21st century, modeling studies must utilize future climate scenarios to understand climate change implications on air quality. Modeling studies must also consider the potential impact of changes in both biogenic and anthropogenic emissions under future climate scenarios (Kumar and Omidvarborna, 2018c).

O_3, PM, and SOA also affect light scattering and interact with terrestrial radiation, leading to climate forcing effects (Horne and Dabdub, 2017; Tsigaridis and Kanakidou, 2018). It is essential that air quality management strategies account for possible climate-driven changes in pollutant levels as well as changes in the frequency or severity of air pollution episodes that result from climate change. The potential air quality benefits of emission control policies and associated reductions in precursor emissions such as NOx and VOCs may be offset by increasing temperatures or other climate change-induced meteorological changes.

The climate also influences the atmosphere's development and the flow of air pollutants. Climate can change the height of atmospheric layers and chemical reaction rates in the air. However, it is difficult to exactly predict how the quality of air will be impacted by climate change. Figure 2 shows the projection of changes in average surface temperature and precipitation because of GHG emissions in Europe, Asia, and North America for 2080 – 2099 versus 1980 – 1999 (Jacob and Winner, 2009). The increment of the Earth surface temperature is the result of the Industrial Revolution in the current century.

In addition to surface temperature, precipitation slightly increases due to evaporation from the oceans, especially in northern parts of North America and Europe; however, relative humidity is not expected to change significantly (Held and Soden, 2000). In general, projection models are in accord with the view that high latitudes will become wetter and subtropical latitudes drier (Jacob and Winner, 2009). However, there is still a debate whether the future climate will be wetter or drier (Christensen et al., 2007).

Both physical and mathematical models have been used for assistance in research of the interaction of air pollution and climate change. Reduction in cloud cover over southern and central Europe is expected in the summer times, which is an important factor promoting O_3 formation (Forkel and Knoche, 2006; Meleux et al., 2007). The prediction models found an increase in the number of lightning, even if it is less frequent, in the future climate (Hauglustaine et al., 2005; Brasseur et al., 2006; Wu et al., 2008a). It isn't clear what effect climate change has on mixing depth since the models showed both increases and decreases in various regions with no consistent patterns (Wu et al., 2008b; Lin et al., 2008).

Figure 2. Differences in annual mean surface air temperatures and precipitation in different continents (Europe, Asia, and North America) for 2080–2099 versus 1980–1999 (Christensen et al., 2007; Jacob and Winner, 2009); from left to right, Europe, Asia, and North America

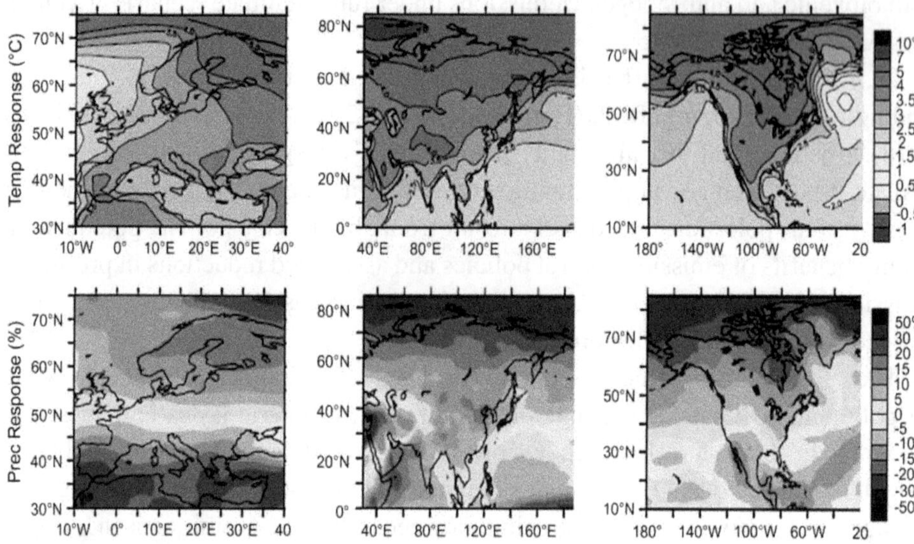

GREENHOUSE EFFECT

The GHG effect is a natural phenomenon that warms the Earth, as shown in Figure 3. The Earth receives a portion of the radiation (55%) from the Sun in different forms (ultraviolet, visible, and near infrared ray), which are absorbed by the land and the ocean. The atmosphere and the cloud reflect (about 26%) or absorb (19%) of the remaining radiation from the Sun. The remaining heat, excluding a small amount, which is bounced back to the atmosphere from the inside, gets absorbed by the GHGs, thus causing a greenhouse effect. The main effect of increased GHG elements, CO_2, CH_4, N_2O, and CFCs, in the Earth's atmosphere is global warming.

This is needed so that Earth's temperature would not be below freezing point. However, Earth's greenhouse effect is strengthening as GHG emissions increase, especially from fossil fuels to the atmosphere. Computer predictions indicate that by 2050 the Earth could have an average rise in temperature by as much as 1.5 – 4°C. That is increasing at an unprecedented rate leading to an overall warming of the planet.

To prepare for this effect, studies are being done on the major GHGs (including both short- and long-lived climate pollutants), which are CO_2, CH_4, CFCs, N_2O, PMs, O_3, BC and water vapor (Seinfeld and Pandis, 2016; California Air Resources

Figure 3. Earth's greenhouse effect
(Climate Change on USEPA website)

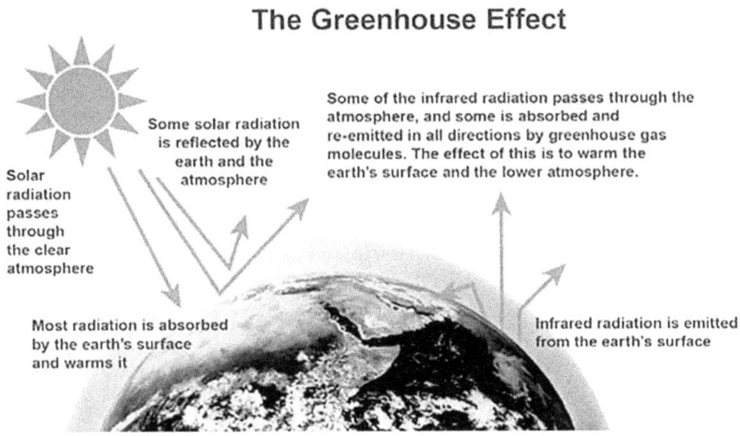

Board website). Studies have shown that over the last two centuries the concentration level of the GHGs in the atmosphere has increased quite drastically, mostly through combustion of fossil fuels and wildfires. For example, 2.8 billion MT of CO_2 (91.3% of the GHGs) was released to the atmosphere in 2015. CH_4, N_2O and fluorinated gases (HFCs, PFCs, SF_6) accounted for 7.2, 0.9, and 0.7% of reported 2015 GHG emissions, respectively (Blok et al., 2017; GHGs on USEPA website; Chen et al., 2017).

Short-lived climate pollutants are especially dangerous pollutants that remain in the atmosphere for less time than that of the long-lived climate pollutants (e.g., CO_2 and N_2O). This characteristic increases their potential as heating trap compounds that can be thousands of times higher than that of the other GHGs even at a short period of time. Thus, reducing these emissions can make an immediate positive impact on climate change. A brief discussion on the major GHGs and their roles on climate change are summarized in Table 3 (GHG on USEPA website; Climate Change indicators on USEPA website; Matyssek et al., 2013). GHG levels are often expressed in terms of mixing ratios. The various volumes for both the molar and mass concentrations are examined. For modeling climate change, it is recommended to use a vertical coordinate system.

GHG effects can be controlled by:

- Reduce the emissions of harmful pollutants from the sources, such as vehicles, by using control devices or implementing alternative and renewable resources.

Table 3. Characteristics of major GHGs

Contributed gases	Molecular formula	Possible sources and impacts
CH_4		• CH_4 is the principal component of natural gas. • CH_4 produces both through natural sources and human activities (coal mining, gas drilling, and decomposition of wastes in landfills, agriculture, and rice cultivation, ruminant digestion and manure management associated with domestic livestock). • It is a far more active GHG than CO_2, but also one which is much less abundant in the atmosphere (20 times more powerful gas than CO_2). • Its production is linked to the world's growing population (need for food and disposable goods).
CO_2	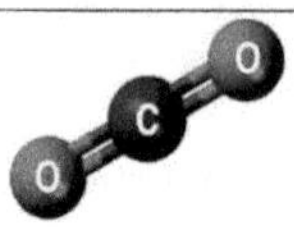	• A minor but very important component of the atmosphere. • It is the most important long-lived "forcing" of climate change (emission rate: 2~3 ppm/year (Hoffman et al., 2006; Butler and Montzka, 2016)). • A by-product of most living things and many commercial processes (burning foods by organisms, combustion of fossil fuels, deforestation, land use changes, etc.). • It is also released through natural processes (e.g., respiration and volcano eruptions). • It can absorb infrared radiation (heat) from the Sun, which increases the temperature of the Earth. • Humans have increased atmospheric CO_2 concentration by more than a third since the industrial revolution.
Water vapor	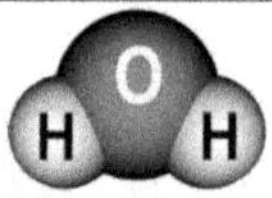	• The most abundant and important GHG (responsible for about 75% of the natural greenhouse effect). • It acts as a feedback to the climate (it increases as the Earth's atmosphere warms, but so does the possibility of clouds and precipitation, making the most important feedback mechanisms to the greenhouse effect). • Unlike most of the other GHGs, human activities impact for water vapor is negligible compared to natural impact. • No opportunities for humans to directly control the amount of water vapor in the lower atmosphere (troposphere) as the Earth is getting warm due to increase of other GHGs.
N_2O		• A powerful GHG in the atmosphere. • It produces by soil cultivation practices (especially the use of nitrogenous fertilizers), combustion of fossil fuel, nitric acid production, biomass burning, de-nitrification by bacteria in soils, groundwater and the oceans. • N_2O is increasing in the atmosphere at a rate of 0.25%/year.
O_3		• About 90% of the O_3 is in the stratosphere, where it protects the Earth's surface from solar ultraviolet radiation. O_3 also controls the temperature structure of the stratosphere. • O_3 is generated in the troposphere by photochemical oxidation of CO, CH_4, and NMVOCs by the OH radical in the presence of reactive NOx (Jacob and Winner, 2009). • O_3 is strongly correlated with temperature (Cox and Chu, 1996); hence, O_3 pollution is a major concern during summer season or hot climates due to the photochemical nature of the source. • Currently O_3 is being depleted in the upper atmosphere; however, in the lower atmosphere, O_3 produced by combustion (e.g., vehicles and power plants) acts as a GHG.

continued on following page

Table 3. Continued

BC (Climate Pollutant Reduction CARB website)		• A fine PM, recognized as a leading environmental risk factor for premature death (Schreiner et al., 2006). • Carbonaceous particles (contain organic and BC), which are abundant in the atmosphere. • It is generated during the incomplete combustion of coal, fossil fuels (Shandilya and Kumar, 2014; Omidvarborna et al., 2014, 2015, 2016), forest fires, etc. • Vehicles running on diesel fuels are the major emitter of BC (Shandilya, 2011, 2012; Omidvarborna, 2017). • It contributes to climate change both directly (absorbing sunlight) and indirectly (depositing on snow and by interacting with clouds and affecting cloud formation).
CFCs	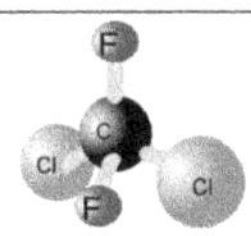	• A part of a larger family of compounds called as halocarbons (containing chlorine, fluorine and bromine). • Non-reactive, non-caustic, non-corrosive, and nonflammable chemicals. • Synthetic compounds entirely of industrial origin (refrigeration, air conditioning, dry cleaning, plastic foams, and aerosols). • They have ability to contribute to destruction of the O_3 layer (break off via ultra violet light and then deplete O_3 in the upper atmosphere). • As O_3 depleted, the O_3 layer gets thinner and more ultraviolet rays can reach the surface of Earth. • They are used in a number of applications, but now are largely regulated in production/release to the atmosphere by international agreement. • The current strategies to reduce CFCs emissions contain replacing them with low GWP compounds (e.g., ammonia).

(CARB website; GHG USEPA website; Climate Change indicators: GHG USEPA website; Climate change NASA website)

- Manufacturing and producing quality products that have a long life and recycling of recyclable materials.
- Reforestation approach, both at government and public level.

TEMPERATURE CHANGE DUE TO CHANGE IN POLLUTION LEVELS

Aw and Kleeman (2003) showed that temperature is the single most important meteorological variable influencing both gas-phase and particulate air quality due to its effect on chemical reaction rates and gas-to-particle partitioning. They showed decreases in $PM_{2.5}$ concentrations when temperatures were increased uniformly. Their results indicated that inland O_3 concentrations are the most affected by temperature changes.

As a secondary air pollutant, O_3 is formed in the atmosphere from a combination of NOx, VOCs, CO, and CH_4 in the presence of sunlight. The well-established O_3

formation chemistry happens via the following set of reactions (Sillman, 1999; Monks et al., 2015; Von Schneidemesser et al., 2015):

Reaction 1:

$$\mathrm{VOC} + \mathrm{OH} \xrightarrow{\mathrm{O_2}} \mathrm{RO_2} + \mathrm{H_2O}$$

Reaction 2:

$$\mathrm{CO} + \mathrm{OH} \xrightarrow{\mathrm{O_2}} \mathrm{HO_2} + \mathrm{CO_2}$$

Reaction 3:

$$\mathrm{RO_2} + \mathrm{NO} \xrightarrow{\mathrm{O_2}} \text{secondary VOC} + \mathrm{HO_2} + \mathrm{NO_2}$$

Reaction 4:

$$\mathrm{HO_2} + \mathrm{NO} \rightarrow \mathrm{OH} + \mathrm{NO_2}$$

Reaction 5:

$$\mathrm{NO_2} + h\vartheta \rightarrow \mathrm{NO} + \mathrm{O}$$

Reaction 6:

$$\mathrm{O} + \mathrm{O_2} + \mathrm{M} \rightarrow \mathrm{O_3} + \mathrm{M}$$

A later study by Steiner et al. (2006) reported the impact of future climate changes and emissions on O_3 air quality. Perturbations to temperature, atmospheric water vapor, and biogenic VOC emissions caused an increase in daily peak O_3 levels. Simultaneous perturbations that combined climate change effects resulted in higher O_3 increase. Carreras-Sospedra et al. (2006) evaluated the sensitivity of the CIT air-shed model to various inputs, including temperature and boundary conditions. They found that simulated O_3 concentrations are sensitive to meteorological conditions, particularly temperature. In addition to affecting chemical reaction rates in the model, high temperatures favor the decomposition of PAN, increasing NOx, an important

Figure 4. Categorization scheme of different methods to study climate change

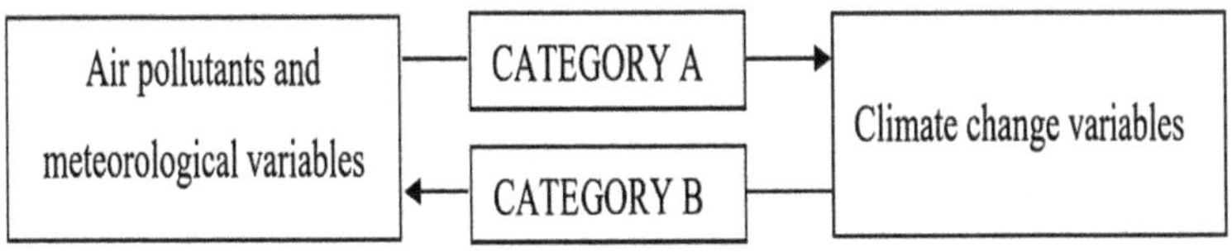

O_3 precursor. They concluded that the impact of temperature on O_3 concentrations is greatest in locations where O_3 levels are already high.

Millstein and Harley (2009) referred to Steiner et al. (2006) to study the effect of future climate and emissions-related perturbations on O_3 air quality. They found rises in peak O_3 levels across the domain due to increased temperatures. The combined climate forcing case described by Millstein and Harley (2009) (with changes to temperature, absolute humidity and biogenic emissions) likewise caused rises in peak O_3.

Although O_3 production rises alongside increasing temperatures in polluted areas, several global studies report O_3 lowers in remote areas due to the destruction of O_3 by water vapor at these higher temperatures (Marland and Rotty, 1984; Millstein and Harley, 2009).

Furthermore, sensitivity to climate-change parameters varies depending on location, local emissions, chemical species of interest, time of day, and other variables. O_3 increases due to increased temperatures are greatest in polluted areas with high anthropogenic emissions (Carreras-Sospedra et al., 2006; Millstein and Harley, 2009).

MODELING

Modeling has been evolving over the last 20 years. Models are now available to model climate change directly related to air quality. All the available models and methods can be divided in two categories as shown in Figure 4. Category A consists of the methods and models applied to study the effects of air pollutant and meteorological variables on the climate system, and Category B includes the methods and models applied to study the effect of climate variables on air pollution scenario.

The regional climate responses, such as atmospheric temperature, precipitation, and circulation patterns are improbable to scale simply to global mean radiative forcing (RF) or global mean surface temperature (GMST). Hence, in general, 3D numerical circulation models (GCMs) are the suggested tools for the investigation of regional climate responses. In the past decade, a new tool for studying the interactions between air pollutants and climate has been developed. Note that the climate models are based on the solution of Navier-Stokes equation along with the equations for other parameters while the pollutant concentrations are obtained using the solution of convective-diffusion equations. The current generation of models is incorporating atmospheric chemistry into the climate/air pollutant modelling systems.

The model of atmospheric chemistry and climate models (CCMs) was established by merging GCMs with chemistry–transport models (CTMs), which directly coupled the climate system with atmospheric composition (Fiore et al., 2015). CH_4 is naturally set to observed abundances in GCMs, CCMs, and CTMs models by fixing a lower boundary condition or concentration (Lamarque et al., 2013). However, for O_3 and PM in GCMs, precursors released in the CCMs and CTMs are subjected to chemical/physical processing prior to final removal in the models (Fiore et al., 2015). The current generation of CCMs and CTMs have been recently assessed with available observations by other studies (Schulz et al., 2009; Lamarque et al., 2013).

According to chemistry-transport model based on radiative transfer model (CTM-RTM) and CCM studies, short-lived species induce spatially heterogeneous RF, roughly following the geographic patterns of emissions. The forcing location (i.e., geographic and vertical) may regionally affect surface temperature, precipitation, and circulation patterns, although the climate responses do not necessarily reflect the spatial patterns of the RF (Fiore et al., 2015). A summary of different modelling methods for each category is provided in Table 4.

A SCREEN CLIMATE MODEL: BASIC CONCEPTS

This section presents a simple screening type climate model to get an idea on the magnitude of climate change because of a given energy consumption scenario. Basic concepts involved in modeling are discussed first.

1. *RF on Climate:* The GHGs induce RF that can affect the climate. As defined by the IPCC and WMO, RF of the surface-troposphere system is the change in net radiative flux at the tropopause after allowing for stratospheric temperatures to readjust to radiative equilibrium (Ramaswamy et al., 2001). RF is the major driving force behind the climate change. It can be used as a common quantity for the greenhouse impact of most GHGs. By using RF as the measure of greenhouse impact, the dynamics of the phenomena can be represented more accurately than if all emissions are converted into CO_2 equivalents (CO_2E). This is well suited for situations where the emissions are changing with time.
2. *Relative Forcing Effects of Various Gases:* Some GHGs are more potent in affecting global temperatures than the others, resulting in comparable increases

Table 4. Summary of methods and models used to study air quality-climate connections

Approach	Focus Area(s)	Features	Limitations
Category A: Methods Applied to Study Air Pollutant Impacts on the Climate System			
RTM forced with distributions of atmospheric constituents from CCM, or a GCTM (Stevenson et al., 2013; Lamarque et al., 2013; Morgenstern et al., 2017)	RF, Changes in atmospheric constituents (emissions)	Enables attribution of RF to changes in individual atmospheric constituents	Inputs like cloud distributions are not necessarily consistent with the simulation of atmospheric constituents. The simulations used to estimate changes in atmospheric distributions of constituents are often limited to a few years of meteorology
Global CCM driven by fixed sea surface temperatures (Hurrell et al., 2008; Garfinkel et al., 2013; Unger et al., 2013)	ERF from a particular atmospheric species	As above and also include processes like aerosol-cloud interactions	Land warming complicates the calculation
Regression of net energy imbalance onto the change in GMST in a transient climate simulation (Boucher et al., 2013; Gregory et al., 2016)	As above	As above	Estimate can be confounded by natural variability or time-varying feedbacks
GCMs: Typical IPCC-class models used to project changes in climate variables (Guilyardi et al., 2009; Hare et al., 2012; Molod et al., 2015)	Climate evolution; Climate response to forcing	Long, multi-ensemble simulations allow separation of climate change signal from variability representation of process	Prescribed O_3 and aerosol fields drive evolution of climate system, introducing inconsistencies in the distributions of chemical and meteorological variables; calculating RF from a single constituent requires the GMC skill at representing large-scale circulation
Global CCMs, either coupled to a full ocean model, or driven by sea surface temperatures and sea ice archived from a GCM (Eyring et al., 2006; Pawson et al., 2008; De Jong et al., 2009; Yamashita et al., 2010; Garfinkel et al., 2013; Morgenstern et al., 2017; Young et al., 2018)	Response of climate system to changes in air quality	Consistent simulation of meteorology and air pollutants; represents at least some two-way interactions between air pollutants and climate	Computational expense from chemistry implies coarser resolution than possible with GCMs; simplistic/absent representations of some processes exists; multiple sensitivity simulations are required for source attribution
Regional CCMs (Giorgi, 2006; Forkel and Knoche, 2006; Pan et al., 2010; Rummukainen, 2010)	Regional climate response to local air quality changes	As above but with higher resolution and/or more complex description of processes like for resolving clouds	As above
Nested grid models (e.g., one-way global through urban scales) (Krol et al., 2005; Willems et al., 2012)	Relating air pollutants and specific sectors from local-to regional climate	Fully consistent simulation that resolves processes at the relevant regional scale	Limited by ability to resolve relevant processes
Reduced-complexity climate (and carbon cycle/Earth system models) (Van Vuuren et al., 2008; Matthews et al., 2009; Unger et al., 2013)	Different scenarios of GMST	Emulates GCMs for rapid calculation of GMST	Simple representation of the climate system may not properly account for couplings in the system such as chemical feedbacks
Analytical formulae (Ramaswamy et al., 2001)	Relating RF with changes in GHG	Simple relationship between equilibrium RF to a change in emissions	As above

continued on following page

Table 4. Continued

Approach	Focus Area(s)	Features	Limitations
Simple box models (Boucher, 2012; Boucher et al., 2012; Flato et al., 2013)	GMST response to different RF scenarios	Can incorporate multiple time scales to estimate transient and steady-state GMST changes for a variety of scenarios	As above; neglects possible dependence of GMST response to the spatial distribution of RF and thus is unlikely to approximate well the response to short-lived air pollutants
Adjoint methods in a GCTM (Kopacz et al., 2010; Henze and Lee, 2012; Young et al., 2018)	Air emissions (species, region, sector) contribution to RF by a particular atmospheric constituent (regional, global)	Receptor-oriented framework allows attribution to multiple factors in a single simulation; complementary to forward modeling source attributions	Assumes a linear system; by design it targets a single quantity ('receptor')
Category B: Methods Applied to Study Climate Impacts on Air Pollution			
Observed relationships between chemical and meteorological variables (Dawson et al., 2007; Jacob and Winner, 2009; Bloomer et al., 2009; Tai et al., 2010; Rasmussen et al., 2012; Tai et al., 2012; Dawson et al., 2014)	Air pollution response to changing weather	Observational constraints on models	Does not directly provide information on how climate will change, and may not correctly identify the underlying drivers of the correlation
Statistical downscaling using observed relationships (above) and climate change projections (Halloway et al., 2008; Pielke and Wilby, 2012)	Air pollution response to changes in regional/global climate	Can be applied to a large suite of GMC physical climate change simulations	Assumes stationary in the air pollutant response to the projected meteorological variable
CTMs driven by perturbed meteorological variables (Steiner et al., 2006; Jacob and Winner, 2009; Rasmussen et al., 2012)	Air pollution response to changes in a meteorological driver	Develops process-level understanding of air pollutant response to changes in a single factor; helps diagnose responses in more complex models	Changes in air pollutants do not feedback on the climate system
Global CTM driven by meteorological from a GCM (or observed for present-day source attributions) (Wu et al., 2008a, 2008b; Jacob and Winner, 2009; Wang et al., 2013; Yu et al., 2018)	Alteration in air quality by climate change	Sensitivity simulations separate role of changes due to meteorology versus emissions/ chemistry	As above, plus often limited to a few years of meteorology from the parent GCM, complicating attribution to climate change
Regional CTM driven by meteorology from a GCM or RCM (Jacob and Winner, 2009; Chang et al., 2014; Lacressonniere et al., 2017; He et al., 2018)	As above, but with a region-specific focus	As above, may also consider region-specific feedbacks/processes	As above but with additional dependence on chemical and meteorological boundary conditions which may not be consistent
Global CCMs (Sheng et al., 2015; Luhar et al., 2017; Morgenstern et al., 2017; Young et al., 2018; Gaudel et al., 2018)	Alteration in air quality by climate change	As above but with a consistent simulation of meteorology and pollutants	Has coarse resolution; uses simplistic/absent representations of biosphere processes
Regional CCMs (Nabat et al., 2015; Scinocca et al., 2016)	As above, but with region-specific focus	As above but with higher resolution and/or more complex description of processes	Boundary conditions for chemical and meteorological variables not necessarily consistent
Global-to-regional nested CCM (one-way) (Bartok et al., 2017; Young et al., 2018)	As above	As above but avoids inconsistent boundary conditions	Limited by ability to resolve relevant processes

(Modified from Fiore et al. (2015))

in the concentrations of different GHGs having vastly different heat-trapping effects. Among those observed, the CO_2 is the least effective as a GHG. The concept of GWP has been used by scientists to study the relative effects of various GHGs in trapping the Earth's heat. The effects of various gases on global warming are difficult to be summarized into a single number due to the following reasons:

a. Each gas absorbs radiation in specific wavelengths or "window" in the spectrum. For some cases where the concentrations of the gas are low and no other gases block radiation in the same window, small emissions of the gas will have a disproportionate absorptive effect. However, if concentration of gas rises over time, a larger portion of the total light passing through the "window" will already have been captured, and the marginal effects of additional emissions will decline. Therefore, the effect of an additional unit of emissions of the gas that is relatively plentiful in the atmosphere, such as water vapor or CO_2, tends to be less than that of a rare gas, such as SF_6. This implies that the concentration of a gas reduces the impact of adding additional quantities of that gas. Thus, the relative impacts of various gases will change as their relative concentrations in the atmosphere change.
b. Many natural processes will decompose the GHGs into other gases, or to be absorbed into the ocean or ground. These processes can be summarized in terms of the "atmospheric lifetime" of a gas, or the period it would take for natural processes to remove a unit of emissions from the atmosphere. CH_4, which decays into CO_2 over a period, has a much larger short-run effect on the global warming than does on the equal amount of CO_2. However, over longer periods of time, from 10 years to 500 years, the differences between the CH_4 and CO_2 will become less significant.
c. Many chemically active gases react in the atmosphere in ways that alter the formation of other GHGs. For example, nitrogen and CO combine to promote the formation of O_3, which is a potent GHG, while CFCs tend to destroy the atmospheric O_3, thus promoting global cooling.

3. *RF on Climate Change:* The incoming solar energy absorbed by the Earth and its atmosphere are balanced by the outgoing long wave radiation. If, for some reason, an additional amount of energy is added to the incoming energy, then the balance will be temporarily upset. Over time, however, the climate system will adjust to that change by either increasing or decreasing the surface temperature of the Earth until a balance is once again attained. It is mathematically represented as:

Initially the balanced system has the energy absorbed equal to the energy radiated:

$$Q_{abs} = Q_{rad} \quad (1)$$

when the system is perturbed by addition of RF, ΔF (W/m^2), to the incoming absorbed energy, a new equilibrium will eventually be established so that

$$(Q_{abs} + \Delta Q_{abs} + \Delta F) = (Q_{rad} + \Delta Q_{rad}) \quad (2)$$

where the deltas refer to the changes in the quantity considered. Subtracting equation (1) from equation (2) gives

$$\Delta F = \Delta Q_{rad} - \Delta Q_{abs} \quad (3)$$

Climate Sensitivity Parameter: The quantity that links the changes in the amount of surface temperatures, (ΔTs), for a given change in RF (ΔF), is called the climate sensitivity factor, (λ). This can be defined by the following relationship:

$$\Delta T_s = \lambda \Delta F \quad (4)$$

By using Equations (3) and (4), we get the expression for λ as shown below:

$$\lambda = \Delta T_s/\Delta F = \Delta T_s/(\Delta Q_{rad} - \Delta Q_{abs}) \quad (5)$$

Rearranging the above expression we get another expression as below:

$$\lambda = ((\Delta Q_{rad}/\Delta T_s) - (\Delta Q_{abs}/\Delta T_s))\text{-}1 \quad (6)$$

If the RF is expressed as a function of concentration as below:

$$F = f(C) \quad (7)$$

where, C is the concentration, then an incremental change in concentration from C_0 to C will yield an incremental change in RF of

$$\Delta F = f(C) - f(C_0) \quad (8)$$

For CO_2, the RF can be expressed as a logarithmic increase:

$$\Delta F = k_3 (\ln C - \ln C_0) = k_3 \ln (C/C_0) \quad (9)$$

The coefficient k_3 is approximately 6.3 (W/m^2) for CO_2 gas. Hence the equation of the change in the RF for the CO_2 gas can be rewritten as follows:

$$\Delta FCO_2 = 6.3 \ln ((CO_2)/(CO_2)_0) \tag{10}$$

Since the concentration is mass divided by volume, one could get an idea on concentration from the emission rate. The fuel consumption data and formulas given by Marland and Rotty (1984) may used to calculate the CO_2 emission rates. For each type of fuel, the annual CO_2 emissions are the product of three terms: the amount of consumed fuel, the fuel fraction that becomes oxidized, and a factor for the carbon content of the fuel. In equation form it can be represented as:

$$CO_{2i} = (P_i)(FO_i)(Ca_i) \tag{11}$$

where, subscript i represents a specific fuel commodity, P represents the amount of fuel i that is consumed each year, FO is the fraction that is oxidized, Ca is the average carbon content for fuel i, and CO_2 is the resulting CO_2 emissions for fuel i expressed in mass of carbon. The effectiveness of the combustion process determines the fraction of the carbon that is oxidized. For liquid fuels and for natural gas, the hydrogen carbon ratio largely determines the heating value, and for solid fuels most of the combustion energy is from oxidation of carbon.

The Conversion factors that are to be considered while calculating the emission rates for the electricity generation data are given below. The Conversion factor that is used to convert from kWh to MJ is 3.6, which is an exact conversion.

From Natural gas production:

CO_{2g} = CO_2 emissions in 10^6 tons of carbon
FO_g = effective fraction oxidized in year of production = 0.98
C_g = carbon content in 10^6 tons per thousand 10^{12} joules = 0.0137

From Crude oil and natural gas liquids production:

CO_{2l} = CO_2 emissions in 10^6 tons Carbon
FO_l = effective fraction oxidized in year of production = 0.918
C_l = carbon content in 10^6 tons per thousand 10^{12} joules = 0.85

From Coal Production:

CO_{2s} = CO_2 emissions in 10^6 tons carbon
FO_s = effective fraction oxidized in year of production = 0.982

C_s = carbon content in 10^6 tons per thousand 10^{12} joules = 0.746

The emission rates (or, concentrations) are used in Equation 10 to determine the RF. ΔFCO_2 is the change in RF for the change in the emission rates of CO_2 in W/m^2. CO_2 and $(CO_2)_0$ are the initial and final emission rates of CO_2 for the period considered, as the units cancel any volumetric units that can be used for the emission rates. The formula given below can be used to find the surface temperature change that is caused by the change in the calculated RF:

$$\Delta T_s = \lambda \Delta F \tag{12}$$

where, ΔT_s is the surface change in temperature in °C. λ is the climate sensitivity factor = 0.57 in °C/W/m² (Change, 1996) and ΔF is the change in the RF W/m² Equations (10), (11) and (12) can now be used to compute the change in surface temperature for a given consumption of fuels over a time period. Equation (12) is basic equation of GCM model as used by Ramanathan (1977).

For the GHGs, the temperature change screening equations from the literature are given below. Apling et al. (1990) have given a summary of the screening equations based on the work of T.M.L. Wigley and others (Change, 1990; Krapivin and Shutko, 2012).

$$CO_2: \Delta T_s = 6.3 \times \ln((CO_2)/(CO_2)_0) \tag{13}$$

$$N_2O: \Delta T_s = 0.057 \times ((N_2O)^{0.5} - (N_2O)_0^{\ 0.5}) \tag{14}$$

$$O_3: \Delta T_s = 0.7 \times ((O_3) - (O_3)_0)/15 \tag{15}$$

$$CH_4: \Delta T_s = 0.19 \times ((CH_4)^{0.5} - (CH_4)_0^{\ 0.5}) \tag{16}$$

$$\text{CFC-11}: \Delta T_s = 0.14 \times ((\text{CFC-11}) - (\text{CFC-11})_0) \tag{17}$$

$$\text{CFC-12}: \Delta T_s = 0.16 \times ((\text{CFC-12}) - (\text{CFC-12})_0) \tag{18}$$

The units of concentrations of CO_2, N_2O, O_3, CH_4, CFC-11, and CFC-12 are: ppmv, ppmv, ppbv, ppbv, ppbv, and ppbv, respectively.

CO_2 LEVELS AND EMISSIONS

On May 10, 2013, for the first time in a million years, the CO_2 level in the Earth's atmosphere passed 400 ppm (Gillis, 2013). Consequently, extreme heat waves, droughts, wildfires, melting glaciers and sea level rise in the future may be due to the accumulation of GHGs (principally CO_2 from the burning of fossil fuels) in the atmosphere. A mathematical model can be helpful in order to understand an introductory global CO_2 model that gives some key numbers, such as CO_2 concentration serving as a function of time for the calendar years 1850 (preindustrial) to 2100 (a modest projection into the future). A basic global warming component is added based on CO_2 buildup in the lower atmosphere while the primary focus is on carbon buildup in the atmosphere and oceans. The ordinary differential equations are carbon balances for seven reservoirs: Upper atmosphere, lower atmosphere, long-lived biota, short-lived biota, ocean upper layer, ocean deep layer and marine biosphere. As stated by the model developers, their intention is to provide a quantitative introduction to the CO_2 problem so that the model is educational and is not a state-of-the-art research model (McHugh et al., 2015). The model will be helpful as a starting point for more detailed study.

Another Climate Model has been developed by the UCAR. The model can be used to explore how the rate of CO_2 emissions affects the amount of CO_2 in the

Figure 5. Sample scenario: continuously rising CO_2 emissions
(developed by Lisa Gardiner and Randy Russell of the UCAR Center for Science Education. Climate model UCAR website)

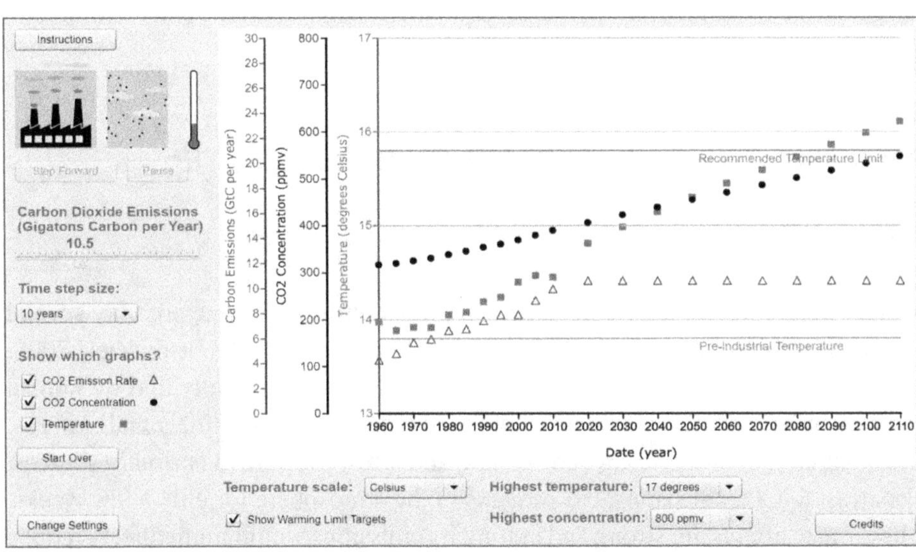

atmosphere and Earth's climate. Additionally, it can be used to explore scenarios for future climate using more detailed instructions. The IPCC uses a similar approach employed in this simple model when preparing their future climate projections based on much more sophisticated, supercomputer-based climate models. The model predicts what will happen to the atmosphere and climate over this century, depending on the amount of CO_2 (Gigatons of CO_2 per year) released into the atmosphere, as shown in Figure 5.

As shown in this model, average global temperature is determined by the atmospheric CO_2 concentration via GHG warming of the atmosphere. The impacts of the Earth's biosphere, land use changes, wind patterns, uptake of CO_2 by the oceans and other factors are not considered by this model. Many of these factors result in lower CO_2 in the atmosphere over time (Climate model UCAR website).

In this model, the starting values for concentration, emission rate, and temperature are right around actual values and in line with predictions scientists had made for the year 2000. In the relationship between atmospheric CO_2 concentration and temperature, the temperature rises about 3° C for each doubling of CO_2 concentration. So, if the concentration goes from 400 ppm to 800 ppm, the model expects the temperature to rise by 3°C. Details of the model are presented elsewhere (UCAR website).

Empirical GHG equivalence factors to measure CO_2 emissions from various emitters in different sectors have been developed over the last decade. Some of the values are presented as CO_2 equivalents (CO_2E) as noted on GWPs. These factors can help individuals as well as public and private sector organizations to understand GHG emissions from normally used units to report reductions into terms that are easier to conceptualize. These factors can be applied to various strategies on climate change. Table 5 summarizes the GHG equivalencies in different units (Butler and Montzka, 2016).

EMERGING TOPICS IN CLIMATE CHANGE DEBATE

Role of BC

BC has emerged as a major contributor to global climate change, possibly second only to CO_2 as the main driver of change (Ramanathan and Carmichael, 2008; Shandilya and Kumar, 2014). BC is also produced both naturally and by human activities because of the incomplete combustion of fossil fuels, bio-fuels, and biomass. Diesel engines, cook-stoves, wood burning and forest fires are the primary sources. In contrast to CO_2 emissions, BC remains in the atmosphere for only a few weeks, so its climate effects are strong and cutting its emissions would immediately reduce

Table 5. GHG equivalence factors

Source	Emission Factor
Electricity Reductions (kWh)	7.03×10^{-4} MT CO_2/kWh
Gallons of gasoline consumed	8.887×10^{-3} MT CO_2/gallon of gasoline
Passenger vehicles per year	4.73 MT CO_2E/vehicle/year
Miles driven by the average passenger vehicle	4.17×10^{-4} MT CO_2E/mile
Therms and Mcf of natural gas	0.005302 MT CO_2/therm 0.054717 MT CO_2/Mcf
Barrels of oil consumed	0.43 MT CO_2/barrel
Tanker trucks filled with gasoline	75.54 MT CO_2/tanker truck
Number of incandescent bulbs switched to light-emitting diode bulbs	2.82×10^{-2} MT CO_2/bulb replaced
Home electricity use	6.772 MT CO_2/home
Home energy use	9.47 MT CO_2/home/year
Number of tree seedlings grown for 10 years	0.039 MT CO_2/urban tree planted
Acres of the USA forests storing carbon for one year	-1.06 MT CO_2 sequestered annually by one acre of average United States forest.[1]
Acres of the USA forest preserved from conversion to cropland	-125.46 MT CO_2/acre/year (in the year of conversion).[2]
Propane cylinders used for home barbecues	0.024 MT CO_2/cylinder
Railcars of coal burned	187.78 MT CO_2/railcar
Pounds of coal burned	9.37×10^{-4}MT CO_2/pound of coal
Tons of waste recycled instead of land-filled	3.15 MT CO_2E/ton of waste recycled instead of land-filled
Number of garbage trucks of waste recycled instead of land-filled	22.06 MT CO_2E/garbage truck of waste recycled instead of land-filled
Coal-fired power plant emissions for one year	3,435,617.88 MT CO_2/power plant
Number of wind turbines installed	3,960 MT CO_2/wind turbine installed

(Butler and Montzka, 2016, and GHG Equivalencies Calculator from EPA website)

[1]Negative values indicate carbon sequestration.

[2]Negative values indicate CO_2 that is not emitted.

the rate of warming. Moreover, reduced exposure to BC provides public health co-benefits, especially in developing countries. BC particles (a carbonaceous aerosol, whereas an aerosol is a suspension of fine solid particles or liquid droplets within a gas, such as smoke, air pollution, smog, oceanic haze, and tear gas) strongly absorb sunlight and give soot its black color (Shandilya and Kumar, 2014; Omidvarborna et al., 2015, 2016). BC warms the climate in two ways (Center for Climate and Energy Solutions (C2ES) website):

- Suspended BC in the air absorbs sunlight and generates heat in the atmosphere, which warms the air and can affect regional cloud formation and precipitation patterns.
- Deposited BC on snow and ice absorbs sunlight, which generates heat that warms both the air above and the snow and ice below, resulting in melting.

Generally, the blacker the soot, the more of a warming agent it is (C2ES website). For example, fossil fuel and bio-fuel soot are blacker than that of soot from biomass burning from forest fires and wood fuel, which is generally more of a brownish color. Based on current information, the United States is responsible for about 6% of global BC emissions; while it has a history of making reductions to improve air quality, further improvements can be made. Most BC emissions come from the developing world: China and India together account for some 25-35% of emissions. Thus, controlling emissions of soot from fuel sources can effectively reduce atmospheric temperatures in the short term (Omidvarborna et al., 2015, 2016). Technologies capable of reducing global BC emissions are retrofitting diesel vehicles with filters to capture BC, fuel switching (e.g., from diesel to natural gas in buses), and replacing inefficient cook-stoves with cleaner alternatives (C2ES website). BC emissions reduction would have an immediate cooling effect on the Earth's climate, delaying temperature increases in the short run and helping reduce irreversible tipping points in the climate system. It would also reduce air pollution, resulting in fewer premature deaths and fewer missed work and school days (C2ES website).

Fine Particulates and Climate Change

Fine PM ($PM_{2.5}$, PM with an aerodynamic diameter of less than 2.5 μm) and climate change share a link (Shandilya et al., 2009; Shandilya and Kumar, 2010, 2011). Fine PM (as an aerosol) cools the climate (Pachauri and Reisinger, 2007; Shandilya et al., 2009; Shandilya and Khare, 2012) and releases from the same source as GHGs, so the mitigation of either $PM_{2.5}$ or GHG emissions will likely also affect the other (ApSimon et al., 2009).

Air pollution depends on weather, demonstrating the impact of climate change on air quality. For example, an increase in precipitation causes a lowering in all $PM_{2.5}$ components through scavenging (Shandilya et al., 2006; Pye et al., 2009). Thus, the uncertain sensitivity to climate change in $PM_{2.5}$ changes reflects the complexity of fine PM components on meteorological variables.

PM and SOA also affect light scattering and interact with terrestrial radiation, leading to climate-forcing effects and contributing to poor visibility (Aw and Kleeman, 2003; Fiore et al., 2015). It is essential that air quality management strategies account for possible climate-driven changes in pollutant levels as well as changes in the frequency or severity of air pollution episodes that result from climate change. The potential air quality benefits of emission control policies and associated reductions in precursor emissions such as NOx and VOCs may be offset by increasing temperatures or other climate change-induced meteorological changes. Steiner et al. (2006) found that although it was not included in their study, particulate chemistry is an important indicator of air quality (Sharma et al., 2003, 2004; Shandilya et al., 2011) and that the impact of future climate changes on the gas phase and particulate phase pollutants will likely be different.

Existing analyses utilize global models whose resolution is unable to provide an accurate picture of the response of PM and SOA to climate change in urban areas (Fiore et al., 2015). Racherla and Adams (2006) simulated a future climate scenario for the year 2050 and showed that some regions show increases in the concentration of fine PM species while others show decreases.

It has also been reported that isolated increases were caused primarily by increased secondary production and reduced regional-scale precipitation despite the global burden in the future climate decreasing by 2-18% because of increased wet deposition loss rates on the global scale. Even in high-resolution modeling studies of local effects on air quality, the effects of climate change on O_3 varies according to location across air basins (Aw and Kleeman, 2003; Steiner et al., 2006; Carreras-Sospedra et al., 2006; Millstein and Harley, 2009). Uncertainty and sensitivity to climate change parameters varies significantly based on location, local emissions, chemical species of interest, time of day, as well as a host of other variables.

Several other studies have been conducted for climate change. Aw and Kleeman (2003) reported that ammonium nitrate concentrations are most sensitive to temperature perturbations in areas that contain comparable molar concentrations of gas-phase ammonia and nitric acid. Conversely, ammonium nitrate particles exhibit little sensitivity to changes in temperature when the ratio of gas-phase ammonia to nitric acid is much greater or much less than unity (Aw and Kleeman, 2003). Dawson et al. (2007) also found significant decline in nitrate and ammonium particulates during summer in response to temperature increases, mostly due to the volatilization of ammonium nitrate.

Dawson et al. (2007) also found that PM concentrations rise with increasing concentration of water vapor. Moreover, they showed that increasing absolute humidity had the greatest impact on ammonium nitrate particulate concentrations due to the ammonia-nitric acid system shifting toward the aerosol phase.

Kleeman (2008) showed that the negative temperature effect on $PM_{2.5}$ concentrations is greater in magnitude than the positive humidity effect. In fact, increasing absolute humidity results in higher concentrations of ammonium nitrate aerosol due to the equilibrium of the ammonia-nitric acid system being shifted toward the aerosol phase (Dawson et al., 2007). This effect partially offsets some of the decrease in ammonium nitrate concentrations that results from increased volatilization at higher temperatures. Such higher temperatures result in decreased gas to particle partitioning in equilibrium partitioning models due to the effects of temperature on Henry's law coefficients and vapor pressure (Griffin et al., 2005). VOCs tend to favor the gas phase as temperatures increase, decreasing SOA yield. These effects are strongest during daytime and afternoon hours when base case temperatures are already high. Higher temperatures also result in decreased gas to particle partitioning in equilibrium partitioning models due to the effects of temperature on Henry's law coefficients and vapor pressure (Kleeman, 2008). PM and SOA levels generally increase slightly with increasing absolute humidity. Increasing humidity results in higher concentrations of ammonium nitrate aerosol due to the equilibrium of the ammonia-nitric acid system being shifted toward the aerosol phase (Dawson et al., 2007).

O_3 and Climate Change

Together, O_3 and PM (particularly $PM_{2.5}$) are of gravest concern for human health (Shandilya and Khare, 2012; Liu and Peng, 2018). O_3, PM, and SOA also affect light scattering and interact with terrestrial radiation, which lead to climate-forcing effects and contributes to poor visibility (Aw and Kleeman, 2003; Fiore et al., 2015). Several modeling studies have examined how climate change-related perturbations affect the O_3 air quality (Aw and Kleeman, 2003; Carreras-Sospedra et al., 2006; Jacob and Winner, 2009; Millstein and Harley, 2009).

Although O_3 production increases with climbing temperatures in polluted areas, several global studies report O_3 decreases in remote areas due to destruction of O_3 by water vapor at these higher temperatures (Millstein and Harley, 2009; Fiore et al., 2015). Even in high-resolution modeling studies of local effects on air quality, the effects of climate change on O_3 differ strongly by location across different air basins (Giorgi and Meleux, 2007; Millstein and Harley, 2009). In an early study, Ebi and McGregor (2009) predicted that surface O_3 background concentrations will reach at least 60 ppb by the years 2050–2060. Such O_3 increases are greatest in polluted areas with high anthropogenic emissions, while O_3 concentrations in remote and coastal areas are generally less impacted by temperature perturbations (Steiner et al., 2006; Dawson et al., 2007; Millstein and Harley, 2009).

This increase is partly due to surroundings. Steiner et al. (2006) and Millstein and Harley (2009) showed that O_3 concentrations increase due to increased atmospheric water vapor under high anthropogenic emissions conditions. They also showed slight O_3 decreases in some regions with lower NOx concentrations, particularly areas that are NOx limited. However, these concentrations are less sensitive to increases in the concentration of water vapor than to increases in temperature.

O_3 precursors such as NOx and VOCs are affected by increases in water vapor concentrations due to increased production of HOx from O_3 photolysis (Millstein and Harley, 2009). As a result, a decrease in NOx levels causes O_3 concentrations to increase in regions that are NOx-saturated and decrease in NOx limited regions. This suggests that air quality in polluted regions is more sensitive to changes in temperature and absolute humidity when they occur simultaneously. Increasing absolute humidity causes O_3 concentrations to increase in areas with high anthropogenic emissions (particularly NOx saturated regions) due to increased production of NOx from O_3 photolysis. These increases in O_3 that result from increases in the concentrations of water vapor are exacerbated when temperatures rise simultaneously due to the effect of temperature on kinetic rate constants.

The presence of greater precursor emissions of NOx and VOCs when using 2005 emissions influences the sensitivity of O_3 to simultaneous temperature and humidity perturbations. As a result, O_3 air quality in most locations shows an increasing sensitivity to temperature and humidity perturbations when the perturbations are performed simultaneously. In polluted regions, simultaneous increases to temperature and absolute humidity generally exhibit a compounding effect on increasing afternoon O_3 concentrations due to the combination of increased reaction rates and increased production of OH radical (Dawson et al., 2007). This is consistent with existing studies that find larger O_3 increases when temperatures are increased with no change in relative humidity than when temperatures are increased with any change in absolute humidity (Steiner et al., 2006; Dawson et al., 2007). Under conditions of lower anthropogenic emissions, the impact of humidity changes on O_3 air quality is reduced, even when temperature is perturbed simultaneously. Kleeman (2008) also indicated that the magnitude of the increase in O_3 is greater when temperature and absolute humidity are increased simultaneously (constant relative humidity), in agreement with the results presented here when using 2005 emissions.

Role of the Use of Cook-Stove in Developing Countries

In place of the stoves used throughout human history, today's modern kitchens use the standardized and specialized cooking devices. However, in many developing countries, traditional fuels (biomass) are being used to meet the household cooking needs. These primitive cook-stoves emit a significant amount of CO_2, CH_4 and other O_3 producing gases such as CO, as well as short-lived but very efficient sunlight-absorbing particles like BC and brown carbon inside the homes. Therefore, burning biomass fuels in traditional ways contributes to a buildup of GHGs (Venkataraman et al., 2010), as well as other climate forcers, such as BC (Ramanathan and Carmichael, 2008). Moreover, the high demand for local biomass energy exceeds the natural re-growth of local resources, which contributes to deforestation, reducing carbon uptake by forests.

In the last five years, as these concerns have arisen with the advent of climate change, it is apparent that the million tons of biomass fuels (including wood, charcoal, agricultural waste, and animal dung) burned annually in developing countries contribute to the buildup of a variety of GHGs in the atmosphere. The burning of residential solid fuel burning accounts for 25% of global BC emissions, about 84% of which are from households in developing countries, such as cook-stoves in South Asia. BC also disrupts the monsoon and accelerates melting of the glaciers in the Himalayan-Tibetan area. Further, water availability and food security are endangered for millions of people (Clean Cook stoves website).

Current studies on the importance of fine PM emphasized the importance of introducing clean-burning biomass cook-stoves and substituting traditional cook-stoves with those that use modern fuels to mitigate climate change and improve air quality simultaneously. Such innovations justify the development of a sophisticated program to deal with both the emissions and health issues resulting from cooking with open fires or traditional biomass cook-stoves.

CLIMATE CHANGE RISK ASSESSMENT

Climate change risk assessment analyzes climate change risks and opportunities every five years. These assessment studies aim to address the identified risks so as to deliver resilience to climate change on the ground. Figure 6 illustrates the major urgency categories used to assess individual climate risks and opportunities considered by the UK Government on climate change (2016). The figure clearly categorized priority tasks that should be considered in climate change risk assessment studies. Among them, “new and stronger policies or implementation activities in the next five years to reduce long-term vulnerability to climate change” and “research to fill major evidence gaps or reduce the uncertainty in the current understanding in order to assess the need for additional action” both have the highest priority.

FUTURE DIRECTIONS

This study has led to the conclusion that the frequency and severity of air pollution episodes depends strongly on both emissions and meteorological conditions. Implementing control strategies requires a detailed understanding of the chemistry-meteorology-climate interactions of both gas-phase and particulate pollutants.

Figure 6. Urgency categories used in the UK Climate Change Risk Assessment 2017 Evidence Report
(UK Climate Change Risk Assessment, 2016)

It may also be the case that the traditional modeling approach of using recent climate conditions is no longer appropriate to estimate future air quality (Steiner et al., 2006; Huang et al., 2017). As 21st century climate change becomes an increasingly pressing issue, modeling studies must incorporate future climate scenarios in order to understand climate change implications on air quality. Modeling studies must also consider how changes impact both biogenic and anthropogenic emissions under future climate scenarios. PM and SOA must be examined in detail to provide a comprehensive assessment of the impact of future climate and emission changes on air quality in addition to gas-phase pollutants such as NOx, SO_2, CO_2, and O_3. Additional modeling work for this is needed due to potential adverse effects on public health from rising O_3 and PM levels (Giorgi and Meleux, 2007). Air quality management strategies must also account for possible climate driven changes in pollutant levels with changes in the frequency or severity of air pollution episodes resulting from the climate change.

Other existing analyses use global models that provide an insufficient picture of the response of O_3, PM, and SOA to climate change in urban areas (Steiner et al., 2006; He et al., 2018). Few regional modeling studies have examined the impact of changes in future climate and emissions on PM and SOA concentrations. Overall, there is uncertainty regarding the dynamic response of PM and SOA to climate change-induced meteorological changes such as increasing temperatures, increasing absolute humidity, and changes in relative humidity and biogenic emissions associated with a rise in temperature (Liu et al., 2017). Sensitivity to climate change parameters varies based on location, local emissions, chemical species of interest, time of day, and a host of other variables. This, along with the highly intertwined nature of variables such as temperature, humidity, and biogenic emissions further stresses the need for further modeling studies to fill current knowledge gaps and confirm the studies that already exist. High-resolution regional air quality models with developed aerosol modules are necessary to study these effects on local air quality. Furthermore, because the effects of climate change-driven parameters on O_3, PM, and SOA vary by location within an air basin, a fine-scale grid with detailed topography and spatially resolved emissions is required. Temporal and spatial variations in the concentrations of gas phases, particulates, and aerosol chemical species need to be captured to better study the response of air quality to climate change in urban areas. So, due to the nonlinearity of air pollution dynamics and the effects of these perturbations, consideration of different combinations of the perturbations is essential to understand the sensitivity of pollutant concentrations to climate change.

CLOSING REMARKS

The use of energy in current decades is a key reason for GHG emissions. Thus, developing a sustainable energy program is an essential part of dealing with this issue. The most apparent actions needed are to cut down on our demand for electricity generation from fossil fuels, replace the less energy efficient appliance with the most efficient one, use fuel-efficient vehicles for our transportation, and recycle products. Accurate measuring of GHGs experienced at any specific location is a challenging scientific task, but estimates are continuing to improve as regional and/or local-scale models advance.

This Chapter reviewed the major components of the climate systems, processes and key interactions among them. A discussion on the parameters related to climate change and the role of air contaminants in climate change were investigated. Finally, characteristics of major GHGs were explored. Scientists are also looking at reducing the impacts of climate change because of reducing air pollutants under current clean air regulations. These steps will allow us to get closer to achieving greater sustainability for the planet.

ACKNOWLEDGMENT

The chapter uses the material available from: i) The web site http://www.eng.utoledo.edu/~akumar/IAP1/mainpage.htm maintained by Dr. Ashok Kumar; ii) A. Kumar and H. Omidvarborna, Role of Air Pollution in Climate Change and Global Warming: Present Global Scenario, Chapter 3.5, pp 123-134, Handbook of Environment Engineering, McGraw-Hill, 2018.

REFERENCES

A blanket around the Earth, Global Climate Change by NASA. (n.d.). Retrieved from https://climate.nasa.gov/causes/

Adaptation Sub-Committee. (2016). *UK Climate Change Risk Assessment 2017: synthesis report: priorities for the next five years*. Committee on Climate Change.

Apling, A. J., Blanchet, J. P., Charlson, R. J., Crommelynck, D., Grassl, H., Husson, N., & Rodhe, H. (1990). *Radiative Forcing of Climate. IPCC, Climate Change. The IPCC Scientific Assessment*. Cambridge, UK: Cambridge University Press.

ApSimon, H., Amann, M., Åström, S., & Oxley, T. (2009). Synergies in addressing air quality and climate change. *Climate Policy*, *9*(6), 669.

Aw, J., & Kleeman, M. J. (2003). Evaluating the first order effect of intraannual temperature variability on urban air pollution. *Journal of Geophysical Research, D, Atmospheres*, *108*(D12).

Bartók, B., Wild, M., Folini, D., Lüthi, D., Kotlarski, S., Schär, C., ... Imecs, Z. (2017). Projected changes in surface solar radiation in CMIP5 global climate models and in EURO-CORDEX regional climate models for Europe. *Climate Dynamics*, *49*(7-8), 2665–2683.

Blok, K., Afanador, A., van der Hoorn, I., Berg, T., van Vuuren, D., & PBL, O. E. (2017). *Sectoral Greenhouse Gas Emission Reduction Potentials in 2030*. Academic Press.

Bloomer, B. J., Stehr, J. W., Piety, C. A., Salawitch, R. J., & Dickerson, R. R. (2009). Observed relationships of ozone air pollution with temperature and emissions. *Geophysical Research Letters*, *36*(9).

Blunden, J., & Arndt, D. S. (2016). State of the Climate in 2015. *Bulletin of the American Meteorological Society, 97*(8), Si-S275.

Boucher, O. (2012). Comparison of physically-and economically-based CO2-equivalences for methane. *Earth System Dynamics, 3*(1), 49–61.

Boucher, O., Halloran, P. R., Burke, E. J., Doutriaux-Boucher, M., Jones, C. D., Lowe, J., ... Wu, P. (2012). Reversibility in an Earth System model in response to CO2 concentration changes. *Environmental Research Letters, 7*(2), 024013.

Boucher, O., Randall, D., Artaxo, P., Bretherton, C., Feingold, G., Forster, P., & Rasch, P. (2013). Clouds and aerosols. In *Climate change 2013: the physical science basis. Contribution of Working Group I to the Fifth Assessment Report of the Intergovernmental Panel on Climate Change* (pp. 571–657). Cambridge University Press.

Brasseur, G. P., Schultz, M., Granier, C., Saunois, M., Diehl, T., Botzet, M., ... Walters, S. (2006). Impact of climate change on the future chemical composition of the global troposphere. *Journal of Climate, 19*(16), 3932–3951.

Butler, J. & Montzka, S.A. (2016). *The NOAA annual greenhouse gas index (AGGI).* Earth System Research Laboratory, Global Monitoring Division Website.

Carreras-Sospedra, M., Dabdub, D., Rodriguez, M., & Brouwer, J. (2006). Air quality modeling in the South Coast Air Basin of California: What do the numbers really mean? *Journal of the Air & Waste Management Association, 56*(8), 1184–1195. PMID:16933651

Chang, H. H., Hao, H., & Sarnat, S. E. (2014). A statistical modeling framework for projecting future ambient ozone and its health impact due to climate change. *Atmospheric Environment, 89*, 290–297. PMID:24764746

Change, I. C. (1996). *The science of climate change*. Working Group I to the Second Assessment Report of the IPCC.

Chen, E., Ng, Y. K., Tan, Y. F., & Toh, J. S. Y. (2017). Environmentally Responsible Happy Nation Index: Refinements and 2015 Rankings. *Social Indicators Research, 134*(1), 39–56.

Christensen, J. H., Hewitson, B., Busuioc, A., Chen, A., Gao, X., Held, R., & Magaña Rueda, V. (2007). Regional climate projections. In *Climate Change, 2007: The Physical Science Basis. Contribution of Working group I to the Fourth Assessment Report of the Intergovernmental Panel on Climate Change* (pp. 847–940). Cambridge University Press.

Climate Change Indicators. (n.d.). *Greenhouse Gases.* US Environmental Protection Agency. Retrieved from https://www.epa.gov/climate-indicators/greenhouse-gases

Climate Change Indicators: Global Greenhouse Gas Emissions, US Environmental Protection Agency. (n.d.). Retrieved from https://www.epa.gov/climate-indicators/climate-change-indicators-global-greenhouse-gas-emissions

Climate Change Measurements - Climate Measurement. (n.d.). *A review of rainfall and temperature measurement standards in Guyana.* National Institute of Standards and Technology. Retrieved from https://www.nist.gov/sites/default/files/documents/iaao/OdessaShako.pdf

Cooking and Climate Change, Global Alliance for Clean Cookstoves. (n.d.). Accessed on March 12, 2018 from http://carbonfinanceforcookstoves.org/about-cookstoves/cooking-and-climate-change/

Cox, W. M., & Chu, S. H. (1996). Assessment of interannual ozone variation in urban areas from a climatological perspective. *Atmospheric Environment, 30*(14), 2615–2625.

Dawson, J. P., Adams, P. J., & Pandis, S. N. (2007). Sensitivity of PM 2.5 to climate in the Eastern US: A modeling case study. *Atmospheric Chemistry and Physics, 7*(16), 4295–4309.

Dawson, J. P., Bloomer, B. J., Winner, D. A., & Weaver, C. P. (2014). Understanding the meteorological drivers of US particulate matter concentrations in a changing climate. *Bulletin of the American Meteorological Society, 95*(4), 521–532.

De Jong, M. F., Drijfhout, S. S., Hazeleger, W., Van Aken, H. M., & Severijns, C. A. (2009). Simulations of hydrographic properties in the northwestern North Atlantic Ocean in coupled climate models. *Journal of Climate, 22*(7), 1767–1786.

Details of the Math behind the Model. (n.d.). University Corporation for Atmospheric Research (UCAR). Retrieved from https://scied.ucar.edu/activity/very-very-simple-climate-model-activity)

Direct Emissions by GHG. (2015). US Environmental Protection Agency. Retrieved from https://www.epa.gov/ghgreporting/ghgrp-reported-data

Ebi, K., & McGregor, G. (2009). Climate change, tropospheric ozone and particulate matter, and health impacts. *Ciencia & Saude Coletiva, 14*(6), 2281–2293. PMID:20069198

Eyring, V., Butchart, N., Waugh, D. W., Akiyoshi, H., Austin, J., Bekki, S., ... Cordero, E. (2006). Assessment of temperature, trace species, and ozone in chemistry climate model simulations of the recent past. *Journal of Geophysical Research, D, Atmospheres, 111*(D22).

Final Environmental Analysis, Revised Proposed Short-Lived Climate Pollutant Reduction Strategy, Air Resources Board, March. (2017). Retrieved from https://www.arb.ca.gov/cc/shortlived/meetings/03142017/appendixe.pdf

Fiore, A. M., Naik, V., & Leibensperger, E. M. (2015). Air quality and climate connections. *Journal of the Air & Waste Management Association*, *65*(6), 645–685. PMID:25976481

Flato, G., Marotzke, J., Abiodun, B., Braconnot, P., Chou, S. C., Collins, W. J., ... Forest, C. (2013). Evaluation of climate models. In: climate change 2013: the physical science basis. Contribution of working group I to the fifth assessment report of the intergovernmental panel on climate change. *Climatic Change*, *5*, 741–866.

Forkel, R., & Knoche, R. (2006). Regional climate change and its impact on photooxidant concentrations in southern Germany: Simulations with a coupled regional climate chemistry model. *Journal of Geophysical Research, D, Atmospheres*, *111*(D12).

Frich, P., Alexander, L. V., Della-Marta, P. M., Gleason, B., Haylock, M., Tank, A. K., & Peterson, T. (2002). Observed coherent changes in climatic extremes during the second half of the twentieth century. *Climate Research*, *19*(3), 193–212.

Garfinkel, C. I., Waugh, D. W., Oman, L. D., Wang, L., & Hurwitz, M. M. (2013). Temperature trends in the tropical upper troposphere and lower stratosphere: Connections with sea surface temperatures and implications for water vapor and ozone. *Journal of Geophysical Research, D, Atmospheres*, *118*(17), 9658–9672.

Gaudel, A., Cooper, O. R., Ancellet, G., Barret, B., Boynard, A., Burrows, J. P., ... Doniki, S. (2018). Tropospheric Ozone Assessment Report: Present-day distribution and trends of tropospheric ozone relevant to climate and global atmospheric chemistry model evaluation. *Elem Sci Anth*, *6*(1).

Gillis, J. (2013). Heat-trapping gas passes milestone, raising fears. *New York Times*, *11*.

Giorgi, F. (2006, December). Regional climate modeling: Status and perspectives. Journal de Physique IV (Proceedings), 139, 101-118.

Giorgi, F., & Meleux, F. (2007). Modelling the regional effects of climate change on air quality. *Comptes Rendus Geoscience*, *339*(11-12), 721–733.

Greenhouse Gases Equivalencies Calculator - Calculations and References. (n.d.). US Environmental Protection Agency. Accessed on July 12, 2018 https://www.epa.gov/energy/greenhouse-gases-equivalencies-calculator-calculations-and-references

Gregory, J. M., Andrews, T., Good, P., Mauritsen, T., & Forster, P. M. (2016). Small global-mean cooling due to volcanic radiative forcing. *Climate Dynamics*, *47*(12), 3979–3991.

Griffin, R. J., Dabdub, D., & Seinfeld, J. H. (2005). Development and initial evaluation of a dynamic species resolved model for gas phase chemistry and size resolved gas/particle partitioning associated with secondary organic aerosol formation. *Journal of Geophysical Research, D, Atmospheres*, *110*(D5).

Guilyardi, E., Wittenberg, A., Fedorov, A., Collins, M., Wang, C., Capotondi, A., ... Stockdale, T. (2009). Understanding El Niño in ocean–atmosphere general circulation models: Progress and challenges. *Bulletin of the American Meteorological Society*, *90*(3), 325–340.

Hare, J. A., Wuenschel, M. J., & Kimball, M. E. (2012). Projecting range limits with coupled thermal tolerance-climate change models: An example based on gray snapper (Lutjanus griseus) along the US east coast. *PLoS One*, *7*(12), e52294. PMID:23284974

Hauglustaine, D. A., Lathiere, J., Szopa, S., & Folberth, G. A. (2005). Future tropospheric ozone simulated with a climate chemistry biosphere model. *Geophysical Research Letters*, *32*(24).

He, H., Liang, X. Z., & Wuebbles, D. J. (2018). Effects of emissions change, climate change and long-range transport on regional modeling of future US particulate matter pollution and speciation. *Atmospheric Environment*, *179*, 166–176.

Held, I. M., & Soden, B. J. (2000). Water vapor feedback and global warming. *Annual Review of Energy and the Environment*, *25*(1), 441–475.

Henze, D., & Lee, H. (2012). GEOS-Chem Adjoint User's Guide (gcadj v32). Academic Press.

Hoffman, D. J., Butler, J. H., Dlugokencky, E. J., Elkins, J. W., Masarie, K., Montzka, S. A., & Tans, P. (2006). The role of carbon dioxide in climate forcing from 1979 to 2004: Introduction of the Annual Greenhouse Gas Index. *Tellus. Series B, Chemical and Physical Meteorology*, *58*(5), 614–619.

Holloway, T., Spak, S. N., Barker, D., Bretl, M., Moberg, C., Hayhoe, K., ... Wuebbles, D. (2008). Change in ozone air pollution over Chicago associated with global climate change. *Journal of Geophysical Research, D, Atmospheres*, *113*(D22).

Horne, J. R., & Dabdub, D. (2017). Impact of global climate change on ozone, particulate matter, and secondary organic aerosol concentrations in California: A model perturbation analysis. *Atmospheric Environment*, *153*, 1–17.

Huang, J., McQueen, J., Wilczak, J., Djalalova, I., Stajner, I., Shafran, P., ... Huang, H. C. (2017). Improving NOAA NAQFC PM2. 5 Predictions with a Bias Correction Approach. *Weather and Forecasting*, *32*(2), 407–421.

Hurrell, J. W., Hack, J. J., Shea, D., Caron, J. M., & Rosinski, J. (2008). A new sea surface temperature and sea ice boundary dataset for the Community Atmosphere Model. *Journal of Climate*, *21*(19), 5145–5153.

Jacob, D. J., & Winner, D. A. (2009). Effect of climate change on air quality. *Atmospheric Environment*, *43*(1), 51–63.

Kahn Ribeiro, S., Kobayashi, S., Beuthe, M., Gasca, J., Greene, D., Lee, D. S., ... Wit, R. (2007). *Transport and its infrastructure. Climate change 2007: mitigation. contribution of working group III to the fourth assessment report of the intergovernmental panel on climate change*. Cambridge, UK: Cambridge University Press.

Kleeman, M. J. (2008). A preliminary assessment of the sensitivity of air quality in California to global change. *Climatic Change*, *87*(1), 273–292.

Kopacz, M., Jacob, D. J., Fisher, J. A., Logan, J. A., Zhang, L., Megretskaia, I. A., ... Buchwitz, M. (2010). Global estimates of CO sources with high resolution by adjoint inversion of multiple satellite datasets (MOPITT, AIRS, SCIAMACHY, TES). *Atmospheric Chemistry and Physics*, *10*(3), 855–876.

Krol, M. S. B. M. A. P. W. F. P., Houweling, S., Bregman, B., Broek, M. V., Segers, A., Velthoven, P. V., ... Bergamaschi, P. (2005). The two-way nested global chemistry-transport zoom model TM5: Algorithm and applications. *Atmospheric Chemistry and Physics*, *5*(2), 417–432.

Kumar, A., & Omidvarborna, H. (2018a). Introduction to Air Pollution. In Handbook of Environmental Engineering. Academic Press.

Kumar, A., & Omidvarborna, H. (2018b). Technology for Air Pollution Control. In Handbook of Environmental Engineering. Academic Press.

Kumar, A., & Omidvarborna, H. (2018c). Role of Air Pollution in Climate Change and Global Warming: Present Global Scenario. In Handbook of Environmental Engineering. Academic Press.

Kunkel, K. E., Andsager, K., & Easterling, D. R. (1999). Long-term trends in extreme precipitation events over the conterminous United States and Canada. *Journal of Climate*, *12*(8), 2515–2527.

Lacressonnière, G., Watson, L., Gauss, M., Engardt, M., Andersson, C., Beekmann, M., ... Nyiri, A. (2017). Particulate matter air pollution in Europe in a+ 2° C warming world. *Atmospheric Environment*, *154*, 129–140.

Lamarque, J. F., Shindell, D. T., Josse, B., Young, P. J., Cionni, I., Eyring, V., ... Dalsoren, S. (2013). The Atmospheric Chemistry and Climate Model Intercomparison Project (ACCMIP): Overview and description of models, simulations and climate diagnostics. *Geoscientific Model Development*, *6*, 179–206.

Lin, J. T., Patten, K. O., Hayhoe, K., Liang, X. Z., & Wuebbles, D. J. (2008). Effects of future climate and biogenic emissions changes on surface ozone over the United States and China. *Journal of Applied Meteorology and Climatology*, *47*(7), 1888–1909.

Liu, J. C., & Peng, R. D. (2018). Health effect of mixtures of ozone, nitrogen dioxide, and fine particulates in 85 US counties. *Air Quality, Atmosphere & Health*, 1–14.

Liu, S., Bond-Lamberty, B., Boysen, L. R., Ford, J. D., Fox, A., Gallo, K., ... Loveland, T. R. (2017). Grand challenges in understanding the interplay of climate and land changes. *Earth Interactions*, *21*(2), 1–43.

Localized Pollution Potentially Plays Large Role in Future Climate Change. (2008). National Oceanic and Atmospheric Administration (NOAA). Retrieved from http://www.noaanews.noaa.gov/stories2008/20080904_pollution.html

Luhar, A. K., Galbally, I. E., Woodhouse, M. T., & Thatcher, M. (2017). An improved parameterisation of ozone dry deposition to the ocean and its impact in a global climate-chemistry model. *Atmospheric Chemistry and Physics*, *17*(5), 3749.

Marland, G., & Rotty, R. M. (1984). Carbon dioxide emissions from fossil fuels: A procedure for estimation and results for 1950–1982. *Tellus. Series B, Chemical and Physical Meteorology*, *36*(4), 232–261.

Matthews, H. D., Gillett, N. P., Stott, P. A., & Zickfeld, K. (2009). The proportionality of global warming to cumulative carbon emissions. *Nature*, *459*(7248), 829. PMID:19516338

Matyssek, R., Clarke, N., Cudlin, P., Mikkelsen, T. N., Tuovinen, J. P., Wieser, G., & Paoletti, E. (2013). Climate Change, Air Pollution and Global Challenges: Chapter 1. In Climate Change, Air Pollution and Global Challenges: Understanding and Perspectives from Forest Research (Vol. 13). Elsevier Inc.

McHugh, A. J., Griffiths, G. W., & Schiesser, W. E. (2015). *An Introductory Global CO2 Model:(with Companion Media Pack)*. World Scientific Publishing Company.

Meleux, F., Solmon, F., & Giorgi, F. (2007). Increase in summer European ozone amounts due to climate change. *Atmospheric Environment*, *41*(35), 7577–7587.

Millstein, D. E., & Harley, R. A. (2009). Impact of climate change on photochemical air pollution in Southern California. *Atmospheric Chemistry and Physics*, *9*(11), 3745–3754.

Molod, A., Takacs, L., Suarez, M., & Bacmeister, J. (2015). Development of the GEOS-5 atmospheric general circulation model: Evolution from MERRA to MERRA2. *Geoscientific Model Development*, *8*(5), 1339.

Monks, P. S., Archibald, A. T., Colette, A., Cooper, O., Coyle, M., Derwent, R., ... Stevenson, D. S. (2015). Tropospheric ozone and its precursors from the urban to the global scale from air quality to short-lived climate forcer. *Atmospheric Chemistry and Physics*, *15*(15), 8889–8973.

Morgenstern, O., Hegglin, M. I., Rozanov, E., O'Connor, F. M., Abraham, N. L., Akiyoshi, H., ... Deushi, M. (2017). Review of the global models used within phase 1 of the Chemistry-Climate Model Initiative (CCMI). *Geoscientific Model Development*, *10*(2), 639.

Nabat, P., Somot, S., Mallet, M., Sevault, F., Chiacchio, M., & Wild, M. (2015). Direct and semi-direct aerosol radiative effect on the Mediterranean climate variability using a coupled regional climate system model. *Climate Dynamics*, *44*(3-4), 1127–1155.

National Drought Mitigation Center. (2016). *Maps and data*. Retrieved from http://droughtmonitor.unl.edu/MapsAndData.aspx

Omidvarborna, H. (2017). *Combustion chemistry of biodiesel for the use in urban transport buses: experiment and modeling* (PhD Thesis). University of Toledo.

Omidvarborna, H., Kumar, A., & Kim, D. S. (2015). Recent studies on soot modeling for diesel combustion. *Renewable & Sustainable Energy Reviews*, *48*, 635–647.

Omidvarborna, H., Kumar, A., & Kim, D. S. (2016). Variation of diesel soot characteristics by different types and blends of biodiesel in a laboratory combustion chamber. *The Science of the Total Environment*, *544*, 450–459. PMID:26657390

Omidvarborna, H., Kumar, A., Kim, D. S., & Shandilya, K. K. (2014). Analysis of particulate matter from the exhaust of biodiesel transit buses under idling conditions. In Air Waste Manage. Assoc. Annu. Meeting Conf. Exhibition (Vol. 106, No. 4, pp. 2712-2725). Academic Press.

Pachauri, R. K., & Reisinger, A. (2007). *IPCC fourth assessment report*. Geneva: IPCC.

Pan, L. L., Bowman, K. P., Atlas, E. L., Wofsy, S. C., Zhang, F., Bresch, J. F., ... Cooper, W. A. (2010). The stratosphere–troposphere analyses of regional transport 2008 experiment. *Bulletin of the American Meteorological Society*, *91*(3), 327–342.

Pawson, S., Stolarski, R. S., Douglass, A. R., Newman, P. A., Nielsen, J. E., Frith, S. M., & Gupta, M. L. (2008). Goddard Earth Observing System chemistry climate model simulations of stratospheric ozone temperature coupling between 1950 and 2005. *Journal of Geophysical Research, D, Atmospheres*, *113*(D12).

Pielke, R. A., & Wilby, R. L. (2012). Regional climate downscaling: What's the point? *Eos (Washington, D.C.)*, *93*(5), 52–53.

Pye, H. O. T., Liao, H., Wu, S., Mickley, L. J., Jacob, D. J., Henze, D. K., & Seinfeld, J. H. (2009). Effect of changes in climate and emissions on future sulfate nitrate ammonium aerosol levels in the United States. *Journal of Geophysical Research, D, Atmospheres*, *114*(D1).

Racherla, P. N., & Adams, P. J. (2006). Sensitivity of global tropospheric ozone and fine particulate matter concentrations to climate change. *Journal of Geophysical Research, D, Atmospheres*, *111*(D24).

Ramanathan, V. (1977). Interactions between ice-albedo, lapse-rate and cloud-top feedbacks: An analysis of the nonlinear response of a GCM climate model. *Journal of the Atmospheric Sciences*, *34*(12), 1885–1897.

Ramanathan, V., & Carmichael, G. (2008). Global and regional climate changes due to black carbon. *Nature Geoscience*, *1*(4), 221.

Ramaswamy, V., Boucher, O., Haigh, J., Hauglustine, D., Haywood, J., Myhre, G., ... Solomon, S. (2001). Radiative forcing of climate. *Climatic Change*, 349.

Rasmussen, D. J., Fiore, A. M., Naik, V., Horowitz, L. W., McGinnis, S. J., & Schultz, M. G. (2012). Surface ozone-temperature relationships in the eastern US: A monthly climatology for evaluating chemistry-climate models. *Atmospheric Environment*, *47*, 142–153.

Reducing Short-Lived Climate Pollutants in California. (n.d.). Air Resources Board, California Environmental Agency. Retrieved from https://www.arb.ca.gov/cc/shortlived/shortlived.htm

Rosenzweig, C., Iglesias, A., Yang, X. B., Epstein, P. R., & Chivian, E. (2001). Climate change and extreme weather events; implications for food production, plant diseases, and pests. *Global Change and Human Health*, *2*(2), 90–104.

Rummukainen, M. (2010). State of the art with Regional Climate Models. *Wiley Interdisciplinary Reviews: Climate Change*, *1*(1), 82–96.

Schreiner, C., Gonyo, E., Shandilya, K. K., Buonocore, J., Zhou, X., Salamero, G., ... Ferro, A. (2006). Children's Exposure to Diesel Exhaust at a Rural Elementary School. *Epidemiology (Cambridge, Mass.)*, *17*, S146–S147.

Schulz, M., Chin, M., & Kinne, S. (2009). The Aerosol Model Comparison Project, Aerocom Phase II: Clearing up Diversity. *IGAC Newslett.*, *41*. Retrieved from http://aerocom.met.no/pdfs/May_2009_IGAC_41.pdf

Scinocca, J. F., Kharin, V. V., Jiao, Y., Qian, M. W., Lazare, M., Solheim, L., ... Dugas, B. (2016). Coordinated global and regional climate modeling. *Journal of Climate*, *29*(1), 17–35.

Seinfeld, J. H., & Pandis, S. N. (2016). *Atmospheric chemistry and physics: from air pollution to climate change*. John Wiley & Sons.

Shandilya, K., Gupta, A. B., & Khare, M. (2011). Formation of atmospheric nitrate under high particulate matter concentration. *World Review of Science, Technology and Sustainable Development*, *8*(2-4), 148–165.

Shandilya, K., & Kumar, A. (2014). Carbon speciation of exhaust particulate matter of public transit buses running on alternative fuels. *Fuel*, *115*, 678–684.

Shandilya, K. K. (2012). *Characterization, speciation, and source apportionment of particles inside and from the exhaust of public transit buses fueled with alternative fuels* (PhD Thesis). University of Toledo.

Shandilya, K. K., & Khare, M. (2012). Particulate matter: Sources, emission rates and health effects. *Advances in Environmental Research*, *23*, 265–322.

Shandilya, K. K., Khare, M., & Gupta, A. B. (2006). Suspended particulate matter concentrations in commercial east Delhi and south Delhi, India. *Indian Journal of Environmental Protection*, *26*(8), 705–717.

Shandilya, K. K., Khare, M., & Gupta, A. B. (2009). Defining aerosols by physical and chemical characteristics. *Indian Journal of Air Pollution Control*, *9*(1), 107-126.

Shandilya, K. K., & Kumar, A. (2010). Morphology of single inhalable particle inside public transit biodiesel fueled bus. *Journal of Environmental Sciences (China)*, *22*(2), 263–270. PMID:20397416

Shandilya, K. K., & Kumar, A. (2011). Physical characterization of fine particulate matter inside the public transit buses fueled by biodiesel in Toledo, Ohio. *Journal of Hazardous Materials*, *190*(1-3), 508–514. PMID:21497439

Shandilya, K. K., & Sitaram, P. (2018). On the Development of a Sustainable Stormwater Management System. *Proceedings for American Society of Engineering Education - North Central Section Spring Conference 2018.*

Sharma, M., Kiran, Y. N. V. M., & Shandilya, K. K. (2003). Investigations into formation of atmospheric sulfate under high PM10 concentration. *Atmospheric Environment*, *37*(14), 2005–2013.

Sharma, M., Kiran, Y. N. V. M., & Shandilya, K. K. (2004). Subset statistical modelling of atmospheric sulfate. *Clean Air and Environmental Quality*, *38*(4), 37.

Sheng, J. X., Weisenstein, D. K., Luo, B. P., Rozanov, E., Stenke, A., Anet, J., ... Peter, T. (2015). Global atmospheric sulfur budget under volcanically quiescent conditions: Aerosol chemistry climate model predictions and validation. *Journal of Geophysical Research, D, Atmospheres*, *120*(1), 256–276.

Sillman, S. (1999). The relation between ozone, NOx and hydrocarbons in urban and polluted rural environments. *Atmospheric Environment*, *33*(12), 1821–1845.

Slater, L. (2016). update and expansion to data originally published in: Mallakpour, I., G. Villarini. 2015. The changing nature of flooding across the central United States. *Nature Climate Change*, *5*, 250–254.

Solomon, S., Qin, D., Manning, M., Averyt, K., & Marquis, M. (Eds.). (2007). *Climate change 2007-the physical science basis: Working group I contribution to the fourth assessment report of the IPCC* (Vol. 4). Cambridge University Press.

Steiner, A. L., Tonse, S., Cohen, R. C., Goldstein, A. H., & Harley, R. A. (2006). Influence of future climate and emissions on regional air quality in California. *Journal of Geophysical Research, D, Atmospheres*, *111*(D18).

Stevenson, D. S., Young, P. J., Naik, V., Lamarque, J. F., Shindell, D. T., Voulgarakis, A., ... Folberth, G. A. (2013). Tropospheric ozone changes, radiative forcing and attribution to emissions in the Atmospheric Chemistry and Climate Model Intercomparison Project (ACCMIP). *Atmospheric Chemistry and Physics*, *13*(6), 3063–3085.

Tai, A. P., Mickley, L. J., & Jacob, D. J. (2010). Correlations between fine particulate matter (PM2. 5) and meteorological variables in the United States: Implications for the sensitivity of PM2. 5 to climate change. *Atmospheric Environment*, *44*(32), 3976–3984.

Tai, A. P., Mickley, L. J., Jacob, D. J., Leibensperger, E. M., Zhang, L., Fisher, J. A., & Pye, H. O. T. (2012). Meteorological modes of variability for fine particulate matter (PM 2.5) air quality in the United States: Implications for PM 2.5 sensitivity to climate change. *Atmospheric Chemistry and Physics*, *12*(6), 3131–3145.

The Effect of Climate Change on Water Resources and Programs, US Environmental Protection Agency. (n.d.). Retrieved from https://cfpub.epa.gov/watertrain/moduleFrame.cfm?parent_object_id=2414#

The Very, Very Simple Climate Model, University Corporation for Atmospheric Research (UCAR). (n.d.). Retrieved from https://scied.ucar.edu/simple-climate-model

Tsigaridis, K., & Kanakidou, M. (2018). The Present and Future of Secondary Organic Aerosol Direct Forcing on Climate. *Current Climate Change Reports*, 1–15.

Unger, N., Zhao, Y., & Dang, H. (2013). Mid-21st century chemical forcing of climate by the civil aviation sector. *Geophysical Research Letters*, *40*(3), 641–645.

USEPA. (2018a). *Climate Impacts on Agriculture and Food Supply.* Accessed on October 10, https://19january2017snapshot.epa.gov/climate-impacts/climate-impacts-agriculture-and-food-supply_.html

USEPA. (2018b). *Sources of Greenhouse Gas Emissions.* Accessed on October 10, https://www.epa.gov/ghgemissions/sources-greenhouse-gas-emissions#t1fn1

Van Aalst, M. K. (2006). The impacts of climate change on the risk of natural disasters. *Disasters*, *30*(1), 5–18. PMID:16512858

Van Vuuren, D. P., Meinshausen, M., Plattner, G. K., Joos, F., Strassmann, K. M., Smith, S. J., ... Den Elzen, M. G. J. (2008). Temperature increase of 21st century mitigation scenarios. *Proceedings of the National Academy of Sciences of the United States of America*, *105*(40), 15258–15262. PMID:18838680

Venkataraman, C., Sagar, A. D., Habib, G., Lam, N., & Smith, K. R. (2010). The Indian national initiative for advanced biomass cookstoves: The benefits of clean combustion. *Energy for Sustainable Development*, *14*(2), 63–72.

Von Schneidemesser, E., Monks, P. S., Allan, J. D., Bruhwiler, L., Forster, P., Fowler, D., ... Sindelarova, K. (2015). Chemistry and the linkages between air quality and climate change. *Chemical Reviews*, *115*(10), 3856–3897. PMID:25926133

Wang, Y., Shen, L., Wu, S., Mickley, L., He, J., & Hao, J. (2013). Sensitivity of surface ozone over China to 2000–2050 global changes of climate and emissions. *Atmospheric Environment, 75*, 374–382.

What Is Black Carbon? Center for Climate and Energy Solutions. (2010). Retrieved from https://www.c2es.org/site/assets/uploads/2010/04/what-is-black-carbon.pdf

Willems, P., Arnbjerg-Nielsen, K., Olsson, J., & Nguyen, V. T. V. (2012). Climate change impact assessment on urban rainfall extremes and urban drainage: Methods and shortcomings. *Atmospheric Research, 103*, 106–118.

Wittneben, B. B., & Kiyar, D. (2009). Climate change basics for managers. *Management Decision, 47*(7), 1122–1132.

Wu, S., Mickley, L. J., Jacob, D. J., Rind, D., & Streets, D. G. (2008a). Effects of 2000–2050 changes in climate and emissions on global tropospheric ozone and the policy relevant background surface ozone in the United States. *Journal of Geophysical Research, D, Atmospheres, 113*(D18).

Wu, S., Mickley, L. J., Leibensperger, E. M., Jacob, D. J., Rind, D., & Streets, D. G. (2008b). Effects of 2000–2050 global change on ozone air quality in the United States. *Journal of Geophysical Research, D, Atmospheres, 113*(D6).

Yamashita, Y., Sakamoto, K., Akiyoshi, H., Takahashi, M., Nagashima, T., & Zhou, L. B. (2010). Ozone and temperature response of a chemistry climate model to the solar cycle and sea surface temperature. *Journal of Geophysical Research, D, Atmospheres, 115*(D3).

Young, P. J., Naik, V., Fiore, A. M., Gaudel, A., Guo, J., Lin, M. Y., ... Tilmes, S. (2018). Tropospheric Ozone Assessment Report: Assessment of global-scale model performance for global and regional ozone distributions, variability, and trends. *Elem Sci Anth, 6*(1).

Yu, K., Keller, C. A., Jacob, D. J., Molod, A. M., Eastham, S. D., & Long, M. S. (2018). Errors and improvements in the use of archived meteorological data for chemical transport modeling: An analysis using GEOS-Chem v11-01 driven by GEOS-5 meteorology. *Geoscientific Model Development, 11*(1), 305. PMID:30420911

APPENDIX

Abbreviations

BC: Black Carbon
CCM: Chemistry-Climate Model
CFC: Chlorofluorocarbons
CH_4: Methane
CIT: The acronym CIT was derived from the two organizations responsible for the development of the model, the California Institute of Technology and the Carnegie Institute of Technology at Carnegie Mellon University.
CO: Carbon Monoxide
CO_2: Carbon Dioxide
CTM: Chemical Transport Model
ERF: Effective Radiative Forcing
GCM: General Circulation Model
GCTM: Global Chemistry-Transport Model
GMST: Global Mean Surface Temperature
GHG: Greenhouse Gas
GWP: Global Warming Potential
HFC: Hydro-Fluorocarbon
IPCC: Intergovernmental Panel on Climate Change
kWh: Kilowatt-Hours
Mcf: A Thousand Cubic Feet
MJ: Mega Joule
MT: Metric Ton
N_2O: Nitrous Oxide
NMVOC: Non-Methane Volatile Organic Compound
NOx: Nitrogen Oxides
O_3: Ozone
OC: Organic Carbon
OH: Hydroxyl
PAN: Per-Oxy-Acetyl Nitrate
PFC: Per-Fluorinated Compound
PM: Particulate Matter
PPB: Parts per Billion
PPM: Parts per Million
RCM: Regional Climate Model
RF: Radiative Forcing

RTM: Radiative Transfer Model
SF_6: Sulfur Hexafluoride
SO_2: Sulfur Dioxide
SOA: Secondary Organic Aerosols
UCAR: University Corporation for Atmospheric Research
VOC: Volatile Organic Compound

Chapter 2
Characterization of Magnetic and Non-Magnetic Components From a Low Carbon Fly Ash:
A Solid Air Pollutant

Ruma Rano
National Institute of Technology Silchar, India

ABSTRACT

Low carbon coal ash—a solid air pollutant from super thermal power plant using pulverized fuel combustor—has been characterized in respect of its physico-chemical, mineralogical, and morphological features. Size-classified fractions with their magnetic and non-magnetic components have also been characterized. Low loss on ignition and particle size distribution profile shows fly ash has high utility. The magic number, 10 µm, is attained by greater % of particles. The particles with diameter 50 µm occupy population density of 93%. SEM-EDS reveals that particles are mostly globular with high surface enrichment of Al/Si indicating that it will act as un-reactive inert fillers. In magnetic components, various types of Fe bearing phase are present with less porous and more regular shape than non-magnetic components. Finer fractions have high content of magnetite which is expected to help in coal beneficiation. The finer non-magnetic fraction is a source of alumino-siliceous material for synthesizing a novel solid acid or base catalyst for catalyzing industrially important organic reactions.

DOI: 10.4018/978-1-5225-7289-3.ch002

INTRODUCTION

Coal – based power plants in India account for about 67% of the total electricity production. Eastern region of the country generates the highest (84%) of the power by coal combustion and in tune, results in huge amounts of fly ash as waste material (Guttikunda and Jawahar,2001). Nearly, 200 million tons of fly ash are produced annually, increasing continuously and estimated to have reached 500 million tons by 2022 (Chatterjee, 2014).

Fly ash is recognized as an environmental pollutant, contains toxic and heavy metals. The ponds where it is usually dumped are poorly managed. Fly ash becomes dry as temperature increases and gets airborne. Thus, it becomes one of the major sources of air and water pollution. Apart from causing various diseases, it also leads to reduction in the recharging of groundwater. Air in areas around coal-fired power plants is polluted with fly ash (Economic times, 2018).Crystalline silica and lime along with toxic chemicals represent exposure risks to human health and the environment. Fly ash contains crystalline silica which is known to cause lung disease, in particular silicosis, if inhaled. Crystalline silica is listed by the IARC and US National Toxicology Program as a known human carcinogen (NTP. Retrieve, 2016) Lime (CaO) reacts with water (H_2O) to form calcium hydroxide [$Ca(OH)_2$], giving fly ash a pH somewhere between 10 and 12, a medium to strong base. This can also cause lung damage if present in sufficient quantities. Material Safety Data Sheets recommend a number of safety precautions be taken when handling or working with fly ash (Headwaters Resources. Retrieved,2016). These include wearing protective goggles, respirators and disposable clothing and avoiding agitating the fly ash in order to minimize the amount which becomes airborne.

The literature data revealed that fly ash can be utilized for the removal of NO_x, SO_x, mercury, and other gaseous pollutants from air and other sources (Ahmaruzzaman, 2012). It can be used as a promising sorbent for the removal of various types of air pollutants; however, further research need to be taken in the laboratory and pilot plant scale. Fly ash also contained approximately 10–12% of unburned carbon and these components have an important role in their removal capacity.

A significant contribution to a large amount of fly ash is made by fourteen TPPs under National Thermal Power Corporation (NTPC), which are spread all over India. There are several other power generation corporation /organization – which are owned by public /private enterprises. However, the ash produce by TPPs under NTPC are of good quality as far as the particle size, shape, carbon contents are concerned (Fly ash-A resource material for construction sector, NTPC Limited, Ash Management Group, Jan 2014). Quite a substantial portion of the ash is generated by TPPs under NTPC are being utilised for various purposes like land development, partial replacement of cement / as fillers in concrete, in the development of roads / embankments, raising of ash dyke, making of bricks and other construction materials (ConsoliRocha, and Saldanha,2014;Escalante-Garcia et al,2008;Marroccoli et al,2010).Presently, fly ash is also used in recovery of highly valued metals (Font et al,2007; Navarro et al,2007) in extraction of alumina (Guoet al,2013) and silica (Font et al,2009), in agriculture (Schumann and Sumner,2000), in water and atmospheric pollution control (Ahmaruzzaman,2009), in dye removal (Mohan et al,2002) and in zeolite synthesis (Atun et al,2011). It has been observed that the materials prepared by using NTPC power plant ash of better quality than those prepared by using ash from other TPPs (Rano,2005).However, a detailed physico-chemical, mineralogical, morphological analysis of the various size classified fractions of the as received ash from the various TPPs of NTPC, India have not been carried out earlier so as to create target oriented utilisation of the various fractions, though such exercise has been carried out for fly ash of different origin. Vassilev et al (2003) has proposed and substantial the need for such characterisation for effective use of the various components of the size classified fractions of as received fly ash. In earlier reports on high carbon ash (Sarkar et al, 2006, 2010) a detailed correlation between characterisation and utilisation of the various components of ash has been presented. However, for a low carbon ash, which is entirely different from that of a high carbon ash as far as their physico-chemical, morphological and mineralogical aspects are concerned, such characterisation has not been done. Such characterisation reveals that the various size-classified fractions of the low carbon ash have better utilisation potential than as received fly ash. A scheme of separation procedures applied, to obtain various size-classified ash and their magnetic and non-magnetic components is presented (Figure 1). The present study was undertaken to conduct a detailed physico-chemical, mineralogical and morphological analysis of the various size-classified fractions as well as their magnetic and non-magnetic components and to suggest suitable potential utilisation directions based on the results of analysis. In this chapter an attempts have been made in this regard and found that fly ash possesses various potential application in the management of air quality.

EXPERIMENTAL SECTION

Fly ash (FA) was collected from mechanical hopper of Farakka Super Thermal Power Station (FSTPS), Dist: Murshidabad(Northern side), State: West Bengal, India.FSTPS is a Super Thermal Power Station controlled by NTPC, uses Pulverised Fuel Combustor (PFC), higher capacity and is better maintained. The relevant combustion data of FSTPS for a unit of 200 MW installed capacity are as follows:

Figure 1. Separation procedures followed to obtain various size-classified ash and their magnetic and non-magnetic components

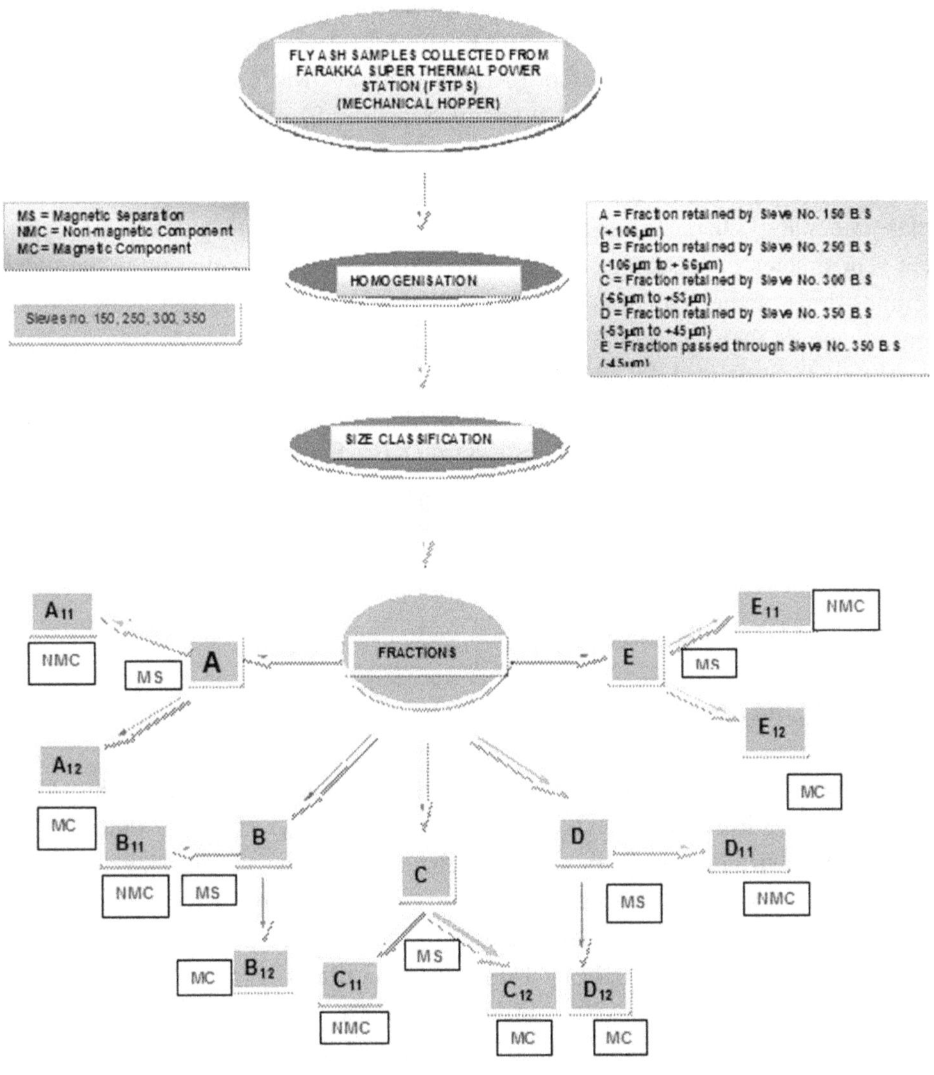

Scheme 1

the combustion temperature is between 800-1100^{0}C at the initial combustion level which increases gradually to the super heater stage up to a temperature range of 1400-1600^0 C. The airflow rate is in the range of 620-670 t/h. The excess oxygen is 3-3.5% (Rano, 2005).

FA was collected, and after proper homogenisation, 10 Kg of the sample was prepared for various experimental investigations. Size-classification of the homogenised ash samples (10Kg) as well as isolation of magnetic and non-magnetic components was carried out according to Figure 1. LOIs, bulk and apparent densities were determined following standard ASTM methods. Morphological analyses of the various fractions were carried out using Scanning Electron Microscope (SEM) Model S-440, LEO Electron Microscopy equipped with an Oxford link-Isis Energy Dispersive X-ray analyser (EDX), Powder X-ray diffraction (XRD) was used for the phase mineralogy characterisation of the various magnetic, non-magnetic components and cenospheres concentrates (Model PW-1710 with CuK_{α} radiation). Chemical compositions of the five sizes-classified fractions were determined using X- ray Fluorescence (XRF), model: PW 2404 X-ray spectrometer. Particle size analysis was carried out using LASER based model Fritsch Particle Sizer Analysette 22. FTIR spectroscopic analyses were carried out on FTIR Spectrometer - Perkin Elmer Model-System 2000 with KBr as dispersant (KBr: Sample ratio of 1:100). After characterization of the samples, various applications were carried out to check their utility potential.

RESULT AND DISCUSSION

Physical Characterization

FSTPS fly ash is in general low carbon ash. LOI of the as received ash is 0.46% the coarser fractions (A, B) have slightly higher LOI than the finer ones (Table -1). LOI data with the weight percentage distribution of the various size classified fractions give a primary clue to the device of the ash fractions for use in cement / construction industries. The finer fraction (C, D, &E) together contribute around 70% to the range of (0.36 -0.47%). Hence, sieving the as received ash with sieve of mesh size BS: 300 will result in production of ash which have lower LOI. The usability of such fractions of ash as partial replacement of cement / fine aggregates has been further substantiated by the morphological assessment of there fractions.

The apparent and bulk densities increase with decrease in particle size (Table-1). The lower densities of the coarser fractions may be due to the presence of either the cenospheres /or char particles (Gurupia et al, 2001) or both. However, the low carbon content in the coarser fractions as well in general in all the fractions precludes the second possibility SEM-EDX analysis of the coarser fraction (A & B) has not revealed any particles with significant enrichment of carbon . Hence density data points towards the presence of ceramic cenospheres in the coarser fractions. Since

Table 1. Weight percentage size distribution, LOI %, apparent / bulk densities of the various size classified fractions of fly ash

TPS	Collection Point	Size Classified Fraction				
		A	B	C	D	E
		Weight % & LOI %				
Farakka Thermal Power Station	Mechanical Hopper	8.62 (0.53)	20.18 (0.52)	44.01 (0.47)	23.53 (0.42)	3.18 (0.36)
		Apparent Density (g/cm³)				
		1.68	1.74	1.85	2.21	2.22
		Bulk Density (g/ cm³)				
		0.82	0.90	0.95	1.00	1.13

(Values in parentheses reflect LOI%)
* The results are average of three experiments
Where,
A= Fraction retained by Sieve No. 150 B.S (+ 106μm)
B= Fraction retained by Sieve No. 250 B.S (-106μm to + 66μm)
C= Fraction retained by Sieve No. 300 B.S (-66μm to +53μm)
D= Fraction retained by Sieve No. 350 B.S (-53μm to +45μm)
E= Fraction passed through Sieve No. 350 B.S (-45μm)

the finer fractions specially the fractions A &B which have higher apparent and bulk densities may lie quite useful as engineering fill materials (Torrey, 1978). The same trend of increasing bulk density with decrease in particle size is observed in both the magnetic and non-magnetic components (Table-2). The various size classified fractions of FSTPS ash. However, the bulk densities of the magnetic components are significantly high as compared with that of non-magnetic ones for reasons that are obvious (Giere et al, 2003).

Particle size analysis (Table-3) of the various size classified fractions reveals that fractions D & E may be quite useful as fine aggregates since their D_{10}, D_{50} and

Table 2. Distribution of magnetic and non-magnetic components in various size-classified fractions of fly ash with their bulk densities

TPS	Collection Point	Size Classified Fraction									
		A_{12}	B_{12}	C_{12}	D_{12}	E_{12}	A_{11}	B_{11}	C_{11}	D_{11}	E_{11}
		Magnetic Weight Percentage (%)					Non-Magnetic Weight Percentage (%)				
Farakka Thermal Power Station	Mechanical Collector	6.1	6.4	6.4	7.2	8.6	93.2	92.2	91.5	88.5	88.3
		Bulk Density (g/ cm³)									
		1.46	1.53	1.67	1.75	1.81	0.73	0.87	0.89	0.95	1.05

Table 3. Particle size distribution profiles of size classified fractions of fly ashes

	Fraction	Undersize (%)			D_{10}	D_{50}	D_{90}	AMD (µm)	SPAN $\frac{(D_{90}-D_{10})}{(D_{50})}$
		<10µm	<50µm	<100µm					
TPS / Collection Point	A	20.65	27.01	38.67	3.25	116.51	192.56	108.13	1.62
	B	2058	31.99	75.85	3.54	78.24	119.57	68.10	1.48
	C	41.36	34.46	98.99	2.30	14.94	62.80	23.34	4.05
	D	57.24	100.0	100.0	1.72	8.08	22.20	10.37	2.53
	E	63.11	100.0	100.0	1.55	7.12	20.29	9.29	2.63

Where,
D_{10}: 10% of the particles is smaller than this diameter (µm)
D_{50}: 50% of the particles is smaller than this diameter (µm)
D_{90}: 90% of the particles is smaller than this diameter (µm)
AMD: Average Mean Diameter
SPAN: $(D_{90} - D_{10})/ D_{50}$

D_{90} values as well as arithmetic mean diameter (AMD) are quite low (< 45µm). Particularly, fraction E is quite useful in this respect with AMD less than 20 µm.The magic number (<10µm) is attained by somewhat greater percentage of particles (~ 37%). Moreover, the particles with diameter <50µm occupy a population density of 93%. Magnetic fractions have higher AMD than non-magnetic fractions. Such analysis of the physical characteristics coupled with other relevant morphological and chemical analysis of various size classified ash fractions, may be quite useful foe identifying their probable uses (Sarkar et al, 2005).

Chemical Composition

The chemical composition of the various size classified fractions of FSTPS is presented in Table-4. It is observed that in general all the size classified ash fractions fall under the class F category (ASTM C 618 94,1994). Hence, size classification has not resulted in any change of classification. Therefore, all the fractions can be used for specific utilisation process which requires use of class F fly ash (www. Geocities.Chemistry). It has been proposed that in certain utilisation processes like inert fillers, effective catalyst support etc., class F fly ash is better than class C. Hence, these size classified fractions may be termed as class F use specific.

Basically all the size classified fractions are alumino-siliceous in nature with SiO_2 / Al_2O_3 ratio for the fraction A-E. Hence these fractions can be used where there is a requirement for alumino-siliceous ash. The occurrence of Ca and Fe are

Table 4. Particle size distribution profiles of As received, Magnetic fraction and Non-magnetic fraction of fly ashes

TPS	Undersize (%)			D_{10}µm	D_{50}µm	D_{90}µm	AMD (µm)	SPAN
	<10µm	<50µm	<100µm					
Farakka Thermal Power Station (Mechanical Collector) As received	37.54	93.67	99.67	2.70	14.75	38.64	18.96	2.44
Magnetic	17.38	39.71	75.79	4.89	66.87	126.86	65.84	1.82
Non-magnetic	36.80	86.31	94.87	2.95	15.73	67.93	26.62	4.13

Where,
D_{10}: 10% of the particles is smaller than this diameter (µm)
D_{50}: 50% of the particles is smaller than this diameter (µm)
D_{90}: 90% of the particles is smaller than this diameter (µm)
AMD: Average Mean Diameter
SPAN: $(D_{90} - D_{10})/ D_{50}$

moderate and that of Mg is low. However, concentrations of Fe in the finer fractions are some what higher than in the finer fractions.

Trace element analysis (Table-5) using XRF reveals that Ba, V, Cr, Mn, Ni, Cu, Zn, Rb, Sr, Pb, Se and Th are present in the various size classified fractions of the fly ash samples. However, their concentration vary widely amongst all the trace elements present it is observed that Ba, V, Sr and Cr are present in high to moderate concentration. Their concentrations are some what similar to the concentrations reported elsewhere (Vassilev and Menendez, 2005) and has been suggested to have potential utilisation. Specifically for metal extraction Ba, V, Cr, Pb show moderate to limited concentration effect with decrease in particle size. Particularly for V, the trend is noteworthy for other trace elements there is not significant change in concentration with decrease in particle size. A preliminary remark can be made i,e., pollution load may not be too high.

Table 5. Chemical Composition (major elements) of different size classified fraction (% w/w) of fly ashes using PW 2404 XRF

Size Fraction	SiO_2	Al_2O_3	Fe_2O_3	CaO	MgO	P_2O_5	K_2O	Na_2O	TiO_2	SO_3
A	48.98	27.97	4.25	2.66	0.66	0.35	0.89	0.45	2.66	0.1
B	49.56	28.11	4.45	1.72	0.61	0.39	0.89	0.45	2.24	0.1
C	50.12	28.11	4.85	1.99	0.65	0.40	0.91	0.25	2.33	0.09
D	50.91	29.72	5.42	2.69	0.68	0.45	0.92	0.13	2.36	0.08
E	51.04	30.48	5.78	2.09	0.55	0.48	0.62	0.15	5.36	0.1

Mineralogical Composition

The relevant information regarding the mineralogical composition of fly ash, X-ray diffractograms of the non –magnetic and magnetic componentswere obtained(figure 2 & figure 3). XRD analysis of the non-magnetic component reveals that mullite and quartz are the main mineralogical phases present in all the size classified fractions. The peaks [100] due to quartz are quite intense in all the non-magnetic fractions. However, peak intensities due to quartz [100] in the various size fractions cannot be taken as a measure of concentration ofα – quartz, since only qualitative analysis of the size classified fractions were carried out by XRD. Mullite is present in all the fractions. Some of the peaks are due to refractory mullite ($\Theta = 16.4^0$) XRDs of the magnetic components reveal that all the size classified magnetic components contain magnetite and hematite. It is observed that there is a general increase in intensity of the peak [100] due to magnetite with decrease in particle size. The peaks [100] for α – quartz also of much lower intensity than those of observed for the non-magnetic components. Another notable observation is that the peak intensities for α – quartz in the magnetic components decrease with decrease in the particle size of the magnetic components and is almost insignificant in finest magnetic components (E_{12}).

FTIR Spectroscopic Analysis

Qualitative FTIR spectra were recorded for both magnetic and non-magnetic components of FSTPS, so as to obtain information regarding nature of the carbon matters present as well as for any other information regarding the mineral matter present (Table-6). No marked difference between the spectra of the magnetic and non-magnetic components was observed. CH_2 bands are weak and no C=O band could be observed. The low carbon content of the fly ash fractions probably prevents the detection of such oxidised form of the carbon matter. Strong mineral bands are observed (1095, 1031 cm^{-1}) in all the size classified fractions of fly ash (magnetic /

Table 6. Trace element contents in different size-classified fractions of fly ashes (ppm)

Size Fraction	Cr	Mn	Co	Ni	Cu	Zn	V	Se	Rb	Sr	Ba	Pb	Th
A	143.6	302.7	21.1	60.3	69.6	61.5	249.1	2.3	70.8	336.3	1019	37.1	15.5
B	156.3	281.7	22.4	62.5	81.9	68.4	281.4	1.6	73.0	334.8	1132	44.3	11.6
C	165.4	310.5	24.6	65.2	92.5	75.1	301.2	1.2	74.5	335.1	1131	51.6	11.8
D	172.2	358.8	26.1	68.2	96.4	96.5	336.6	0.9	75.0	336.8	1129	66.0	11.9
E	181.7	366.5	30.3	74.9	118.2	129.3	421.8	1.11	96.3	393.8	1275	112.0	nd

*ND: Not detected

Figure 2. XRD patterns of non-magnetic fractions (A_{11}, B_{11}, C_{11}, D_{11}, E_{11}) of Farakka fly ash. Mineral abbreviations: Q-quartz; Mu- mullite

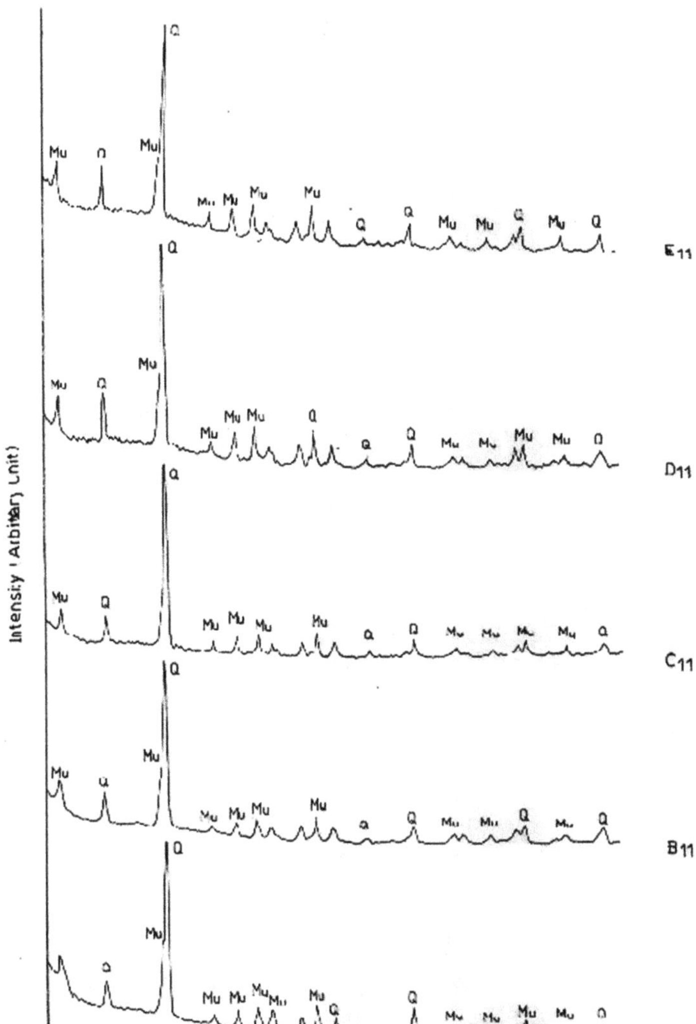

Figure 3. XRD patterns of magnetic fractions (A_{12}, B_{12}, C_{12}, D_{12}, E_{12}) of Farakka fly ash. Mineral abbreviations: Q-quartz; Mu- mullite ; M-magnetite ; H- hematite

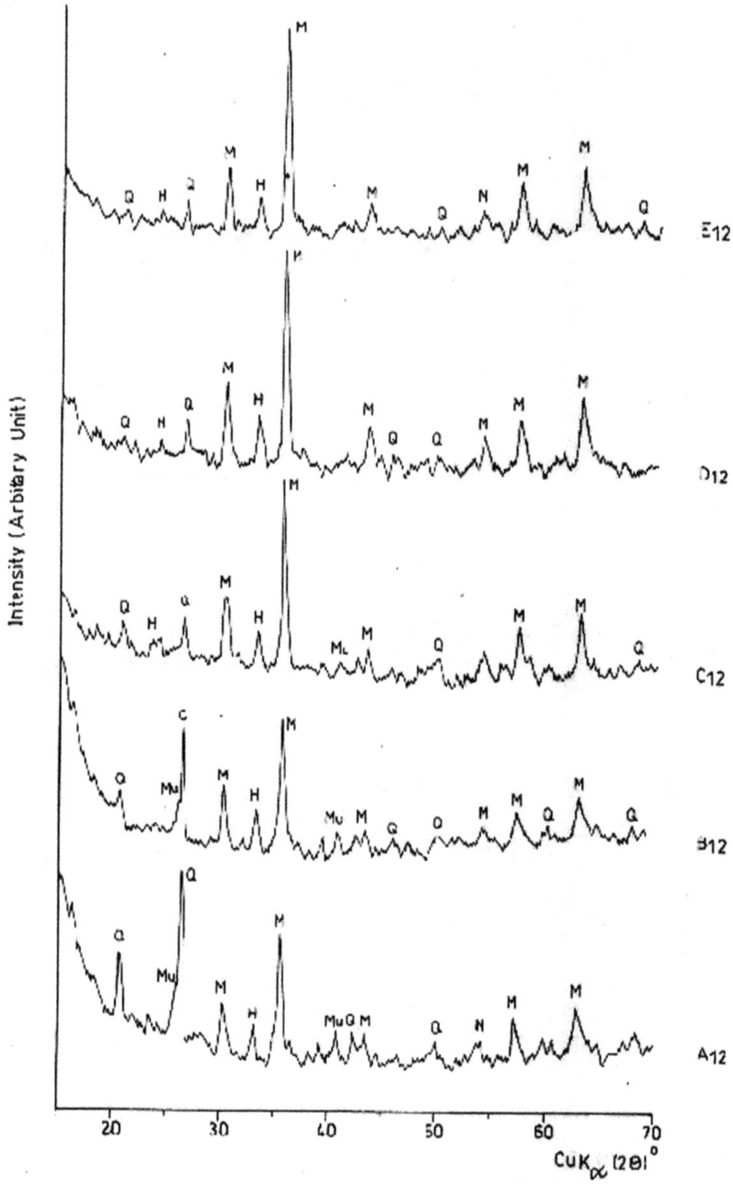

non-magnetic). Such bands Si –O –Si / Si –O stretching) point towards the presence of kaolinite, quartz and mullite.

Quantitative FTIR analysis for the various size-classified fractions (A-E) was carried out to determine the percentage α – quartz ion the various size classified fractions of fly ash (Table-7). The content of α – quartz in each fraction was determined by determining the ratio of peak area due to the peaks (798.16 and 799.33

Table 7. Peak assignment for the FTIR spectra of FSTPS fly ash

Wave Numbers (cm^{-1})	Characteristics	Magnetic Fractions					Non-Magnetic Fractions				
		A	B	C	D	E	A	B	C	D	E
3697	ν_1-OH attached to kaolinite(1:1 layer silicate)	Sh	sh	sh	sh	sh	s, w_i	x	w	w	sh
3450-3300	ν_1-OH of H_2O, H bonded	b,w_i	b,m_i	b,m_i	b,m_i	b,m_i	b, w_i	b,m_i	b,s_i	b,s_i	b,s_i
2950	ν_1-CH_3 asymm. Str.	X	x	x	x	x	sh	sh	sh	sh	sh
2923	ν_1-CH_2 asymm. Str.	sh, w_i	shw_i	s,w_i	s,w_i	s,w_i	m,m_i	m, w_i	m, w_i	s, w_i	s, w_i
2850	ν_2-CH_2 symm. Str.	X	x	w_i	w_i	w_i	m,w_i	m,w_i	m,w_i	m,w_i	mw_i
1729/1710	$\nu_{1,}$C=O	X	x	x	x	x	x	x	x	x	x
1630-34	ν_2-OH bending	b,w_i	b,w_i	b,w_i	sh	sh	b, w_i	b, w_i	b, w_i	b	b
1430	ν_1-CH_3 bending	Sh	sh	sh	sh	sh	b w_i	b, w_i	sw_i	b	b
1091-1096	Si-O str. in kaolinite	b,s_i	b,s_i	s,s_i	s,s_i	s,s_i	s,s_i	s,s_i	s,s_i	s,s_i	s,s_i
1031-1036	Si-O-Si Stretching vib.	X	x	x	x	x	s,s_i	m,w_i	x	w_i	b,s_i
750,796-98	Aromatic C-H Bending	s,w_i	s,w_i	s,w_i	s_i,w_i	s_i,w_i	s_i,w_i	x	s,m_i	b,w_i	s_i,w_i
537-539	$\nu_{1,}$ Si-O bending	X	x	x	x	x	s,s_i	x	x	x	x
415	FeS- pyritic band	X	x	x	x	x	x	x	x	x	x

Where,
sh - shoulder w_i- weak intensity s_i - strong intensity
m_i- medium intensity b - broad s - sharp
m - moderately sharp x – not observed

cm^{-1})It was observed the concentration of α – quartz decreases, with decrease in particle size (A-E) except for one aberration for the size classified fraction B_{11} the moderately coarser fraction.

Morphological Analysis

Scanning electron microscopic study (SEM) of the fly ash fractions was carried out to get useful informations regarding the morphological features of the magnetic and non-magnetic components. It is observed that there is a distinct difference between the magnetic and non-magnetic components, specially the coarser fractions (A_{11}&B_{11}) [Figure 4(a-e)] are somewhat irregular in shape, while the magnetic components are more spherical in nature (Figure 6). EDX analysis of non-magnetic fractions reveals that there is high surface enrichment of Al/Si and O with no trace of C, signifying that the surface is essentially alumino-siliceous in nature. Some of the particles remain fused. The next coarser non-magnetic fraction C_{11} (Figure 4) also contain smaller irregular particles which are also aluminosiliceous in nature. The finer fractions contain particles [Figure 4(h,i)], which are largely spheroidal in character and of much smaller size. These appear to be hollow and porous in nature (Figure 4). EDX analysis reveals that these may have very high surface enrichment of Si or Al/Si both Fraction E_{11} contains particle which are essentially very small with smooth surface and therefore expected to have good ball bearing effect.Coarser magnetic component (A_{12}) reflects that the particles have globular shape,(Figure 5).Some particles have flower like,needle like crystallites on the surface.In other fraction B_{12}, crystallites appear to be irregular and has criss cross pattern(Figure 5).EDX study reveals these particles have moderately high surface enrichment of Fe along with Al/Si and next finer fractions C_{12},D_{12} and E_{12} contain particles that are more uniform and spheroidal with high enrichment of Fe.

Potential Applications Directions

There are distinct differences between magnetic and non-magnetic components of different fractions of fly ash in respect of various physico-chemical, mineralogical and morphological properties. The finer fractions are more regular in shape and size, these particles will be quite useful for using as fillers in cement-mortar preparation and suitable for use in slurry transfer for the following reasons:

1. Spheroidal particle improves the flowability of the slurry by increasing the ball bearing effect.

Figure 4a. (a) SEM image of a general view –non-magnetic component coarser fraction -A_{11} (b) SEM image of particles from A_{11} at higher magnification. (c) SEM image of particles from A_{11}- higher magnification (d) SEM image of a general view –non-magnetic component coarser fraction (B_{11}) (e) SEM image of particles belonging to fraction B_{11} (f) SEM image of a general view –non-magnetic component finer fraction (C_{11}). Presence of alumino-siliceous particles can be observed.

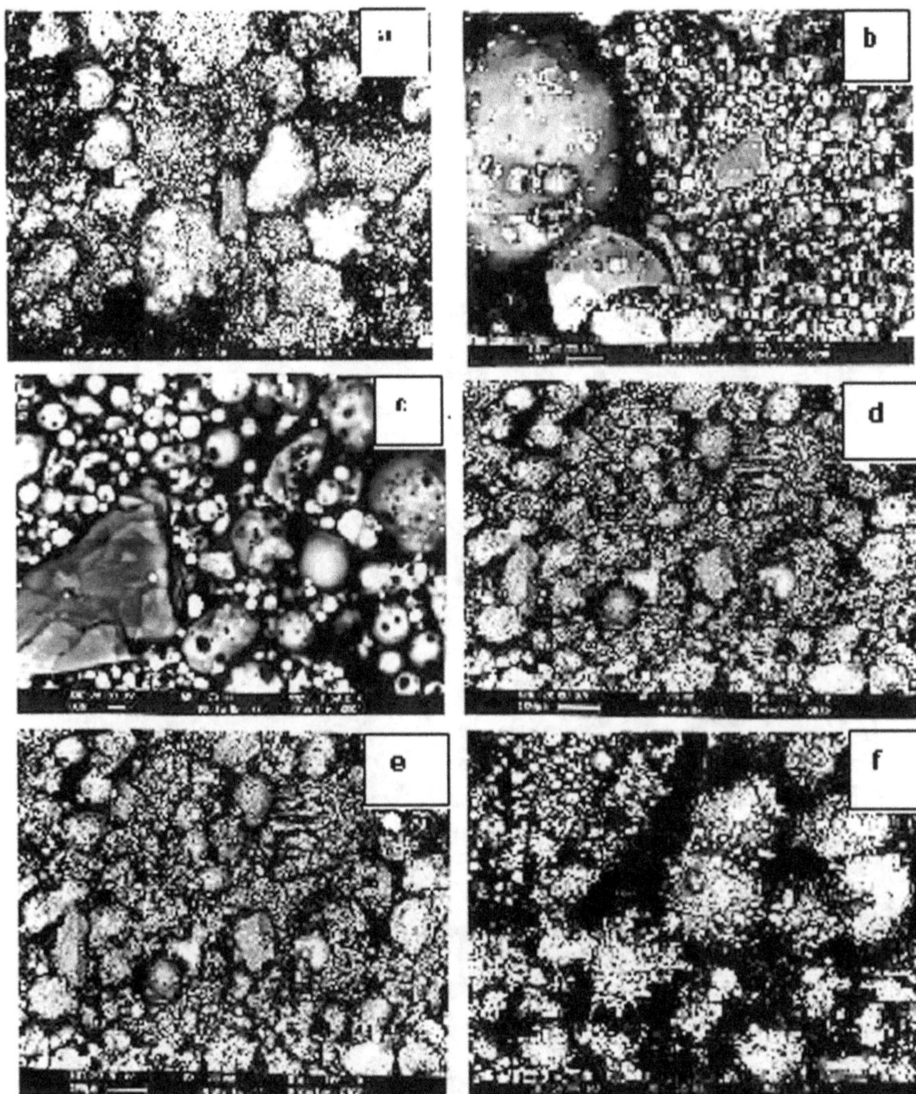

Figure 4b. (g) SEM image of particles from fraction C_{11} (h) SEM image of a general view – non-magnetic component finer fraction (D_{11}). Particles become more globular in shape (i) SEM image of particles from fraction (D_{11})-porous in nature (j) SEM image of a general view, non-magnetic component finer fraction (E_{11}). Particles are highly globular in nature (k) SEM image of particles from E_{11} –alumino-silicate spheres (l) SEM image of particles from E_{11} at higher magnification.

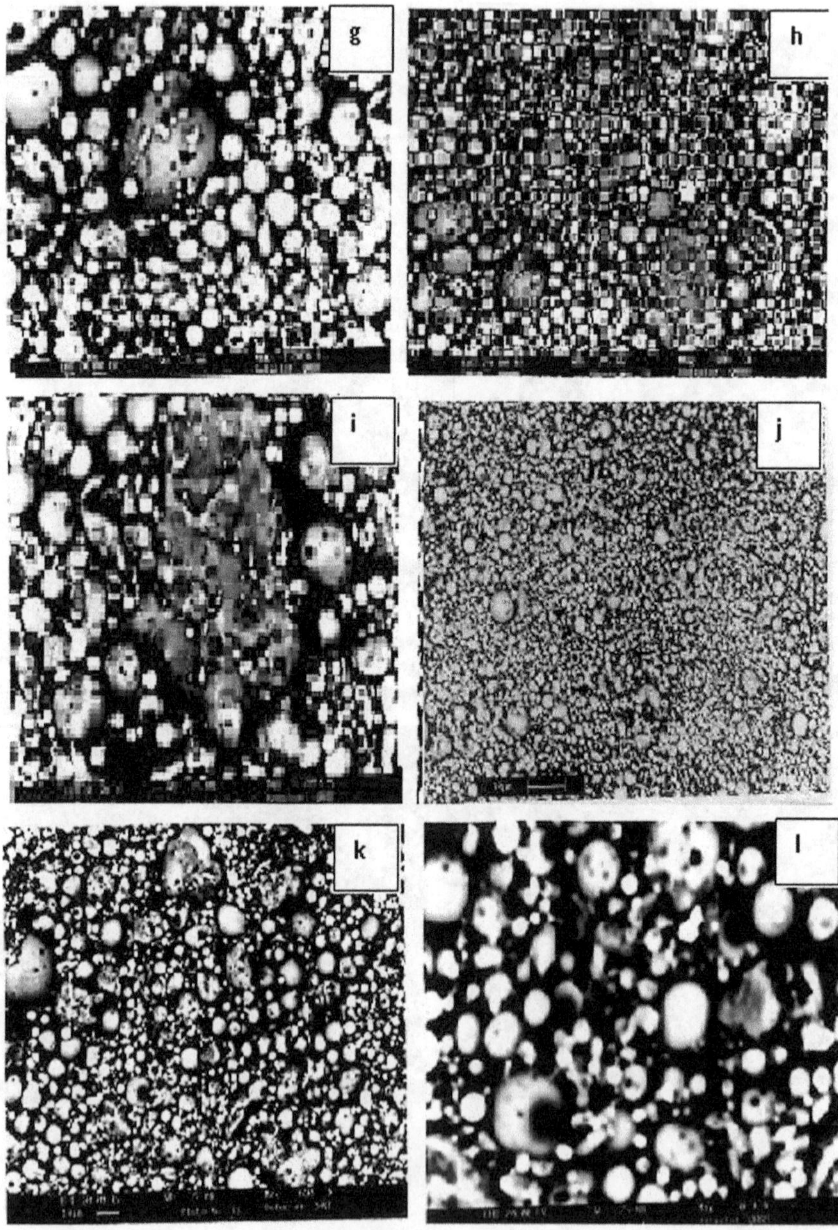

Figure 5a. (a) SEM image of a general view –magnetic component coarser fraction (A_{12}), (b) SEM image of particles from fraction A_{12}. some ferrospheres are observed. (c) SEM image of a general view magnetic component finer fraction (B_{12}), -Globular Particles are observed (d) SEM image of a spheroidal magnetic particle from fraction B_{12} (e) SEM image of particles from B_{12} - criss cross pattern observed (f) SEM image of a general view magnetic component finer fraction (C_{12}).

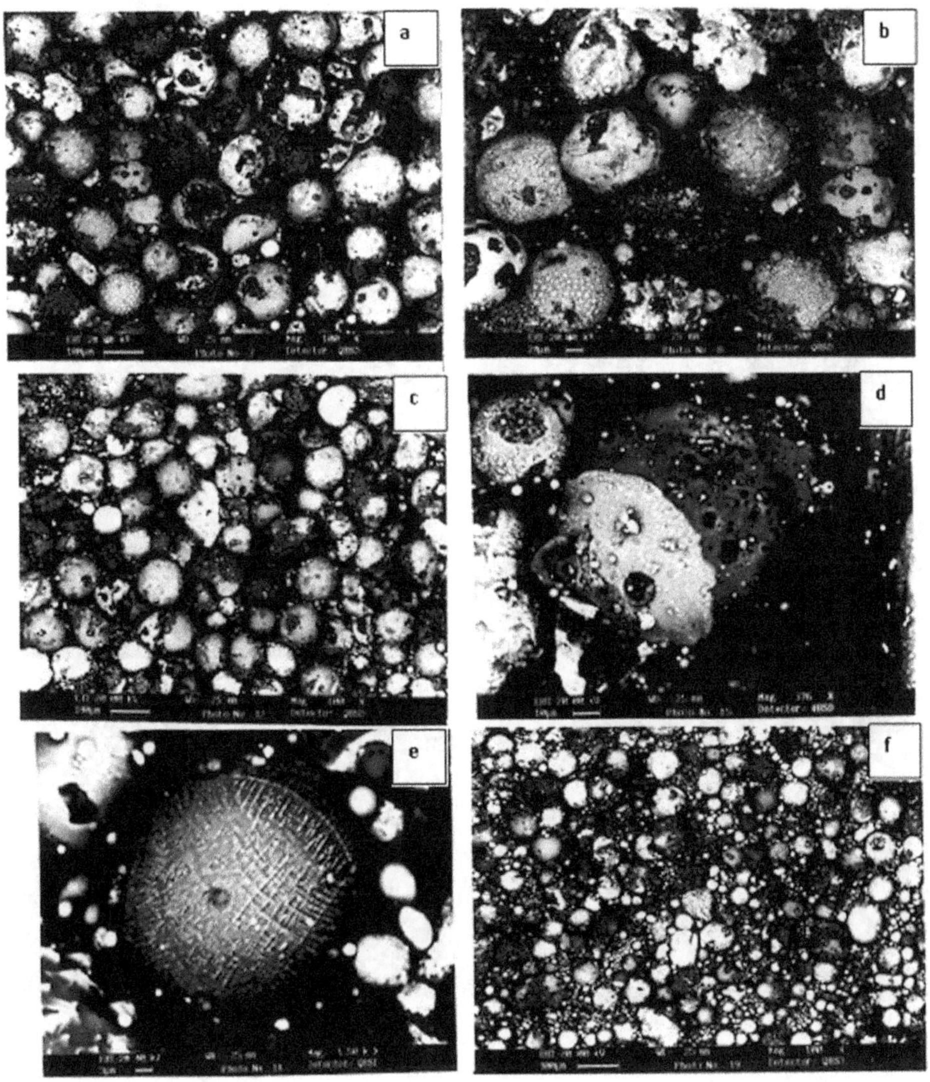

Figure 5b. (g) SEM image of particles from C_{12}-crystallites of different patterns are observed (h) SEM image of particles from C_{12}-ferrospheres observed (i) SEM image of a general view –magnetic component D_{12} (j) Particles become more spheroidal in nature (k) SEM image of particles from D_{12} (l) SEM image of a general view – magnetic component E_{12}

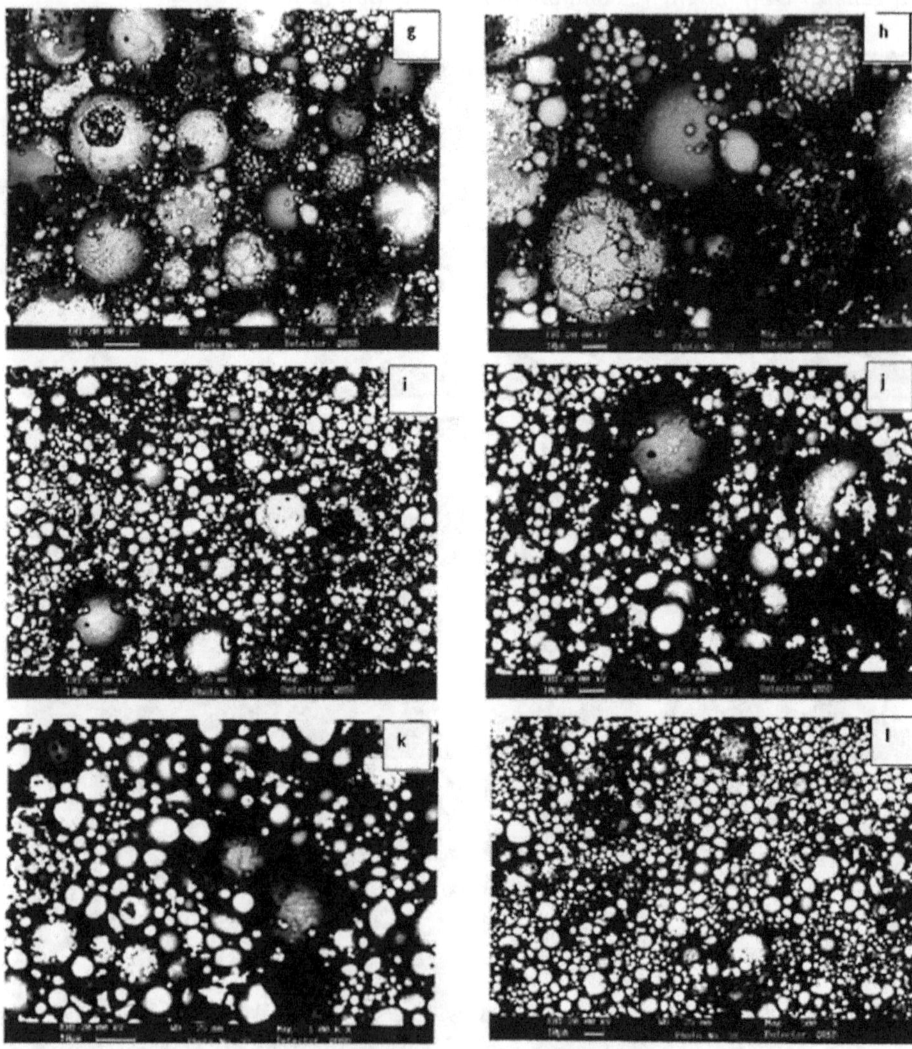

2. It improve the water impermeability of the mass, hence water requirement for slurry preparation will be less.
3. Improves the binding capacity since these contain particles with <10 μm in diameter, which is expected to improve pozzolanic property.
4. The particles being spheroidal is expected to form a compact mass once transported to the void spaces.

Magnetic components find applications in coal beneficiation (Sarkar et al, 2006),ferrosilicon, dense concrete and pigments (Vassilev et al, 2004) because from XRD analysis it appears that possibly the finer fractions have high content of magnetite. Moreover, the finer fractions are spheroidal in nature. These two factors when combined are expected to help in the use of finer magnetic components of fly ash in coal beneficiation. It is recommended that the finest fraction of the magnetic component should be tried for coal beneficiation after determining the concentration of magnetite.

On the other hand, authors find that the finer non-magnetic fraction is an economic source of alumina siliceous material for synthesizing an efficient, novel solid base catalyst system for catalyzing various industrially important organic reactions (Mazumder and Rano, 2015).

An efficient catalytic methods (Figure 6) has been established for synthesis of dibenzylideneacetone DBA(94% yield) by crossed aldol condensation reaction between acetone and benzaldehyde using a solvent free environmentally greener solid base catalyst (SBC). Enhanced increase in surface area and percentage of alumina are the important factors for the increase in surface hydroxyl groups responsible for basicity and higher catalytic activity of SBC. Recyclability up to four reaction

Figure 6. Crossed Aldol condensation of acetone with benzaldehyde over solid base catalyst (SBC)

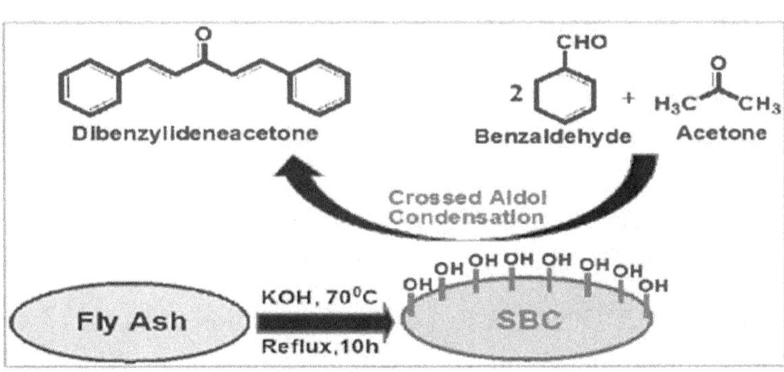

Scheme 2

cycles indicates the extraordinary stability and capability of the catalyst. This results in the production of a low cost, effective, environment friendly and stable SBC.

Various value added components such as cenospheres and ferrospheres can be extracted which have nano-sized active centers on their surfaces, which if properly modified may leads to inspiring developments in the field of photocatalysis. Photocatalyst synthesized by surface modification of fly ash associated with excellent photocatalytic activity towards the degradation of various organic dyes in aqueous solution under direct sunlight irradiation. These may be used for purification of polluted water released from various leather, color, textile, rubber, cosmetics, aqua-culture, food and pharmaceutical industries for environmental remediation (Huo et al, 2010).

CONCLUSION

The magnetic and non-magnetic components can be easily isolated by sieving and magnetic separation from fly ash. Less LOI content coupled with better particle size distribution profile implies that the combustion efficiency is good and it is suitable for various applications. Thus thorough characterisation reveals that the various size-classified fractions of a low carbon fly ash have better utilisation potential than as received fly ash. It has been also found that fly ash possesses potential application in the management of air quality.

ACKNOWLEDGMENT

The authors are thankful to the Director, Indian school of Mines, Dhanbad for providing the facilities for carrying out the research.Also we are grateful to Mr. SabyasachiShome, Geologist, Paleontology Divisionof GSI, Kolkata for performing morphological analysis using SEM-EDX.

REFERENCES

Ahmaruzzaman, M. (2009). Role of fly ash in the removal of organic pollutants from wastewater. *Energy & Fuels*, *23*(3), 1494–1511. doi:10.1021/ef8002697

Ahmaruzzaman, M., & Gupta, V. K. (2012). Application of Coal Fly Ash in Air Quality Management. *Industrial & Engineering Chemistry Research*, *51*(47), 15299–15314. doi:10.1021/ie301336m

ASTM C 618 94. (1994). Specification for coal fly ash and raw calcined natural pozzolan for use as a mineral admixture in Portland cement concrete. *Annual Book of ASTM Standards, 4*(2), 296–298.

Atun, G., Hisarlı, G., Kurtoglu, A. E., & Ayar, N. (2011). A comparison of basic dye adsorption onto zeolitic materials synthesized from fly ash. *Journal of Hazardous Materials, 187*(1-3), 562–573. doi:10.1016/j.jhazmat.2011.01.075 PMID:21300434

Chatterjee, A. (2014). *Technological initiatives and status of fly ash in India. In Fly ash: Sources, applications and potential environmental impacts* (pp. 315–359). New York: Nova Science Publishers.

Consoli, N. C., Rocha, C. G., & Saldanha, R. B. (2014). Coal fly ash–carbide lime bricks: An environment friendly building Product. *Construction & Building Materials, 69*, 301–309. doi:10.1016/j.conbuildmat.2014.07.067

Escalante-Garcia, J. I., Rios-Escobar, M., Gorokhovsky, A., & Fuentes, A. F. (2008). Fluorgypsum binders with OPC and PFA additions, strength and reactivity as a function of component proportioning and temperature. *Cement and Concrete Composites, 30*(2), 88–96. doi:10.1016/j.cemconcomp.2007.05.015

Fly ash-A resource material for construction sector. (2014). NTPC Limited, Ash Management Group.

Font, O., Moreno, N., Díez, S., Querol, X., López-Soler, A., Coca, P., & Peña, F. G. (2009). Differential behaviour of combustion and gasification fly ash from Puertollano Power Plants (Spain) for the synthesis of zeolites and silica extraction. *Journal of Hazardous Materials, 166*(1), 94–102. doi:10.1016/j.jhazmat.2008.10.120 PMID:19097700

Font, O., Querol, X., Juan, R., Casado, R., Ruiz, C. R., López-Soler, Á., ... Peña, F. G. (2007). Recovery of gallium and vanadium from gasification fly ash. *Journal of Hazardous Materials, 139*(3), 413–423. doi:10.1016/j.jhazmat.2006.02.041 PMID:16600480

Giere, R., Carleton, L. E., & Lumpkin, G. R. (2003). Micro- and nanochemistry of fly ash from a coal-fired power plant. *The American Mineralogist, 88*(11-12), 1853–1865. doi:10.2138/am-2003-11-1228

Guo, Y., Li, Y., Cheng, F., Wang, M., & Wang, X. (2013). Role of additives in improved thermal activation of coal fly ash for alumina extraction. *Fuel Processing Technology, 110*, 114–121. doi:10.1016/j.fuproc.2012.12.003

Gurupia, T., Jones, C. L., Howard, A., Lockert, C., Wandell, T., & Stencel, J. M. (2001). *International Ash Utilization Symposium, Centre for Applied Energy Research*. University of Kentucky, Paper#44.

Guttikunda, S. K., & Jawahar, P. (2014). Atmospheric emissions and pollution from coal-fired thermal power plants in India. Atmospheric environment. *Atmospheric Environment*, *92*, 449–460. doi:10.1016/j.atmosenv.2014.04.057

Headwaters Resources Class F fly ash safety data sheet, (2016). Headwaters Resources.

Huo, P., Yan, Y., Li, S., Li, H., Huang, W., Chen, S., & Zhang, X. (2010). H2O2 modified surface of TiO2/fly ash cenospheres and enhanced photocatalytic activity on methylene blue.Desalination. *Desalination*, *263*(1-3), 258–263. doi:10.1016/j.desal.2010.06.067

Marroccoli, M., Pace, M. L., Telesca, A., Valenti, G. L., & Montagnaro, F. (2010). Utilization of coal combustion ashes for the synthesis of ordinary and special cements. *Combustion Science and Technology*, *182*(4-6), 588–599. doi:10.1080/00102200903466210

Mohan, D., Singh, K. P., Singh, G., & Kumar, K. (2002). Removal of dyes from wastewater using flyash, a low-cost adsorbent. *Industrial & Engineering Chemistry Research*, *41*(15), 3688–3695. doi:10.1021/ie010667+

Navarro, R., Guzman, J., Saucedo, I., Revilla, J., & Guibal, E. (2007). Vanadium recovery from oil fly ash by leaching, precipitation and solvent extraction processes. *Waste Management (New York, N.Y.)*, *27*(3), 425–438. doi:10.1016/j.wasman.2006.02.002 PMID:16563726

Nurul, M., & Ruma, R. (2015). An efficient solid base catalyst from coal combustion fly ash for green synthesis of dibenzylideneacetone. *Journal of Industrial and Engineering Chemistry*, *29*, 401–414.

Rano, R. (2005). *Evaluation of the Physico-Chemical and Morphological Properties of Flyash obtained from Different Thermal Power Plants for their Potential Utilisation* (PhD thesis).

Sarkar, A., Rano, R.& Kumar A (2010). Characterization of high carbon ash from thermal power station using pulverized fuel combustor. *Energy. Source Part- A*, *32*, 607-609.

Sarkar, A., Rano, R., Mishra, K. K., & Sinha, I. N. (2005). Particle size distribution profiles of some Indian fly ash-A comparative study to assess their possible uses. *Fuel Processing Technology*, *86*(11), 1221–1238. doi:10.1016/j.fuproc.2004.12.002

Sarkar, A., Rano, R., Udaybhanu, G., & Basu, A. K. (2006). A comprehensive characterization of fly ash from a thermal power plant in Eastern India. *Fuel Processing Technology*, *87*(3), 259–277. doi:10.1016/j.fuproc.2005.09.005

Schumann, A. W., & Sumner, M. E. (2000). Chemical evaluation of nutrient supply from fly ash-biosolids mixtures. *Soil Science Society of America Journal*, *64*(1), 419–426. doi:10.2136ssaj2000.641419x

Schumann, A.W. & Sumner, M.E. (n.d.). *Chemical Evaluation of Nutrient Supply From Fly Ash- Substances listed in the thirteenth report on Carcinogens.* NTPC.

Somoano, D. M. (2003). Phase-minerql and chemical composition of fly ashes as a basis for their multicomponent utilisation.1. Characterization of feed coal and fly ashes. *Fuel*, *82*(14), 1793–1811. doi:10.1016/S0016-2361(03)00123-6

Torrey, S. (1978). Coal Ash Utilization: Fly Ash, Bottom Ash and Slag. Noyes Data.

Vassilev, S. V., & Menendez, R., Borrego, A.G., Somoano, M.D., Tarazona, M., & Rosa M. (2004). Phase-minerql and chemical composition of fly ashes as a basis for their multicomponent utilisation.-2Characterization of ceramic chemosphere and salt concentrates. *Fuel*, *83*, 1563–1583. doi:10.1016/j.fuel.2004.01.010

Vassilev, S. V., & Rosa, M. (2005). Phase-minerql and chemical composition of fly ashes as a basis for their multicomponent utilisation. 4.Characterization of heavy concentrates and improved fly ash residues. *Fuel*, *84*(7-8), 260–231. doi:10.1016/j.fuel.2004.11.021

Chapter 3
Indexing Method for Assessment of Air Quality:
A Case Study for Dharamshala City in India

Rajiv Ganguly
https://orcid.org/0000-0001-5356-1635
Jaypee University of Information Technology, India

ABSTRACT

Rapid urbanization and globalization has led to severe degradation of existing air quality in the majority of Indian cities. In this context, the general public has been aware of their exposure to ambient air quality and the effects of such air pollutants on human health. Hence, the concept of Air quality indices (AQI) is often used by regulatory authorities in conveying the status of existing ambient air quality to the general public. The chapter presents the application of air quality indices for assessing the existing air quality standards in an Indian city, Dharamshala, a tourist location in Himachal Pradesh, for the period of 2016-2017. Two different methods of determining AQI have been used wherein one method is used as the revised Central Pollution Control Board (CPCB), India with different sub-indices for Indian conditions based on the United States Environmental Protection Agency (USEPA) methodology and another alternative method utilizes contribution from all criteria pollutants.

DOI: 10.4018/978-1-5225-7289-3.ch003

INTRODUCTION

Air pollutants generated from vehicular exhausts are one of the major sources of pollution in cities of India and worldwide (Ganguly et al., 2014, 2015; Kumar et al., 2013, 2015, 2016). The important criteria air pollutants generated from traffic sources are nitrogen oxides (NO_x), carbon monoxide (CO), carbon dioxide (CO_2), volatile organic compounds (VOCs), particulate matter (PM) and several metals (Ganguly and Broderick, 2008). Criteria pollutants are those pollutants whose limiting concentrations in atmosphere has been predefined as per the National Ambient Air Quality Standards (NAAQS) to minimize their effects on human health. Continuous and long-term exposure to such pollutants can lead to severe health impacts including asthma and respiratory problems (Anderson et al., 2011; Clark et al., 2010; Heal et al., 2012), inadequate development of lung functions in infants and small children (Gauderman et al., 2007), cardiovascular diseases (Lindgren et al., 2010) and may lead to cancer (Buffler et al., 2005; Langholz et al., 2002). In this context, ambient air quality standards have been fixed for these criteria pollutants in most of the developing and developed countries. In an Indian context, the Central Pollution Control Board (CPCB) has specified the NAAQS for such pollutants as shown in Table 1.

The major difficulty associated with the representation of these prescribed standards is unifying them over a reference scale for representing the health effects of air pollutants to the citizens (Nagendra et al., 2007). Further, increased air pollution concentrations and their recurrence intervals often pose challenges to general citizens of understanding of the existing ambient air quality. As such, Air Quality Index (AQI) are often used by regulatory authorities of informing the common public of the existing nature of the ambient air quality.

In principle, an AQI is a single number determined on the principle of sub-indices of different criteria pollutants and which defines the nature of the ambient air quality in the context of human health effects (Bortnick et al., 2002; Murena, 2004). The pollutants generally considered for the determination of AQI involves the monitored data of all the criteria pollutants (SO_2, NO_x, SPM and RSPM) in India. The methods of analysis of AQI using different methods primarily depend on pollutants considered, sampling period and the considered breakpoints (Monteiro et al., 2016). Different methodologies are used for determination of AQI with different categorizations and there exists no common single technique which has been universally endorsed. The determination of AQI is a simple yet an effective technique in summarizing the potential heath effects of the surrounding ambient air quality (Kowalska et al., 2009; Monteiro et al., 2016). The AQI provide the status of the ambient air quality to the general public without delving into the computational details and alerts the

Table 1. National ambient air quality standards

Pollutant	Time Weighted Average	Concentration in Ambient Air (CPCB, 2009)			European Union (EU) Air Quality Directive	World Health Organization (WHO) Guidelines
		Industrial Areas	Residential, Rural and Other Areas	Sensitive Areas		
Sulfur dioxide (SO_2) $\mu g/m^3$	Annual[a]	80	60	15		
	24 h[b]	120	80	30		
Oxides of Nitrogen (NO_2) $\mu g/m^3$	Annual[a]	80	60	15	40	40
	24 h[b]	120	80	30	200 (not to exceed 18 times in a year)	200
Suspended Particulate Matter (SPM) $\mu g/m^3$/ PM10	Annual[a]	360	140	70	40	20
	24 h[b]	500	200	100	50 (Not to be exceeded on more than 35 days per year)	50 (99th percentile, 3 days per year)
Respirable Suspended Particulate Matter (RSPM) / PM2.5$\mu g/m^3$	Annual[a]	120	60	50	25	10
	24 h[b]	150	100	75		25 (99th percentile, 3 days per year)
Lead (Pb) mg/m^3	Annual[a]	1.0	0.75	0.50		
	24 h[b]	1.5	1.00	0.75		
Ammonia (NH_4) mg/m^3	Annual[a]	0.1	0.1	0.1		
	24 h[b]	0.4	0.4	0.4		
Carbon Monoxide (CO) mg/m^3	8 h[b]	5.0	2.0	1.0		
	1 h	10.0	4.0	2.0		
Ozone $\mu g/m^3$	8 h[b]				120	100

However, 2% of the time, it may exceed but not on two consecutive days.

[a]Annual arithmetic mean of minimum 104 measurements in a year taken twice a week 24 hourly at a uniform interval.

[b]24 hourly/8 hourly values should be met 98% of the time in a year.

citizens to take pertinent precautionary steps to prevent themselves from the harmful effects of such air pollutants (Monteiro et al., 2016). The AQI is a single number and is often associated with the preordained categorizations of air quality like *'good', 'moderate', 'poor' and 'severe'*.

In general context, about 80% of the global population living in urban areas experience high doses of pollutant concentrations exceeding the norms as prescribed by World Health Organization (WHO). The most affected population demographics are the residents living in middle to low income countries wherein about 98% of such cities having a population greater than 100000 do not meet the WHO air quality guidelines. The control of urban air pollution is often beyond the capable of the citizens and needs a concerted effort from governmental and international organizations in measures to reduce the concentrations of air pollution. In such aspects, air quality monitoring in cities is an important tool in determining the existing pollutant concentrations in the city and hence its possible health effect to the residents of the city.

As observed from the WHO report that air pollution is a serious environmental health hazard in developing countries. As per the report (CAI, 2013) more than 100 million people were exposed to contaminated air exceeding the standards. Some of the salient features of the report were that majority of Latin American countries failed in meeting the criteria of particulate pollutants whereas ozone concentrations were exceeded in in cities of Santiago (Chile), Mexico City and Quito. Another important finding was that seven of the countries failed to meet WHO's recommendation on nitrogen dioxide. Major cities in Latin America like Monterrey, Mexico City, Santiago, Lima, Bogota etc all exceeded the specified pollutant concentrations as per WHO limits. In this context, several of the Latin American countries like Chile, Mexico, Brazil are already implementing strict regulations to curb the menacing effect of air pollution. Other countries like Peru, Costa Rica, panama etc are also making holistic effects to reduce the effects of air pollutants. Similarly, the air pollutant concentrations in China are also extremely severe. In this context, Air Quality Index China (AQICN) collects data of air pollutant concentrations and maps them in real time showing the air quality concentrations in those locations. It uses a six-point classification system making easy to discern the existing air quality in the chosen location (Climate and Environment, 2016).

Several reported literature exists on the use of AQI being successfully used for signifying the state of the existent ambient air quality condition. For example, in a study carried out in Athens by Kassomenos *et al.,* (1999) classified the air quality in seven different categories based on AQI determined on the monitoring data between the periods of 1983 to 1995. Other such reported literature has been presented by Longhurst (2005), Landulfo *et al.* (2007), Mayer and Kalberlah (2008), and Elshout *et al.* (2008). Several AQI studies have been carried out in Indian cities including for Delhi (Bishoi et al., 2009; Kumar and Goyal, 2009; Kumar and Goyal, 2011), Bangalore (Nagendra et al., 2007), Kanpur (Sharma et al., 2003), Mumbai (Sharma, 1999), Coimbatore (Saravanakumar et al., 2016) and Keonjhar (Dash and Dash, 2015) using different techniques in assessing and classifying the air quality.

The main aim of the work presented in the book chapter is to determine and analyze the AQI for two newly operational monitoring stations located in Dharamshala city in Himachal Pradesh utilizing the revised CPCB and an alternative methodology for a study period of two years 2016 and 2017 and compare the efficiency of both the selected methods in categorizing the existing ambient air quality.

MATERIAL AND METHODS

Site Location

Dharamshala, one of the major cities of Himachal Pradesh lies at an elevation of 1457 m above the MSL with a population count of about 53543 as per the National Census Record of 2015. The city lies with in the latitude and longitude locations of 32.218° N to 76.320ºE covering an area of 27.6 km^2. The city can be classified as experiencing mild temperate climate (*Cwa*) as per the *Koppen climate classification.* Figure 1 shows the location of the study site.

There exist two new monitoring stations in Dharamshala namely Municipal Corporation (MC) Office (Station I) and Office Building (Station II) functioning under the National Ambient Monitoring Program (NAMP) governed by the CPCB and became operational in 2016. The monitoring data of the criteria pollutants for the study period of 2016 and 2017 were obtained from the website (http://xgn.hp.nic.in/naaqm.aspx) of Himachal Pradesh Pollution Control board (HPPCB) which is maintained in collaboration with National Ambient Air Quality Monitoring (NAAQM) of India.

Monitoring Details

The monitoring of the different pollutants at both the stations of the study location conforms to the method as prescribed by CPCB. To summarize, determination of SO_2 concentrations are carried out using the modified *West and Gaeke Method* (Ganguly and Thapa, 2016; Saravanakumar et al., 2016). The process involves passing a known volume of air in a solution of *sodium tetrachloromercurate* to form a stable compound *dichlorosulphitomercurate* which is then reacted with *formaldehyde solution.* The colour intensity of the final end product compound so formed is then determined using spectrphotometer analysis for analyzing the SO_2 concentrations.

Figure 1. Geographical location of Dharamshala City

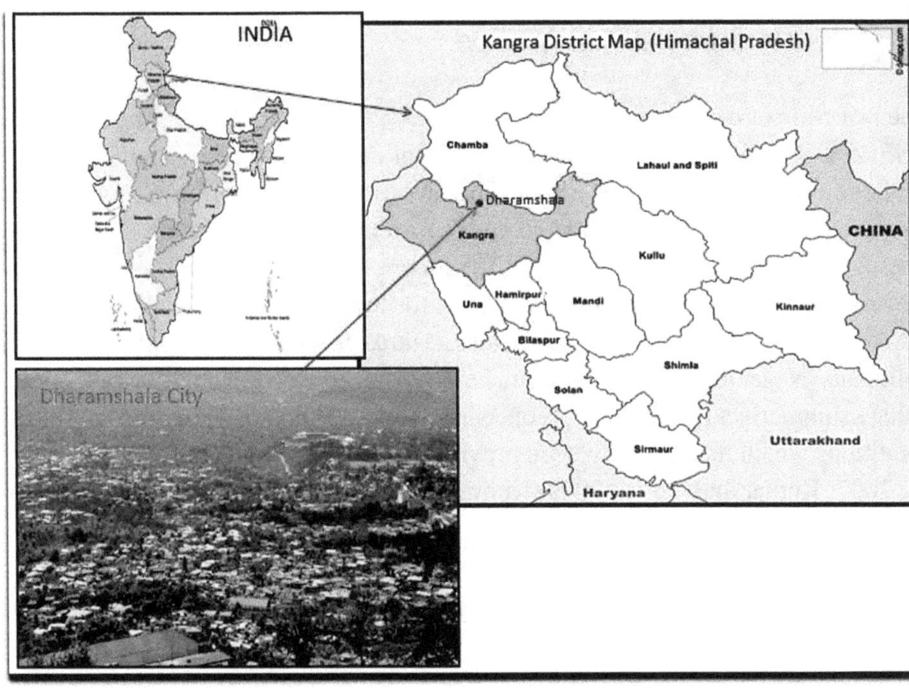

Similarly, the pollutant NO_x is analyzed using the *modified Jacobs and Hochheiser method* (Ganguly and Thapa, 2016; Saravanakumar et al., 2016). This methodology involves passing of a fixed volume of ambient air in an initial mixture of a solution of *sodium hydroxide* and *sodium arsenite*. The resulting nitrite ion concentration is then determined on the basis of color intensity formed due to the reaction of nitrite ion with phosphoric acid, sulfanilamide, and N-(1-naphthyl)- ethylenediamine dihydrochloride (NEDA) based on spectrophotometer analysis at 540 nm

Measurement of Respirable Suspended Particulate Matter (RSPM) are carried out using Respirable Dust sampler (RDS) which functions on the principle of air suction via cyclonic connector for collection of the suspended particulate matter (>10 μm). The average flow rate to be maintained during the collection process is 1.1 m^3/min (Ganguly and Thapa, 2016). Thereafter, the collected particulate matter on a filter paper is tested in the laboratory to determine the mass of RSPM which when divided by the air volume sampled leads to the concentration of RSPM denoted in units of μg/m^3 or ppm (Saravanakumar et al., 2016).

Air Quality Index

Air Quality Index by CPCB Method

The principle of determination of AQI by the CPCB methodology is based on the USEPA technique but using distinct breakpoint indices for Indian conditions. The methodology in principle for an AQI involves the determination of sub-indices for the criteria pollutants considered and their relative agglomeration (Kumar and Goyal, 2009; Kumar and Goyal, 2011). In contrast to the USEPA methodology the breakpoint indices for criteria pollutants for Indian conditions are based on the NAAQS standards and the potential health hazards posed by the considered criteria pollutants (Nagendra et al., 2007; Kumar and Goyal, 2009; Kumar and Goyal, 2011). Table 2 summarizes the breakpoint concentration and proposed sub-indices for Indian conditions which have already been previously reported in literature (Nagendra et al., 2007; Kumar and Goyal, 2009; Kumar and Goyal, 2011)

Table 2. Proposed sub-index and breakpoint pollutant concentration for Indian-AQI (Nagendra et al., 2007)

SI. No.	Index Values	Descriptor	SO_2 (24-h avg.) (µg/m^3)	NO_2 (24-h avg.) (µg/m^3)	SPM (24-h avg.) (µg/m^3)	RSPM (24-h avg.) (µg/m^3)
1.	0–100	Good	0–80	0–80	0–200	0–100
2.	101–200	Moderate	81–367	81–180	20–260	101–150
3.	201–300	Poor	368–786	181–564	261–400	151–350
4.	301–400	Very poor	787–1572	565–1272	401–800	351–420
5.	401–500	Severe	>1572	>1272	>800	>420

[a]Good: Air quality is acceptable; however, for some pollutants, there may be a moderate health concern for a very small number of people.

[b]Moderate: members of sensitive groups may experience health effects.

[c]Poor: Members of sensitive groups may experience more serious health effects.

[d]Very Poor: Triggers health alert, everyone may experience more serious health effects.

[e]Severe: Triggers health warnings of emergency conditions

Table 3. Air quality categories based on air quality index as per alternative methodology guidelines

Sl. No.	Index Values	Descriptor
1.	0–50	Good
2.	51–100	Moderate
3.	101–150	Unhealthy for sensitive groups
4.	151–200	Unhealthy
5.	≥ 200	Hazardous

(Rao and Rao, 1986)

The formulae used for determination of AQI using the USEPA method is as follows

$$I_p = \left[\frac{\left(I_{HI} - I_{LO}\right)}{\left(BP_{HI} - BP_{LO}\right)}\right]\left(C_p - BP_{LO}\right) + I_{LO} \tag{1}$$

Where,

I_p = the AQI for pollutant p
C_p = the actual ambient concentration of pollutant p
BP_{HI} = the breakpoint given in Table 2 that is greater than or equal to C_p
BP_{LO} = the breakpoint given in Table 2 that is less than or equal to C_p
I_{HI} = the sub-index value corresponding to BP_{HI}
I_{LO} = the sub-index value corresponding to BP_{HI}

The highest value amongst the AQI determined for the criteria pollutant is considered to be the overall AQI (Nagendra et al., 2007; Kumar and Goyal, 2009; Kumar and Goyal, 2011).

2.3.2 Air Quality Index by alternative method (AQI_{am})

The determination of AQI using the alternative methodology uses the following expression (Dash and Dash, 2015; Saravanakumar et al., 2016).

$$I_p = \frac{1}{4}\left(\frac{I_{PM10}}{S_{PM10}} + \frac{I_{PM2.5}}{S_{PM2.5}} + \frac{I_{SO2}}{S_{SO2}} + \frac{I_{NO2}}{S_{NO2}}\right) \times 100 \tag{2}$$

Where I_p is the overall AQI values. I_{values} is the actual ambient concentrations of pollutants PM_{10}, $PM_{2.5}$, SO_2 and NO_2. S_{values} is the NAAQS of pollutants PM_{10}, $PM_{2.5}$, SO_2 and NO_2 as per CPCB.

The interrelationship between the AQI determined on the alternative methodology (AQI_{am}) and the particular the air quality divisions associated with using this methodology has been summarized in Table 3 (Rao and Rao, 1986). A higher value of AQI indicates higher impacts of air pollutants on human health.

RESULTS AND DISCUSSION

AQI Values Using CPCB Methodology

The AQI for the two monitoring stations at the study locations have been determined on the basis of monitored concentrations of the important criteria pollutants for the study period of two years of 2016 and 2017. In this context, the determined AQI parameter for both the monitoring stations using the revised CPCB methodology has been presented in Table 4. It is observed from Table 4 that the annual average AQI for monitoring station I is 40 and 35 and for monitoring station II is 42 and 38 respectively for the study years of 2016 and 2017 and hence classified as *'good'* category as per the breakpoint indices value shown in Table 2. In fact, for both the years, the annual AQI value was less than 50 at both the monitoring locations. The highest annual average AQI values for both the monitoring stations were reported in the year 2016. Further, averages of the AQI indicated that the air quality was '*good*' irrespective of seasonal variations at both the monitoring sites. The highest AQI was reported for winter in 2016 (AQI = 43) and spring in 2017 (AQI = 41) for monitoring station I and spring in 2016 (AQI = 42) and summer in 2017 (AQI = 39) for monitoring station II. The annual and seasonal variation of the AQI for the modified CPCB methodology is shown in Figure 2. The most critical pollutant for determining the AQI values at both the monitoring locations was RSPM and is identical to earlier reported literature for Delhi (Kumar and Goyal, 2011; NAMP 2016) and Bangalore (Nagendra et al., 2007). This also acts a corollary to earlier studies based on exceedence analysis wherein it has been reported that exceedence factors for NO_x and SO_2 are often low whereas it may vary from high to low for RSPM (Ganguly and Thapa, 2016). The results of the AQI analysis determined using this technique show compliance with results from other Tier-II cities in India wherein the air quality is generally classified as *good to moderate* levels. The air quality for metropolitan and Tier-I cities are generally classified as *poor to severe* levels.

Table 4. Air quality index values for both monitoring stations for the study period (2016-2017) using CPCB methodology; note that the symbol ("-" refer to the unavailability of the data)

Months	Monitoring Station –I		Monitoring Station –II	
	2016	2017	2016	2017
January	42	39	48	41
February	40	38	51	39
March	32	36	37	34
April	32	37	39	39
May	47	49	49	41
June	49	48	36	42
July	37	-	39	-
August	28	27	27	36
September	38	28	35	37
October	36	26	36	36
November	52	31	54	-
December	52	31	55	37
Annual Average	40	35	42	38

Figure 2. Annual and seasonal variation of AQI over the study period: (a) Monitoring Station –I; (b) Monitoring Station -II

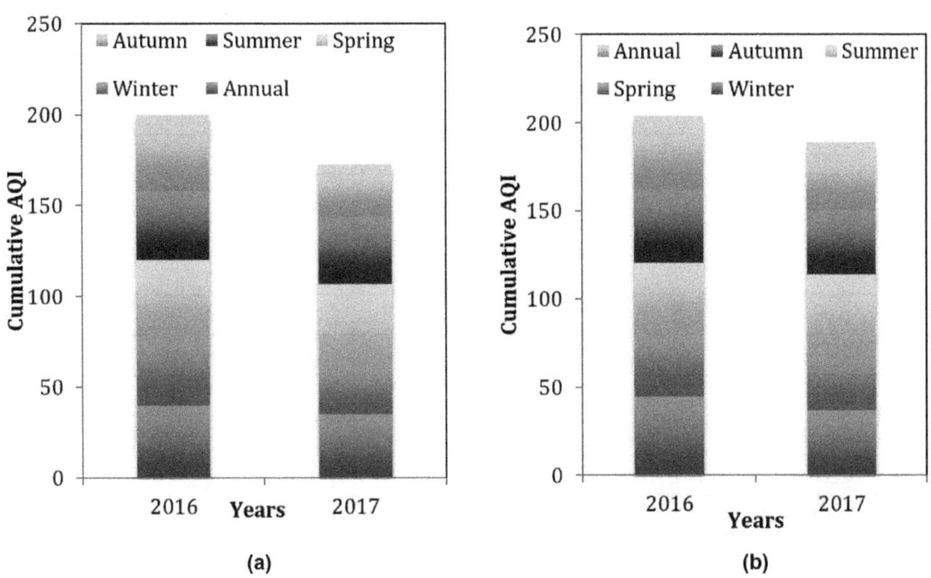

AQI Values Using Alternative Methodology

Similar to the above analysis carried out for determining the AQI using the CPCB methodology, the AQI_{am} for both the monitoring locations of the study area for the study period was determined using the alternative methodology. The results for the AQI_{am} analysis using this technique has been summarized in Table 5. It is observed from Table 5 that the annual averaged AQI values were determined to be 14 and 12 for monitoring station I and 13 and 12 for monitoring station II for the year 2016 and 2017 respectively. A comparative analysis of these AQI values with the air quality descriptors as shown in Table 3 shows that the air can be classified as *'good'* category. The highest annual averages of AQI_{am} for both the monitoring locations were observed for year 2016 similar to the AQI using the modified CPCB methodology. Further, seasonal variations of the AQI_{am} showed that the highest value was observed for winter (AQI_{am} = 14) in 2016 and summer for 2017 (AQI_{am} = 13). The AQI determined using the alternative method uses the limit values of all the criteria pollutants to determine the AQI_{am} = 14. It can be mentioned from the

Table 5. Air quality index values for both monitoring stations for the study period (2016-2017) using alternative methodology; note that the symbol ("-"refer to the unavailability of the data)

Months	Monitoring Station –I		Monitoring Station –II	
	2016	2017	2016	2017
January	14	13	15	13
February	14	13	15	12
March	11	12	12	11
April	11	12	12	13
May	15	16	16	13
June	16	15	12	13
July	13	-	13	-
August	11	9	10	12
September	13	9	12	12
October	13	9	12	12
November	17	10	16	-
December	17	10	17	12
Annual Average	14	12	13	12

methodology that there is no single definitive pollutant, yet the maximal fraction for determination of AQI_{am} is from RSPM. The annual and seasonal variation of the AQI_{am} is shown in Figure 3.

Comparative analysis of AQI calculated on the alternative methodology with other Tier-II cities in India showed that the AQI_{am} was *maxim* near industrial sources and *less* in residential areas in a study conducted in Coimbatore (Saravanakumar et al., 2016). In a similar study carried out in Biliepeda (Dash and Dash, 2015) the AQI_{am} was classified as *moderate* with all major sources having relative contributions to the existing air quality. In contrast to the earlier reported literature, wherein the studies were conducted in cities having both industrial and vehicular sources affecting the air quality the monitoring stations in Dharamshala are primarily hilly location having almost no industries and comparatively less traffic flow (even though it is a tourist destination). Further, the meteorological conditions at the study location vary from previously reported literature sources which also have an influence on the existing air quality in Dharamshala.

Figure 3. Annual and seasonal variation of AQI over the study period at station I: (a) Monitoring Station –I ; (b) Monitoring Station -II

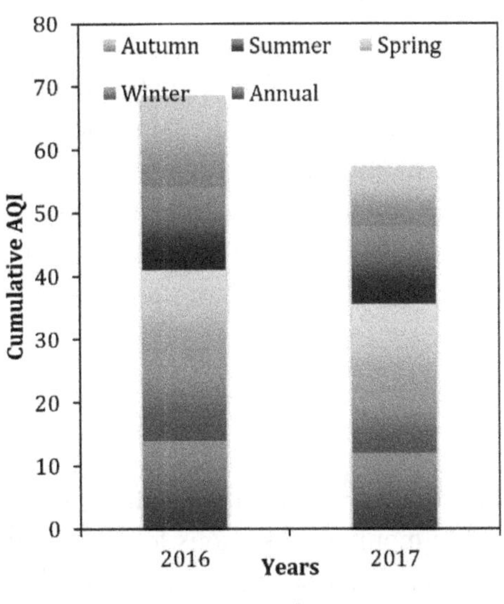

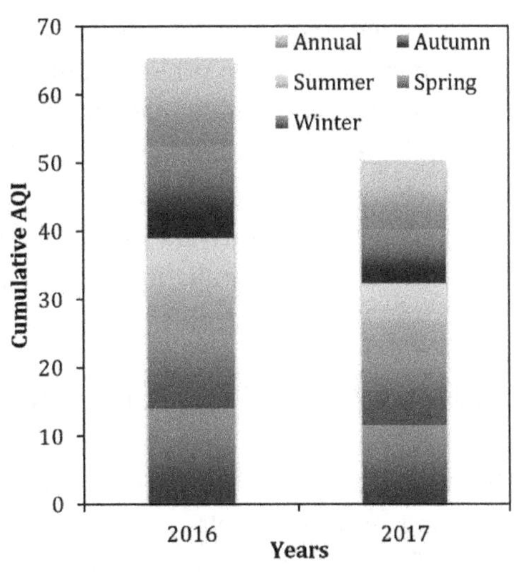

Comparison of AQI Values Using CPCB and Alternative Methodology

A detailed analysis of the methodology of determination of AQI using both the methods show that while the CPCB methodology uses the highest value of AQI determined from the concentrations of the criteria pollutants based on the sub-indices for Indian conditions the alternative methodology determines the AQI_{am} based on the contribution of all the relevant criteria pollutants. This can be noted from both *Equations 1 and 2*. In this context, the different methods have distinct and separate AQI categorizations as observed from Tables 2 and 3. Further, a review of literature shows that while the revised CPCB methodology for determination of AQI is more prevalent for metropolitan and Tier-I cities in India (Delhi, Bangalore) (Nagendra et al., 2007; Kumar and Goyal, 2011), the alternative methodology has been well used for predicting AQI for Tier-II and lower category of cities of India (Dash and Dash, 2015; Saravanakumar et al., 2016).

The output results of the comparison of the AQI values of both the monitoring locations in the study area shows that while the annual average AQI using the CPCB methodology is higher for monitoring site –II (AQI = 42 for 2016; AQI = 38 for 2017) the AQI based on the alternative methodology (AQI_{am} = 14 for 2016; AQI_{am} = 12 for 2017) is higher for monitoring site –I. Detailed information regarding the location of the actual monitoring stations are presently a little skewed but can be adjudged to be set in urban locations as per the names of the monitoring sites assigned. Evaluation of annual and seasonal variations of AQI by using both the methodologies showed similar trends for both the monitoring locations for the study period. However, most importantly the AQI values obtained using both the methods were in complete accord and surmised the air quality to be of *good quality* as per their own individual air quality categorization criteria for the entire study period.

SUMMARY AND CONCLUSION

This book chapter presents the evaluation of the existing air quality in Dharamshala city using AQI methodology at both the quality monitoring stations for the study period of 2016 and 2017. Two different methodologies were utilized for determination of AQI. The modified CPCB methodology based on the USEPA method but using different break point indices relative Indian conditions calculates the AQI based on the highest value of the AQI determined for the criteria pollutants whereas the alternative methodology determines the AQI based on the relative fractions of all the considered criteria pollutants. The computed AQI for both the methods showed that the air quality can be categorized as '*good*' irrespective of any seasonal variations. RSPM was the most significant criteria pollutant affecting the determination of AQI for both the methods. Hence, as such based on the categorizations of AQI there is no need of immediate apprehension in the context of air quality but a detailed environmental plan should be created as slight change in in the air quality can significantly affect the biodiversity of the study location. Till date, continuous monthly air quality monitoring data is available for both the sites at the study location and needs to be continued so as to develop existing baseline concentrations of criteria pollutants in Dharamshala and any significant changes in their concentrations. This is particularly important because the city has been proclaimed as the second capital of the state which could significantly increase vehicular population leading to increased pollutant concentrations. Further, it is a popular tourist spot and also being considered as smart city, regular monitoring of pollutants and computation of AQI is important to keep track of any abrupt changes in air quality in the city.

REFERENCES

Anderson, H. R., Favarato, G., & Atkinson, R. W. (2011). Long-term exposure to air pollution and the incidence of asthma: Meta-analysis of cohort studies. *Air Quality, Atmosphere & Health*, *6*(1), 47–56. doi:10.100711869-011-0144-5

Batterman, S., Ganguly, R., Isakov, V., Burke, J., Arunachalam, S., Snyder, M., ... Lewis, T. (2014). Dispersion Modeling of Traffic-Related Air Pollutant Exposures and Health Effects among Children with Asthma in Detroit, Michigan. *Transportation Research Record: Journal of the Transportation Research Board*, *2452*(1), 105–113. doi:10.3141/2452-13

Bishoi, B., Prakash, A., & Jain, V. K. (2009). A Comparative Study of Air Quality Index Based on Factor Analysis and USEPA Methods for an Urban Environment. *Aerosol and Air Quality Research*, *9*(1), 1–17. doi:10.4209/aaqr.2008.02.0007

Bortnick, S. M., Coutant, B. W., & Eberly, S. I. (2002). Using Continuous PM2.5 Monitoring Data to Report an Air Quality Index. *Journal of the Air & Waste Management Association*, *52*(1), 104–112. doi:10.1080/10473289.2002.1047076 3 PMID:15152670

Buffler, P. A., Kwan, M. L., Reynolds, P., & Urayama, K. Y. (2005). Environmental and Genetic Risk Factors for Childhood Leukemia: Appraising the Evidence. *Cancer Investigation*, *23*(1), 60–75. doi:10.1081/CNV-46402 PMID:15779869

Central Pollution Control Board (CPCB). (2009). *National Ambient Air Quality Standards for India*. Available online at http://cpcb.nic.in/National_Ambient_Air_Quality_Standards.php

Clark, N. A., Demers, P. A., Karr, C. J., Koehoorn, M., Lencar, C., Tamburic, L., & Brauer, M. (2010). Effect of early life exposure to air pollution on development of childhood asthma. *Environmental Health Perspectives*, *118*(2), 284–290. doi:10.1289/ehp.0900916 PMID:20123607

Clean Air Institute. (2013). *La Calidad del Aire en América Latina: Una Visión Panorámica*. Retrieved from http://www.cleanairinstitute.org/calidaddelaireamericalatina/cai-report-spanish.pdf

Climate and Environment. (2016). Retrieved from https://www.triplepundit.com/2016/04/world-air-quality-index-brings-data-life/

Dash, S. K., & Dash, A. K. (2015). Determination of Air Quality IndexStatus near Bileipada, Joda Area of Keonjhar, Odisha, India. *Indian Journal of Science and Technology*, *8*(35). doi:10.17485/ijst/2015/v8i35/81468

Elshout, S. (2008). Comparing Urban Air Quality in Europe in Real Time A Review of the Existing Air Quality Indices and the Proposal of a Common Alternative. *Environment International*, *34*(5), 720–726. doi:10.1016/j.envint.2007.12.011 PMID:18234340

Ganguly, R., Batterman, S., Isakov, V., & Breen, M. (2015). Effect of geocoding errors on traffic-related air pollutant exposure and concentration estimates. *Journal of Exposure Science & Environmental Epidemiology*, *25*(5), 490–498. doi:10.1038/jes.2015.1 PMID:25670023

Ganguly, R., & Broderick, B. (2008). Performance evaluation and sensitivity analysis of the general finite line source model for CO concentrations adjacent to motorways: A note. *Transportation Research Part D, Transport and Environment*, *13*(3), 198–205. doi:10.1016/j.trd.2008.01.006

Ganguly, R., & Thapa, S. (2016). An assessment of ambient air quality in Shimla city. *Current Science*, *111*(3), 509–516. doi:10.18520/cs/v111/i3/509-516

Gasana, J., Dillikar, D., Mendy, A., Forno, E., & Ramos Vieira, E. (2012). Motor vehicle air pollution and asthma in children: A meta-analysis. *Environmental Research*, *117*, 36–45. doi:10.1016/j.envres.2012.05.001 PMID:22683007

Gauderman, W. J., Vora, H., McConnell, R., Berhane, K., Gilliland, F., Thomas, D., ... Peters, J. (2007). Effect of exposure to traffic on lung development from 10 to 18 years of age: A cohort study. *Lancet*, *369*(9561), 571–577. doi:10.1016/S0140-6736(07)60037-3 PMID:17307103

Heal, M. R., Kumar, P., & Harrison, R. M. (2012). Particles, Air Quality, Policy and Health. *Chemical Society Reviews*, *41*(19), 6606–6630. doi:10.1039/c2cs35076a PMID:22660420

Kassomenos, P., Skouloudis, A. N., Lykoudis, S., & Flocas, H. A. (1999). Air-Quality Indicators for Uniform Indexing of Atmospheric Pollution Over Large Metropolitan Areas. *Atmospheric Environment*, *33*(12), 1861–1879. doi:10.1016/S1352-2310(98)00355-0

Kowalska, M., Osrodka, L., Klejnowski, K., Zejda, J. E., Krajny, E., & Wojtylak, M. (2009). Air quality index and its significance in environmental health riskcommunication. *Archives of Environmental Protection*, *35*(1), 13–21.

Kumar, A., & Goyal, P. (2011). Forecasting of daily air quality index in in Delhi. *The Science of the Total Environment*, *409*(24), 5517–5523. doi:10.1016/j.scitotenv.2011.08.069 PMID:21962560

Kumar, A., & Goyal, P. (2011). Forecasting of air quality in Delhi using principal component regression technique. *Atmospheric Pollution Research*, *2*(4), 436–444. doi:10.5094/APR.2011.050

Kumar, P., Andrade, M. F., Ynoue, R. Y., Fornaro, A., & de Freitas, E. D. (2016). New Directions: From biofuels to wood stoves: the modern and ancient air quality challenges in the megacity of São Paulo. *Atmospheric Environment*, *140*, 364–369. doi:10.1016/j.atmosenv.2016.05.059

Kumar, P., Jain, S., Gurjar, B. R., Sharma, P., Khare, M., Morawska, L., & Britter, R. (2013). Can a "Blue Sky" return to Indian megacities? *Atmospheric Environment*, *71*, 198–201. doi:10.1016/j.atmosenv.2013.01.055

Kumar, P., Khare, M., Harrison, R. M., Bloss, W. J., Lewis, A., Coe, H., & Morawska, L. (2015). New Directions: Air pollution challenges for developing megacities like Delhi. *Atmospheric Environment*, *122*, 657–661. doi:10.1016/j.atmosenv.2015.10.032

Landulfo, E., Matos, C. A., Torres, A. S., Sawamura, P., & Uehara, S. T. (2007). Air Quality Assessment Using a Multi-Instrument Approach and Air Quality Indexing in an Urban Area. *Atmospheric Research*, *85*(1), 98–111. doi:10.1016/j.atmosres.2006.11.008

Langholz, B., Ebi, K. L., Thomas, D. C., Peters, J. M., & London, S. J. (2002). Traffic Density and the Risk of Childhood Leukemia in a Los Angeles Case-Control Study. *Annals of Epidemiology*, *12*(7), 482–487. doi:10.1016/S1047-2797(01)00317-9 PMID:12377426

Lindgren, A., Bjork, J., Stroh, E., & Jakobsson, K. (2010). Adult asthma and traffic exposure at residential address, workplace address, and self-reported daily time outdoor in traffic: A two-stage case-control study. *BMC Public Health*, *10*(1), 716. doi:10.1186/1471-2458-10-716 PMID:21092159

Longhurst, J. (2005). 1 to 100: Creating an Air Quality Index in Pittsburg. *Environmental Monitoring and Assessment*, *106*(1-3), 27–42. doi:10.100710661-005-0758-x PMID:16001710

Mayer, H., & Kalberlah, F. (2008). Two Impact Related Air Quality Indices as Tools to Assess the Daily and Long-Term Air Pollution. *International Journal of Environment and Pollution*, *32*, 232–243.

Monteiro, A., Viera, M., Gama, C., & Miranda, A. I. (2016). Towards an improved air quality index. *Air Quality, Atmosphere & Health*. doi:10.100711869-016-0435-y

Murena, F. (2004). Measuring Air Quality over Large Urban Areas: Development and Application of an Air Pollution Index at the Urban Area of Naples. *Atmospheric Environment*, *38*(36), 6195–6202. doi:10.1016/j.atmosenv.2004.07.023

Nagendra, S. M. S., Venugopal, K., & Jones, S. L. (2007). Assessment of air quality near traffic intersectionsin Bangalore city using air quality index. *Transportation Research Part D, Transport and Environment*, *12*(3), 167–176. doi:10.1016/j.trd.2007.01.005

Rao, M. N., & Rao, H. V. N. (1986). Air pollution. TATA McGraw-Hill Publishing Company.

Saravanakumar, R., Sivalingam, S., & Elangovan, S. (2016). Assessment of Air Quality Index of Coimbatore City in Tamil Nadu. *Indian Journal of Science and Technology*, *9*(41). doi:10.17485/ijst/2016/v9i41/99185

Sharma, M., Pandey, R., Maheshwari, M., Sengupta, B., Shukla, B. P., Gupta, N. K., & Johri, S. (2003). Interpretation of Air Quality Data Using an Air Quality Index for the City of Kanpur, India. *Journal of Environmental Engineering and Science*, *2*, 453–462. doi:10.113903-047

Sharma, V. K. (1999). Development of Air Quality Indices for Mumbai, India. *International Journal of Environment and Pollution*, *11*(2), 141–146. doi:10.1504/IJEP.1999.002254

Zlauddin, A., & Siddiqui, N. A. (2006). Air quality index (AQI) A tool to determine ambient air quality. *Pollution Research*, *25*(4), 885–887.

Chapter 4
Climate Changes and Atmospheric Pollution:
Global and Regional Impacts

Gustavo Marques da Costa
Universidade Feevale, Brazil

Darlan Daniel Alves
Universidade Feevale, Brazil

Danielle Paula Martins
Universidade Feevale, Brazil

Katiucia Nascimento Adam
Universidade Feevale, Brazil

Sabrina Antunes Vieira
Universidade Feevale, Brazil

Daniela Muller de Quevedo
Universidade Feevale, Brazil

Daniela Montanari Migliavacca Osório
Universidade Feevale, Brazil

ABSTRACT

The objective of this chapter is to present the central concepts, parameters, and methods for the monitoring of climate changes, with a focus on air pollution, and the possible global and regional impacts of climate changes as well. There are plant species used as bioindicators that have a high sensitivity or ability to accumulate environmental pollutants. Another method that this chapter will present is the use of receiver models that employ both mathematical and statistical approaches to quantify the individual contribution of a given number of emission sources in the composition of a sample. The data presented in this chapter will provide reliable bases and methodologies for environmental control, supporting the adoption of more restrictive policies.

DOI: 10.4018/978-1-5225-7289-3.ch004

INTRODUCTION

Monitoring of the impacts of climate changes introduces some questions about its effectiveness, timing, intervention levels, the uncertainty of data, available financial resources, among others, which need to be considered.

Air pollution can be defined as the presence of harmful substances in the atmosphere in an amount sufficient to affect its composition or equilibrium, damaging the environment and the most varied life forms. This type of pollution causes severe impacts not only on human life, but also on animal and plant life, as well as the deterioration of cultural leisure goods, and the disablement or depletion of natural resources. In this context, bioindicators are organisms that react to environmental changes by modifying their vital functions, and their chemical composition, providing information about the environmental scenario. A bioindicator should be taxonomically defined and easily recognizable by non-specialists; present a wide geographical distribution; be abundant or easy to collect; have well-known ecological characteristics and be able to be used in laboratory studies. Indicator organisms are still a tool of great importance to diagnose the quality of ecosystems, besides contributing to the identification of reference areas and critical areas in a given region. Morphological, physiological or genetic changes in microorganisms, plants, and animals including humans may be a consequence of potentially toxic or genotoxic chemicals present in the atmosphere.

The global impact of air pollutants refers to those that can affect the planet as a whole. The most critical problem caused by the use of fossil fuels is gases, such as carbon dioxide (CO_2), chlorofluorocarbons (CFCs), methane (CH_4), nitrous oxide (N_2O), etc. These gases are known as greenhouse gases and capture part of the infrared radiation that the Earth returns to space, causing the increase of the atmospheric temperature due to climatic changes. The most important greenhouse gas is carbon dioxide which is the main compound resulting from the complete combustion of fuels. When in large quantities, carbon dioxide and other pollutants end up forming a filter in the atmosphere, during the day the Earth is heated by the sun, and at night it loses the stored heat, resulting in a reduction in temperature. However, with the layer of pollutants present, the heat is retained in the Earth, causing an increase in the average temperature. The transport sector has a considerable stake in climate issues as it contributes 22% to the global emission of greenhouse gases, employing practically 100% of fossil fuels. Brazil has one of the highest growth rates of its CO_2 emissions due to the industrialization that has taken place, and the growth of the existing vehicle fleets. There is, therefore, a great need for effective public policies in this sector, in order to change this framework in which, in addition to climate change, there are direct damages to human health and the environment.

The Metropolitan Region of São Paulo, considered the largest Brazilian urban area, with about 18 million inhabitants, today presents one of the most critical tables in the country regarding the change in temperature due to the increase of carbon dioxide (CO_2). This scenario is due to a large number of individual transport users, which ends up producing more than half of all the city's air pollution. Besides, CO_2 generates the greenhouse effect, which aggravates the environmental situation, compromising the quality of the environment for its inhabitants. Areas of heavy traffic accumulate complex compounds of pollutants that contribute about 90% to carbon dioxide emissions. This complex mixture of pollutants can cause adverse effects on organisms and the environment due to the synergistic and antagonistic impact of pollutants and their association with weather conditions.

Atmospheric air quality is commonly characterized and monitored by monitoring networks, using physicochemical methods for the qualitative and quantitative detection of pollutants. Based on analyzes of this nature, limit values are established for concentrations of pollutants in the air, recommended by government agencies. There are also automatic networks and manual pollutant monitoring. However, there are plant species used as bioindicators that have a high sensitivity or ability to accumulate environmental pollutants and can, therefore, be used for the qualitative and quantitative assessment of atmospheric air, indicating the risks of exposure of organisms to pollutants. Plants are generally more sensitive than animals, have a faster developmental stage and reproductive cycle, and are still able to respond to environmental conditions in a short period. Active biomonitoring is a tool used to evaluate the air quality of environments with different magnitudes of environmental impacts and consists of the systematic analysis of the responses of organisms introduced in a standardized way in the natural environment for a specified period.

Identification of pollutant emission sources is a fundamental requirement in the design process of effective air pollution control strategies. The identification and characterization of the emission rates of the sources allow the elaboration of emission inventories, which can be used in the development of control strategies, including the possibility of reviewing existing regulations or the formulation of more restrictive rules. The distribution of pollution is the process of quantifying the individual contribution of each emission source to the mass, and chemical composition of the samples, and might be obtained through the use of receiver models.

Another method that will be presented in this chapter is the use of receiver models that employ mathematical and/or statistical approaches to quantify the individual contribution of a given number of emission sources in the composition of a sample, based on the structure or characteristics of the source profiles of emission. The fundamental principle of receiver modeling is that mass conservation can be assumed, and mass balance analysis can be used to identify and distribute emission sources.

There is a variety of receiver models available that can be used for studies of the identification of particulate matter emission sources. The United States Environmental Protection Agency (EPA) offers the EPA Chemical Mass Balance (EPA CMB), EPA Unmix and EPA Positive Matrix Factorization (EPA PMF 5.0) models.

Thus, it is essential to understand how climate changes can impact life on the planet, as the results of the report published by the Brazilian Panel on Climate Change warn of the average increase in temperatures between 2.5ºC and 3.0ºC until 2100, in the South Atlantic Forest (Brazil). This increase could be even higher if the population and economy continue to grow rapidly and the consumption of fossil fuels is maintained. The Paris Accord, an international pact that should guide the fight against climate change during this century, has set the objective that the average temperature increase of the planet by 2100 is less than two degrees Celsius and that as far as possible is limited to 1.5 degrees. It, therefore, points to the need for a "balance between anthropogenic emissions" and "absorption" of greenhouse gases by "drains" in the second half of the century.

The objective of this chapter will be to present the main concepts, parameters, and methods for the monitoring of climate changes, with a focus on air pollution, and the possible global and regional impacts of climate changes as well.

AIR POLLUTION

The Concept of Pollution and Contamination

Atmospheric air in regions with urban agglomeration usually presents a variety of pollutants from stationary and mobile sources (Guimarães et al., 2000, Teixeira et al., 2008 & Teixeira et al., 2012), mainly due to the increasing use of motor vehicles (Crispim et al., 2012). In this sense, the growth of the vehicular fleet leads to a lower fluid flow capacity, which leads to more time spent in congestion and, consequently, increases the combustion of fuels, generating more pollutant gases into the atmosphere (Guimarães et al., 2000, Meireles and Cerqueira, 2011 & Alves et al., 2011) with different physicochemical characteristics and toxicities (Carreras et al., 2013).

Clean air is considered an essential requirement for the quality of life and consequently human well-being and health. However, in recent years air pollution has significantly threatened global health (WHO, 2005). According to WHO (WHO, 2005), air pollution is the spatial and temporal variation of substances in the atmosphere. These substances can be considered as pollutants when concentrations found in the environment can render atmospheric air harmful or offensive to health, inconvenient to public welfare, harmful to materials, fauna, and flora or detrimental to safety, use and enjoyment of property, and the normal activities of the community.

Air pollution encompasses a range of activities, phenomena, and substances that contribute significantly to the imbalance and deterioration of air quality (Dallarosa et al., 2008). These chemical constituents present in the air can affect fauna, flora, and materials, as well as cause changes in the quality of life. The impact of air pollution on receptors, especially on flora, has led to several studies in regions potentially affected by air pollution (Klumpp et al., 2004).

The World Health Organization recommends that current limits on the emission of pollutants should be revised since even meeting the standards established by current legislation, studies indicate that concentrations affect human health. For example, in a study by Pope et al. (2009) found a significant association between the low concentrations of $MP_{2.5}$ and life expectancy. The effects of long-term exposure on mortality have been reported by several studies (Miller et al., 2007, Lipsett et al., 2011 & Crouse et al., 2012). In Europe study conducted by Beelen et al. (2008) indicated a mortality risk for MP2.5 of 6%.

The World Health Organization (WHO) reports alarming results, where it is estimated that more than 80% of people living in urban areas are exposed to levels of air quality that exceed the limits of the World Health Organization. Also, it was identified that 98% of cities in low- and middle-income countries with more than 100,000 inhabitants do not meet the WHO air quality guidelines, while in high-income countries this percentage drops to 56%.

MAIN AIR POLLUTANTS

Indicators of Air Pollution

The main air pollutants include sulfur, nitrogen and carbon compounds, often added to atmospheric particulate matter, either in the form of aerosols or wet and dry precipitation, as well as organic compounds and ozone. In industrialized regions, we can find additional pollutants, gaseous by-products, incomplete burning of fossil fuels and typical industrial emissions (CO_2, NH_3, SO_4, SO_2, NOx, trace metals and metal oxides). These chemicals may be incorporated into the particulate matter and cause changes in the elementary cycles of the geochemical processes of the exposed ecosystems, including relating damage to the health of living organisms and materials.

WMO in its publication on the update of the Global Air Quality Guidelines (WHO, 2015) categorized the air pollutants into four groups, presented in Table 1 so that a reflection on inclusion and broadening the discussion of these pollutants at global levels.

The pollutants mentioned in the groups in Table 1 were recommended for a reassessment of the updates of the Global Air Quality Guidelines (WHO, 2015), as there is evidence that they may pose a health risk to human populations, and, therefore, should be carefully studied and regulated by WHO. But, for this to occur, it is necessary to monitor these pollutants in the atmospheric air and also standardization of collection and analysis for the pollutants mentioned in Groups 2, 3 and 4 should occur.

Thus, in this chapter, we will mention only the Group 1 atmospheric pollutants, which include: Particulate Matter, Ozone, Nitrogen dioxide, Sulfur dioxide and Carbon monoxide.

Table 1. Proposition of pollutant groups by WHO (2015)

Groups	Air Pollutants	Observations
Group 1	Particulate Matter Ozone Nitrogen dioxide Sulfur dioxide Carbon monoxide	Pollutants to be considered of importance in the process of updates of the WHO Global Air Quality Guidelines AQGs. Health-related evidence for these pollutants justifies the need for systematic reassessment, especially for the interactions between pollutants.
Group 2	Cadmium Chromium Lead Benzene PCDDs & PCDFs PAHs*	Pollutants which, due to their vast presence in the ambient air and a significant amount of newly available evidence on adverse health effects, should have a systematic review of the existing evidence as they may lead to changes in current guideline values.
Group 3	Arsenic Manganese Platinum Vanadium Butadiene Trichloroethylene Acrylonitrile** Hydrogen sulfide Vinyl chloride Toluene Nickel	The inclusion of these poles for a systematic reassessment of the new evidence, with less urgency than Group 1 and 2, though. Some of the pollutants in this group need to be considered as part of the PM mix.
Group 4	Mercury Asbestos Formaldehyde Styrene Tetrachloroethylene Carbon disulfide Fluoride PCBs 1,2-dichloroethane Dichloromethane	The pollutants in this group, according to experts, do not justify the imminent need for their reassessment in the context of updating WHO environmental AQGs and should, therefore, be retained for future consideration. Also, some of these pollutants are currently addressed in occupational environment guidelines and/or through water guidelines or other types of management processes.

Source: Adapted from WHO (2015)

PARTICULATE MATTER

Atmospheric particles, or also known as particulate matter, come from natural sources such as dust, seaspray (marine spray), volcanic eruptions (volcanoes) and anthropogenic sources, especially the burning of fossil fuels (Seinfeld, J. H. & Pandis, 2006). This particulate material can be classified according to its aerodynamic diameter, and the atmospheric particles generally occur in size range of 1 to 3 µm and are distinguished in thick and thin. By conversion, the coarse fraction of the atmospheric particles consists of an aerodynamic diameter larger than 2.5 µm and these can be further divided into super coarse and coarse particles.

Coarse particles have a diameter of 2.5 µm to 10 µm and are generally produced by mechanical processes such as wind erosion, sea salt particles, pollen, and spores, as well as plant fibers and leaves. As the coarse particles are large, they stay in the atmosphere usually for a few hours after their emission or formation. These particles can still be deposited in the upper airways of the human respiratory system, being excreted (eliminated) by the human body through the nose, cough or swallowing. However, the super-coarse particles have an aerodynamic diameter greater than 10 µm and because they are very large, they cannot be inhaled and have no relevance from the point of view of human health.-

The fine particulate material ($PM_{2.5}$) can be subdivided into particles with aerodynamic diameter of 0.1 to 10 µm, resulting from combustion processes of diesel engines or direct injection gasoline, but can also be formed by natural processes, by coagulation of particle nuclei or by condensation of gas/vapor molecules on the surface of the particles (Kumar et al., 2010). Thus, the difference in the aerodynamic diameter of these particles is directly related to their toxic potential, because the thinner the particle, the greater distance it can travel, either in the environment or when inhaled. In the present day, we have the term nanoparticles, which is used for atmospheric particles in bands of size below 100 nm, 50 nm, 10 nm or occasionally even for particles smaller than 1 µm (Morawska et al. 2008).

Therefore, the concern to evaluate the chemical composition of the particulate matter, mainly the inhalable particles, particle size smaller than 10 µm, is associated with important adsorption characteristic of chemical substances on its surface and that can be transported to the lower respiratory tract, the alveoli. These particles can also affect the climate, vegetation and also reduce visibility (Basha et al., 2010).

Concentrations of PM 10 worldwide are below 100 $\mu g.m^{-3}$, as shown in Figure 1 and Figure 2, but in some places, these values reach concentrations above 250 $\mu g.m^{-3.}$ According to the WHO air quality standards for PM 10, the maximum concentration may be 50 $\mu g.m^{-3}$ (24-hour mean) and 20 $\mu g.m^{-3}$ (annual mean)

Figure 1. PM10 levels by region and city size (annual mean), for available cities and towns latest year in the period 2008- 2015

Afr: Africa; Amr: America; Emr: Eastern Mediterranean; Eur: Europe; Sear: South-East Asia; Wpr: Western Pacific; LMI: Low- and middle-income; HI: high-income. PM10 values for the world are regional urban population-weighted. (Source: WHO's Urban Ambient Air Pollution database - Update 2016)

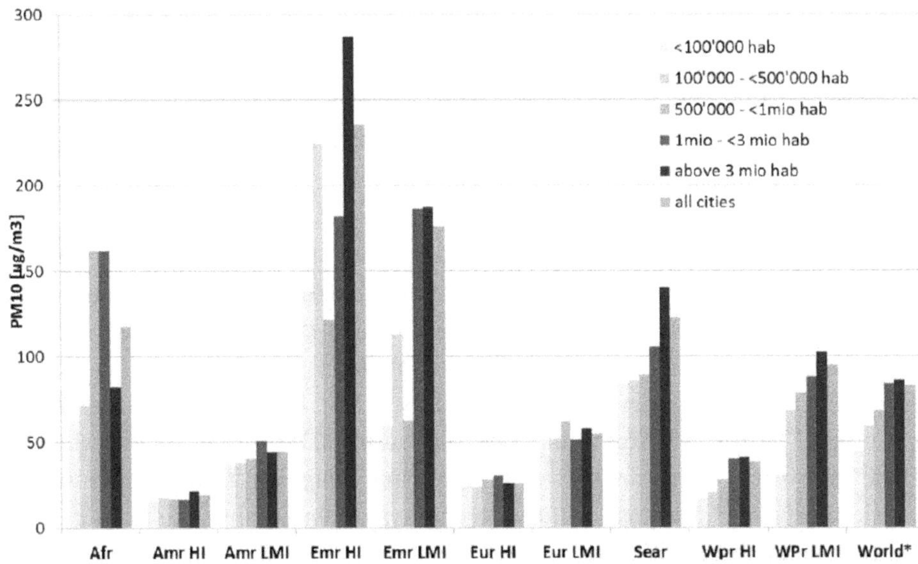

Figure 2. PM10 levels for available mega-cities of more than 14 million habitants for the last available year in the period 2011-2015

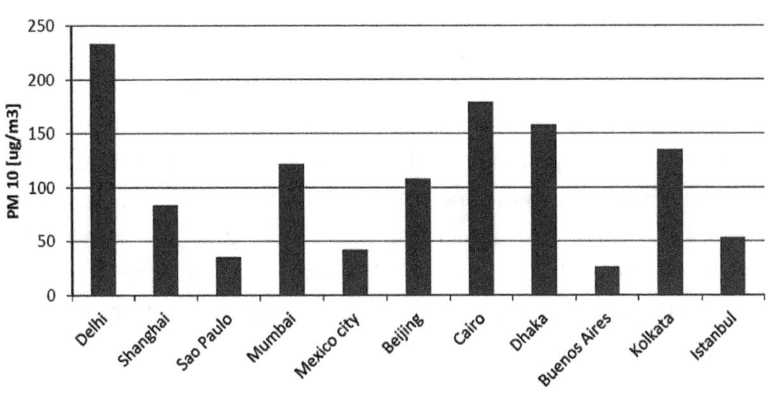

Ozone

Ozone is a secondary air pollutant, being formed from photochemical reactions with other pollutants such as NOx, CO, and CVs. Thus, its main sources are:

- Industrial emissions, such as chemical manufacturers, and sources of combustion, such as power plants burning fossil fuels;
- Small industry emissions such as gasoline distribution facilities, autobody paint shops, and printing shops;
- cars, trucks and buses;
- and off-road engines such as aircraft, locomotives, construction equipment, and gas-powered lawn and garden equipment.

The formation of tropospheric ozone is potentiated due to the climatic conditions, because on hot and sunny days, with little or no wind, the concentrations of ozone can reach unhealthy levels, for these conditions favor the formation of ozone and, if there are no changes in the climatic conditions, these episodes can cause Smog, which is a mixture of pollutants, mainly ozone. High ozone concentrations near ground level can be harmful to people, animals, crops and other materials. Ozone can irritate the respiratory system, causing coughing, throat irritation and/or uncomfortable chest sensation. It also aggravates chronic lung diseases such as emphysema and bronchitis, and reduces the ability of the immune system to fight bacterial infections in the respiratory system. Finally, ozone can cause permanent damage to the lungs. These effects may be worse in children and exercising adults.

Thus, due to the damages O_3 may cause to health, EPA, in 2015, reinforced keeping the 8-hour National Environmental Quality Standard (NAAQS) to 0.070 ppm (~ 150 $\mu g.m^{-3}$) (EPA, 2018). And WHO (2015) suggests encouraging long-term ozone exposure research, even if the concentration levels do not exceed air quality standards, which for ozone is 100 $\mu g / m^3$ for an average exposure of 8 hours.

Figure 3 shows the mean annual concentrations of O_3, 8-hour mean, for Europe from 2003 to 2012, which ranged from 59.6 to 85.8 $\mu g.m^{-3}$ (8-hour mean), well below air quality standards. Even meeting air quality standards, the WHO report (2015) says that European studies reported that daily mortality increased by 0.3%, and heart disease by 0.4%, by an increase of only 10 $\mu g.m^{-3}$ on exposure to ozone.

The results presented by the EEA (2015) also point out that by the end of the century the effects of climate change can increase ozone concentrations up to 8 $\mu g.m^{-3}$.

Figure 3. Mean concentration, 8-hour, in Europa, 2003 to 2012
Source: European Environment Agency (2015)

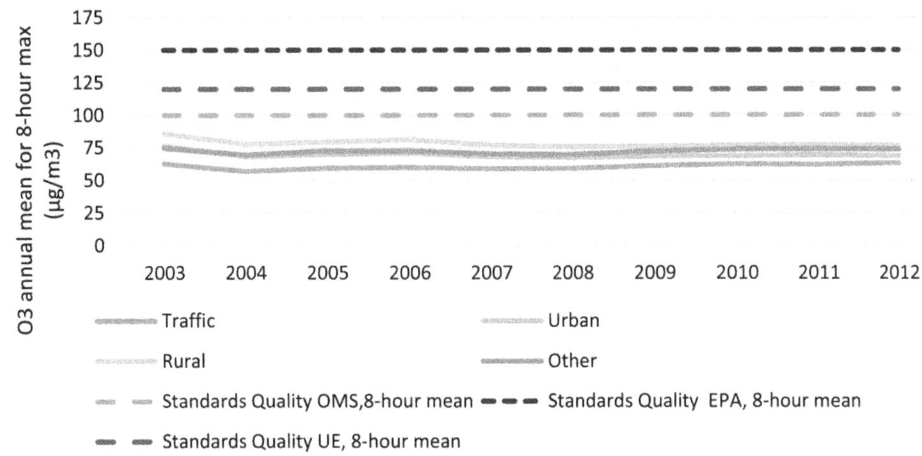

Nitrogen Dioxide: NO_2

Nitrogen dioxide (NO_2) is a group of highly reactive gases known as nitrogen oxides (NOx). Its main legislating pollutant is the NO_2 that has as its source the burning of fuel, vehicular emissions, industrial emissions, etc. Among the environmental effects of NOx, we can mention acid rain, due to NOx reaction with water and oxygen, damaging sensitive ecosystems such as lakes and forests, as well as damage to materials. Nitrate particles (a secondary NOx pollutant) may increase cloudiness, and another significant effect is the increase in nutrient pollution in coastal waters. Because it is a highly reactive gas group, there is current scientific evidence relating to short-term NO_2 exposures, ranging from 30 minutes to 24 hours, with a range of adverse respiratory effects, including increased asthma symptoms and increased disease and respiratory symptoms.

The quality standards for NO_2 for WHO and the EU are 40 $\mu g.m^{-3}$ (annual mean) and 200 $\mu g.m^{-3}$ (1-hour mean); for EPA, the values are 50 ppb (“ 94 $\mu g.m^{-3}$), 100 ppb (“ 188 $\mu g.m^{-3}$) for annual mean, and 1-hour mean, respectively.

NO_2 is the main source of nitrate aerosols, which make up a significant fraction of PM2.5 and, in the presence of ultraviolet light, ozone. Reduced lung function growth is also linked to NO_2 at concentrations currently measured (or observed) in cities in Europe and North America.

Sulfur Dioxide: SO_2

Sulfur dioxide (SO_2) is an important air pollutant that has significant impacts on the environment and climate at a global level at regional and local scales. It oxidizes to form sulfate aerosols that reduce visibility, affect the formation of clouds and can also lead to acid rain. The main anthropogenic sources of SO_2 are the burning of fossil fuels (Fioletov et al., 2015), metal smelting and gas refining (McLinden et al., 2014), which contribute 70% of the global anthropogenic emissions of SO2. The volcanic eruptions, degassing and sea spray representing SO_2 emissions come from natural sources (Faloona et al., 2010).

Environmental levels of SO_2 have declined over time in some regions of the world but have increased in others. Figure 4 shows global annual mean concentrations for the period 1980 to 2016, and a decrease above 80% is observed.

The experts concluded that, according to the results of the REVIHAAP Project, and considering the high levels measured in some countries, the available evidence for SO_2 should be analyzed again in relation to a very short time of 10 min and 24 hours of exposure, but also likely for a long-term exposure, for which there is currently no reference value (with subsequent impacts on a global scale, such as

Figure 4. EPA Data - For each pollutant, the trend statistic is directly related to the level and averaging time of the National Ambient Air Quality Standard (NAAQS). For more information regarding the levels and averaging times for the NAAQS statistics.

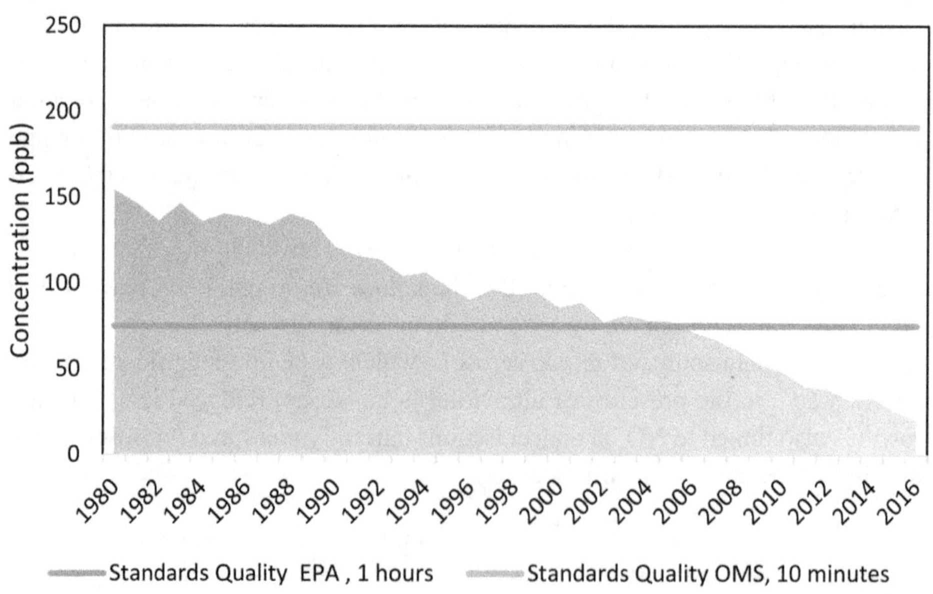

environmental acidification). This systematic review may benefit from recent/future information from recent national US and Canadian assessments, which should provide additional evidence of exposure to very short/short duration SO_2 and respiratory effects based on chamber studies, and suggest mixed evidence for a long deadline, long-term exposure, and adverse health effects. In addition, newly available studies conducted in regions of the world that exhibit higher concentrations of SO_2 should be considered.

Carbon Monoxide: CO

Carbon monoxide is widely found in ambient air, particularly in developing countries where emissions can be high. CO is commonly monitored in ambient air, and there is new evidence on associations between long-term exposure to low concentrations and a wide variety of health effects. There is also a growing understanding that the mechanisms of action of CO are not based only on the hypoxia induced by the formation of carboxyhemoglobin (on which the current guideline is based). When CO levels are elevated outdoors, they can be particularly worrisome for people with some heart disease. These people already have a reduced ability to bring oxygenated blood to their hearts in situations where the heart needs more oxygen than usual. They are especially vulnerable to the effects of CO when exercising or under increased stress. In such circumstances, short-term exposure to elevated CO levels may result in reduced oxygen to the heart, accompanied by chest pain, also known as angina.

GENERAL IMPACTS OF AIR POLLUTION: BIOINDICATORS: METHOD FOR EVALUATION

The emission sources of the various air pollutants can be classified as natural and anthropogenic. Anthropogenic sources, such as combustion processes, are responsible for the increasing emission of particles in the atmosphere, but the influence of natural sources (continental dust, marine spray, and biogenic particles) cannot be ignored. Some of these pollutants, such as heavy metals and polycyclic aromatic hydrocarbons (PAHs), are at the forefront of environmental research because of their toxicity and their carcinogenic and/or mutagenic potential (Ares et al., 2011 & Murakami et al., 2012).

The impact of atmospheric pollution can also be evaluated through the use of indicator species, so active biomonitoring can be used as an instrument to assess the degradation of the environmental quality of certain sites (Oliva and Figueiredo, 2005 & Divan Junior et al., 2008). These living organisms, through biological responses, may indicate the accumulation of certain substances present in the atmosphere.

Lolium multiflorum (ryegrass) is already recognized as a bioaccumulating species for certain groups of substances, among which we can mention fluorine, sulfur and heavy metals (Illi et al., 2017).

Exposure to air pollutants can have serious consequences on the metabolism of individuals and their populations, ultimately leading to loss of viability. Considering that atmospheric contaminants tend to be precipitated by rains, and through their percolation, they present themselves as a potential source of diffuse pollution, the evaluation of the environmental risks associated with its launching is essential. Thus, toxicity bioassays are an important tool for environmental monitoring, allowing the evaluation of complex mixtures to which populations may be exposed.

Bocchi et al. (2016) evaluated the mutagenicity (*Salmonella* test) and genotoxicity (comet and micronucleus test in cell line A549) in the fine particulate matter ($MP_{2.5}$ and $MP_{1.0}$) from Bologna in Italy, also quantifying HPAs. The results show a higher presence of mutagenicity in samples of particulate matter collected in winter and fall. The results with the comet test, however, did not reflect seasonality. Thus, the use of different mutagenicity and genotoxicity assays applied to particulate matter proved to be relevant.

Palacio et al. (2016) evaluated the soluble fraction of total suspended particles and particulate matter (MP10um) in different samples collected in the state of São Paulo. The results of the mutagenicity tests using the *Salmonella* assay showed positive results for all the samples evaluated.

Receptor Models: Method for Evaluation

Receptor models use mathematical and/or statistical approaches to quantify the individual contribution of a given number of emission sources in the composition of a sample, based on the composition or characteristics of the emission source profiles. The basic principle of the receptor modeling is that mass conservation can be assumed and a mass balance analysis can be used to identify and distribute emission sources (Tasic et al., 2010).

There are a variety of receptor models that can be used for studies of identification, for instance, of particulate matter (PM) emission sources. The United States Environmental Protection Agency (EPA) has developed and provides the EPA Chemical Mass Balance (EPA CMB), EPA Unmix and EPA Positive Matrix Factorization (EPA PMF 5.0) models free of charge on its website (EPA, 2017). The EPA PMF 5.0 allows the individual weighting of samples or variables. This feature allows missing data and/or concentrations below the method detection limit (MDL) to be kept in the modeling, by assigning uncertainty values to those data (REFF et al., 2007).

EPA PMF 5.0

EPA PMF 5.0 is a multivariate factorial analysis tool that decomposes a matrix of data into two matrices: factor contributions (G) and factor profiles (F), which must be interpreted by the researcher to identify the types of sources that may be contributing to a particular sample. EPA PMF 5.0 incorporates the uncertainties of the variables, often associated with analyzes of environmental samples, as well as forcing all values in the solutions of the profiles and contributions to be non-negative (REFF et al., 2007).

EPA PMF 5.0 has been used in a variety of environmental studies, mainly related to air quality, for the distribution of pollutants quantified in MP (Brown et al., 2007; Gao et al., 2014; Kim and Hopke, 2007) and, to a lesser extent, for the distribution of pollutants quantified in rainwater (Xiao et al., 2013, Xiao, 2016 & Zhang et al., 2014) and in surface waters (LI et al., 2015).

Modeling of Atmospheric Dispersion

The modeling of atmospheric dispersion is the mathematical simulation of how air masses disperse in the atmosphere, executed by means of softwares that solves the mathematical equations and algorithms of dispersion simulation. The dispersion models are used to estimate or predict concentrations of pollutants in favor of the displacement of air masses emitted by sources such as industrial plants, vehicular traffic or accidental chemical release, among others (Stein et al., 2015). The dispersion models differ according to the mathematical approaches used in their development, but most models require one or more data inputs, as described below:

- Weather conditions: wind speed and direction, amount of atmospheric turbulence, air temperature, solar radiation, etc.
- Source: pollutants concentration and temperature.
- Emissions or emission parameters: location and height of the source, type of source and output speed, output temperature and emission rate.
- Elevation of terrain at the source location and receiver location.
- The location, height and width of any obstructions (such as buildings or other structures) in the path of the emitted gaseous plume.

CLIMATE CHANGE AND GLOBAL SCENARIO: GENERAL ASPECTS

The climate system consists of five main components that interact with one another: the atmosphere, the hydrosphere, the cryosphere, the lithosphere, and the biosphere. This system changes over time by the influence of its own internal dynamics and by external natural and anthropogenic forcing. When variations of this system remain constant, statistically significant and persist for an extended period, typically decades or more, climate change is characterized (IPCC, 2013).

In a climatic state of equilibrium, the average global solar energy absorbed by the Earth's system is balanced by the average global long-wave radiation (infrared radiation) that is emitted into space, thus having a radiative equilibrium at the top of the atmosphere. Radiation imbalances may occur from natural processes (volcanic eruptions, changes in total solar production, among others) and human influences such as the burning of fossil fuels and consequent increase in the concentration of greenhouse gases (GHG) such as CO_2.

Different studies have related the increase in temperature to the concentration of CO_2 in the atmosphere. According to Marcott et al. (2013) in the pre-industrial period (1750) the CO_2 concentration was 275 ppm, going up to 350 ppm in 1988, which would be increasing the average global surface temperature by 1° C. Figure 5 shows the evolution, in terms of global mean, of the concentration of greenhouse gases in ppm from the pre-industrial period to the present.

According to Jalota et al. (2013), the CO_2 level in the planet's atmosphere is 390 ppm with a projection of reaching 700 ppm in different future scenarios of climate change, while in the last 100 years the average global temperature has risen by 0.74 ± 0.18° C (Trenberth and Jones, 2007). Models of climate projections presented by the Intergovernmental Panel on Climate Change (IPCC, 2013) indicate that, under a more pessimistic scenario, the Earth's surface temperature could rise from 1.1 to 6.4° C during the 21st century.

The global climate response to external disturbances can be quite complex. Climatic forcing represents disturbances imposed on the global climate system (Neeling, 2011), with natural origins (changes in solar radiation and volcanic eruptions) and anthropic changes (changes in concentrations of greenhouse gases and aerosols). Positive forcing induce an increase in global mean surface temperature, and negative forcing induce a reduction. According to the IPCC (2014), the effects of human activities must be seen within the global radiative forcing model, which refers to the amount by which a factor changes the radiation balance at the top of the atmosphere expressed in units of W/m^2.

The magnitude of the global temperature response to the climate forcing depends on the corresponding feedbacks. Feedback mechanisms act as internal controls of

Figure 5. Observed changes in atmospheric greenhouse gas concentrations. Atmospheric concentrations of carbon dioxide (CO_2, green), methane (CH_4, orange), and nitrous oxide (N_2O, red). Data from ice cores (symbols) and direct atmospheric measurements (lines) are overlaid.
Source: IPCC (2014).

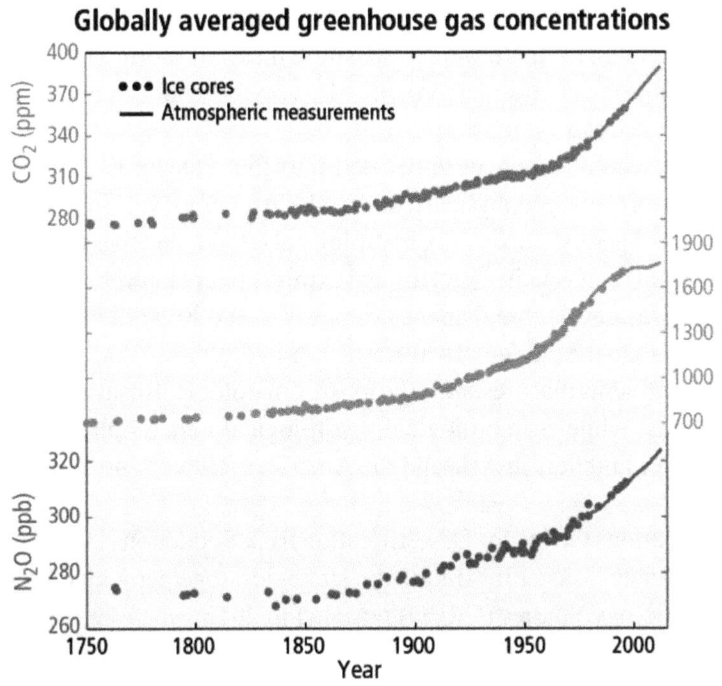

the climate system and result from a special coupling or adjustment between two or more subsystems.

Another way to assess the climatic response to radiative forcing is the concept of climate sensitivity that represents the change in equilibrium in the annual global average of the surface air temperature after a doubling in the equivalent of atmospheric carbon dioxide. That is, climate sensitivity is, basically, the answer to the following question: How much does the temperature increase when the amount of CO_2 doubles?

The doubling in CO_2 concentration is equivalent to a radiative forcing of approximately 4 W/m^2. In response to this duplication, the surface and atmosphere would warm up. At a certain point, the radiative equilibrium would be restored with new, higher surface temperature. Estimates of climate sensitivity obtained with the current generation of global climate models range from 2.1 to 4.4° C, with the best estimate at 3.0° C (Forster et al., 2013).

Consequences of Climate Change

There are several consequences due to climate change, and some changes have already been observed, such as the melting of the polar ice caps and the occurrence of more frequent extremes. According to Guha-Sapir, Hoyos, and Below (2016), in a survey conducted for the Center for Research in Epidemiology and Disasters (CRED), in the year 2015, there were 376 natural disasters worldwide, with a high number of climatological and meteorological disasters, with the greatest number of climatological disasters since 2005.

In this survey, the disasters were divided into four categories: climatological (drought and natural fires), geophysical (earthquakes/tsunamis, volcanic activities and mass movements of geological origin), hydrological (floods and mass movements of hydrological origin) and weather (storms and extreme temperatures). There were 45 climatological disasters, 6 more than the average of 2005-2014, 127 meteorological records, 10 more than the 2005-2014 average, 175 hydrological and 29 geophysical disasters. The continents that presented the most climatological disasters were Africa and the Americas, while meteorological, hydrological and geophysical disasters occurred in greater numbers in Asia and the Americas (Guha-Sapir, Hoyos, Below, 2016).

The United Nations Office for Desertification Risk Reduction (UNISDR), a UN strategy to coordinate disaster risk reduction strategies, together with the Center for Disaster Epidemiology Studies (CRED) revealed in 2015, when 1995 to 2015 was considered, that 90% of the world's major catastrophes and disasters were caused by floods, storms, hot flashes, droughts and other climate-related events, highlighting the close relationship between climate change and disasters.

In addition to these consequences, an increase in the average level of the seas is predicted, through the acceleration of the melting of the polar caps, more and more frequent occurrence of droughts and floods, storms, windstorms, intense heats, among others. Water resources will also be affected by climate change, with an increase in hot days, affecting rainfall and river flow, which could lead to more frequent epidemics. Also, the predicted scenarios indicate that there may be a lack of water for several regions, which will lead to several conflicts to access this resource (Herring et al., 2016 & Marengo et al., 2011).

Climate change will also affect the functioning of the terrestrial ecosystem, as changes in temperature and precipitation would lead to a radical change in the original characteristics, leading to a redistribution of species and a change in the composition of ecological communities, affecting human well-being and production of natural resources (Pecl et al., 2017).

Global Frameworks for Adapting Climate Change

The first event to focus on global warming was the Stockholm Conference in 1972. In 1988 the World Meteorological Organization (WMO) and the United Nations Environment Program (UNEP) established the Intergovernmental Panel on Climate Change (IPCC), which is a group of scientists from several countries to study and publish reports on climate change. Also noteworthy was Eco-92, which resulted in the signing of important environmental agreements, such as Agenda 21 and the Climate Convention; the Kyoto Conference in 1997 that generated the Kyoto Protocol; Rio + 10; Rio + 20; and COP 21.

In 1992, the United Nations Framework Convention on Climate Change was opened for signature at the United Nations Conference on Environment and Development (UNCED), which became known as Eco-92. The Convention came into force in 1994 and, currently, approximately 200 countries and the European Community are Parties to the Convention. Parties annually meet at the Conference of the Parties (COP) to review implementation and continue discussions on climate change (Lago, 2013).

From this, in 1997, in Japan, the Kyoto Protocol was adopted, which establishes greenhouse gas emission reduction targets (GHGs) and implementation mechanisms to achieve these goals. This Protocol will be in force until 2020 (Moreira, Giometti, 2008).

In November 2015, the 21st Conference of the Parties (COP-21) to the United Nations Framework Convention on Climate Change (UNFCCC) and the 11th Meeting of the Parties to the Kyoto Protocol (MOP- 11), took place in Paris.

As a result of this Conference, a new global agreement was adopted that seeks to combat the effects of climate change, in addition to reducing greenhouse gas emissions. The document has been ratified by 195 countries, in addition to the European Union. This agreement was seen as a historic agreement, decisive for reducing the risks of climate change, seeking progress in eradicating poverty, strengthening peace and guaranteeing a life of dignity and opportunity for all (UNFCCC, 2015).

The main points approved were that countries should work to keep warming below 2° C, rich countries should guarantee financing of US $ 100 billion per year, revision of the agreement every 5 years. However, this agreement does not determine how much greenhouse gas emissions need to stop rising or cutting percentage emission of these gases.

In Brazil, the National Adaptation to Climate Change Plan, dated 2016, is an example of strategies built to meet the recommendations of the COP 21, since it set targets to encourage sectors to adapt actions aimed at mitigating vulnerabilities and mitigate natural hazards, caused by the imbalance of nature and climate change.

The last Conference of the Parties, the 23rd Conference, was held in November 2017 in Germany. The main discussion was focused on mechanisms and targets to be met from 2020 when the Paris Agreement comes into force.

It is important to highlight other agreements that demonstrate the need to contemplate and include climate change, such as the Sendai Framework for Disaster Risk Reduction, with global disaster strategies as the horizon between 2015 and 2030. Recommendations of this agreement signal that a mechanism to consider is the construction of measures to adapt to climate change, considering the scenarios and their impacts to reduce the risk of disasters, since many of the disasters are related to the climate.

Also at the Sendai Framework, reaffirm the need to follow the mandate of the United Nations Framework Convention on Climate Change, as it "represents an opportunity to reduce the risk of disasters in a meaningful and coherent way in all inter-governmental interrelated processes."(p.5 Sendai Framework, 2015)

Finally, in the Sustainable Development Objectives composed of 17 goals that compose a new global agenda, which originated from the eight Millennium Goals, the need for global action against climate change is addressed, where in the thirteenth objective three targets aim to take urgent measures to combat climate change and its impacts (UNITED NATIONS IN BRAZIL, 2015).

Climate Change Projections and the IPCC Gas Emission Scenarios

The whole process of obtaining future changes in global terrestrial temperature (and other climatic variables) begins with the construction of scenarios, hence their importance. A scenario can be defined as a description of the future, based on a set of assumptions, about key relationships and generating mechanisms, which must be coherent and consistent. In this case, the results are called climate projections, which according to IPPC (2014) is the response of the climate system to a scenario of emission and/or concentration of greenhouse gases and aerosols, or scenarios of radiative forcing often based on simulations of climatic conditions.

In 2007, in its fourth report (AR4) on climate change, the Intergovernmental Panel on Climate Change (IPCC) adopted a set of climate scenarios, known as SRES (Special Report on Emissions Scenarios) scenarios, which were grouped according to four different projections of greenhouse gas emissions for the future related to social, economic and technological development, population growth, concern for the environment and regional differences. These scenario groups are known as scenarios A1, A1B, A2, B1, and B2. Variations of these groups also occur such as scenarios A1T and A1F.

Briefly, the A2 and A1F scenarios represent a pessimistic perspective, with a forecast of higher GHG emissions in the atmosphere over the coming decades, while scenarios B1, B2, and A1T represent a more optimistic perspective, with a lower GHG emission in the coming decades. Among these sets of scenarios, A1B (called *business as usual* or BAU) is what has been investigated more broadly (Nakicenovic and Swart, 2000, Wehner, 2013) and presents an emission projection of intermediate GHGs in relation to the other scenarios.

In the Fifth Report (AR5), published in 2014 by the IPCC, new scenarios, the so-called RCPs (Representative Concentration Pathways) were established. RCPs are named from radiative forcing levels for the time horizon until the end of this century (2100) in relation to the year 1750. Thus, RCP-X implies in a scenario in which the radiative forcing of stabilization or peak or at the end of the 21st century corresponds to XW/m^2 (Silveira et al., 2013).

Compared to projections of total radiative forcing (RF) for the year 2100 the CPR and SRES scenarios, the estimates of CPRs are more comprehensive. However, there are similarities between the groups. Figure 6 shows the comparison between the overall mean temperature increase predicted by the SRES and RCPs scenarios.

The RCP4.5 scenario is close to SRES B1, RCP6 is close to SRES A1B (more after 2100 than during the 21st century), and RCP8.5 is slightly larger than A2 at

Figure 6. Global temperature change (mean and standard deviation, indicated by shading) relative to 1986-2005, projected by GCMs of the CMIP3 using SRES scenarios (left), and by GCMs of the GMIP5 using RCPs (right). The number of MCGs used for each scenario is indicated in parentheses.
Source: Knutti and Sedlácek (2012).

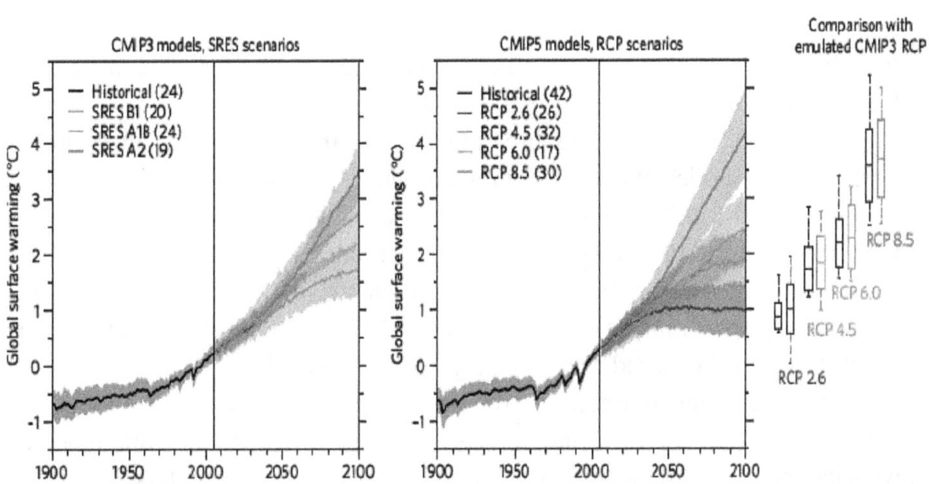

2100 and close to the SRES A1FI scenario. RCP2.6 is smaller than any of the SRES scenarios (Moss et al., 2010 & Van Vuuren et al., 2011).

The IPCC is currently in its Sixth Assessment cycle. During this cycle, the Panel will produce three Special Reports, a Methodology Report on national greenhouse gas inventories and the Sixth Assessment Report (AR6).

The estimated AR6 Synthesis Report will integrate and synthesize the contributions from the three Working Groups that will be rolled out 2022

Climate Models: Method for Evaluation

Climate models are extremely sophisticated computer programs that simulate the climate system and its complex interactions between atmosphere, ocean, land surface, snow, ice and the global ecosystem, as well as a variety of chemical and biological processes (IPCC, 2014). In the General Circulation Models (GCM), popularly known as global climate models, the climate system is described regarding physical, chemical and biological principles through a series of equations that express these laws.

In general circulation models, dynamic and thermodynamic processes and radiation and mass changes are modeled using five basic sets of equations (Barry and Chorley, 2013):

1. Three-dimensional equations of motion, i.e., conservation of moment;
2. Continuity equation (mass conservation or hydrodynamic equation);
3. Continuity equation for atmospheric water vapor (conservation of water vapor);
4. Energy conservation equation, i.e., the thermodynamic equation derived from the first law of thermodynamics;
5. and the equation of state for the atmosphere.

In addition, additional conservation equations for other atmospheric constituents, such as sulfur aerosols, can be applied in more complex models.

The GCMs used in climate research range from simple energy balance models to more complex models that require high computational performance.

The Coupled Models of General Circulation of the Atmosphere and Oceans (CMGCAO) have had great emphasis among the models evaluated by the IPCC. Its main function is to understand and represent the dynamics of the physical components of the climate system (atmosphere, ocean, land and ice sea), generating future climate estimates based on increased greenhouse gases and aerosols.

In order to generate confidence in the performance of global climate models, a project called Coupled Model Intercomparison Project (CMIP) was implemented to evaluate and compare models from various climate research centers around the world. This evaluation is performed using standard procedures and standardized

Figure 7. Mean seasonal cycle of (a) temperature (°C) and (b) precipitation (mm day–1). The average is taken over land areas within the indicated regions, and over the period 1980–1999. The red line is the average over 45 CMIP5 models; the blue line is the average over 22 CMIP3 models. The standard deviation of the respective data set is indicated with shading. The different line styles in black refer to observational and reanalysis data: Climatic Research Unit (CRU) TS3.10, ECMWF 40-year reanalysis (ERA40) and ERA-Interim for temperature; CRU TS3.10.1, Global Precipitation Climatology Project (GPCP), and CPC Merged Analysis of Precipitation (CMAP) for precipitation. Note the different axis-ranges for some of the sub-plots. The 15 regions shown are: Western North America (WNA), Eastern North America (ENA), Central America (CAM), Tropical South America (TSA), Southern South America (SSA), Europe and Mediterranean (EUM), North Africa (NAF), Central Africa (CAF), South Africa (SAF), North Asia (NAS), Central Asia (CAS), East Asia (EAS), South Asia (SAS), Southeast Asia (SEA) and Australia (AUS).
Source: IPCC (2014).

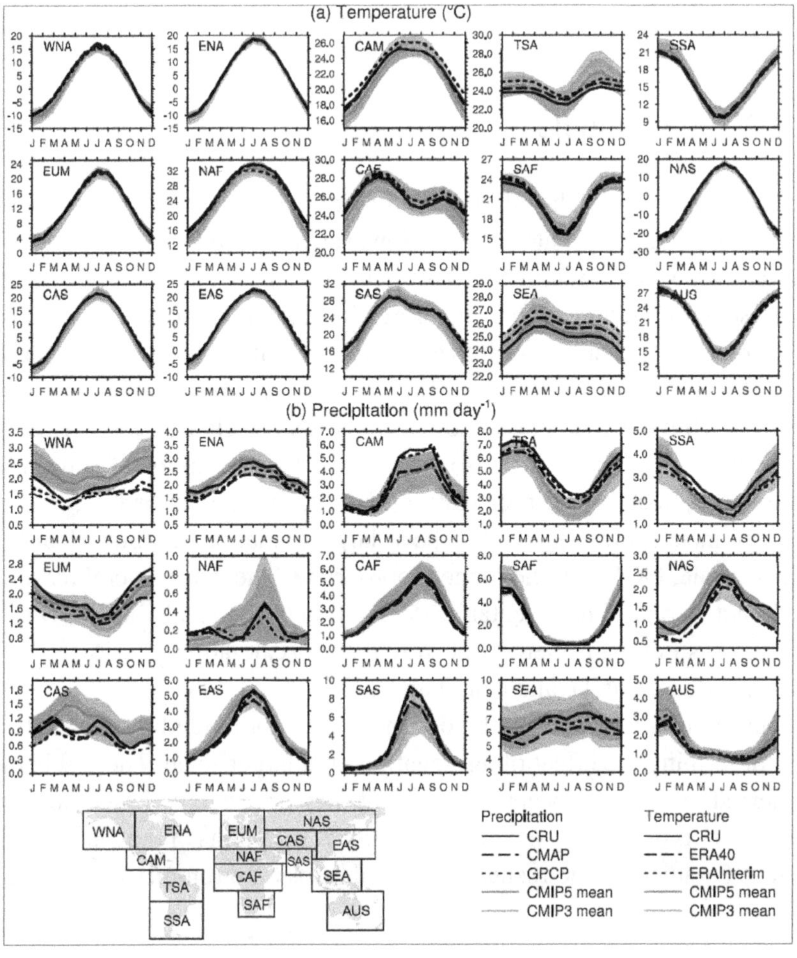

data, such as sea surface temperature, thus providing ample documentation on the structure of the models and the details of their parameterizations. The models cited by the fourth IPCC climate change assessment report, AR4, comprise the CMIP Phase 3 (CMIP3) while the AR5 models comprise the CMIP phase 5 (CMIP5).

Current climate models are incorporating more relevant processes compared to AR4, but confidence in the representation of these processes remains weak (IPCC, 2013). Despite the progress that has been made, scientific uncertainty about the details of many processes remains, such as the representation of aerosol particles and their interaction with clouds and radiative transfer.

A comparison between the CMIP3 and CMIP5 models is shown in Figure 7 for seasonal cycles of temperature and precipitation in different regions. It is observed that the simulation of the temperature is generally better than the precipitation regarding amplitude and phase of the seasonal cycle. The average between models is closer to observations than most individual models. The systematic difference between the CMIP5 and CMIP3 ensembles is small in most regions. For South America (SSA) the projections of the different groups of climate models are quite similar, for both temperature and precipitation.

Global Climate Models

The climate projections of the different available GCMs are widely used in scientific research on impacts of climate change. However, such models are the main source of uncertainty in studies of climate change and its impacts, for example, in the field of water resources (Adam, 2016). Therefore, the assessments of climate projections simulated by GCMs should take into account the limitations of these models.

General Circulation Models (GCMs) can provide a reasonable representation of a regional climatic characterization such as ENSO, Hadley cells, trade winds, and jets in the atmosphere. They also offer a realistic description of patterns called teleconnections, such as the propagation of waves in the atmosphere and ocean. These phenomena, however, tend to have large spatial scales, but when brought to the local scale, GCMs are no longer appropriate. There are several reasons why GCMs do not provide a representative description of the local climate. One is that discretization is only sufficient to portray macro scale atmospheric processes on the Earth's surface (Dibike and Coulibaly, 2005).

The subsample processes of the dot mesh are represented by parametrization schemes describing the effect of aggregation on a larger scale. These schemes are often referred to as the 'physical model' but they are actually based on physical-statistical models describing the mean quantity at the points of the mesh of relevant parameters. Parameterization schemes are usually based on empirical data (e.g., field measurements or in-situ observations), and a typical example of a parameterization scheme is the representation of clouds (Stone and Knutti, 2010).

Climate models need boundary conditions describing surface conditions (e.g., energy and steam flows) to provide a realistic representation of the climate system. Often simple parameterization schemes are employed to provide a reasonable description, but these do not capture the details of the variations associated with small spatial scales (Bouwer et al., 2004).

Among the main limitations of these GCM's are the following:

1. discretization is sufficient only to portray macro scale atmospheric processes on the Earth's surface;
2. there are several parameters and processes that present strong limitations in the representation of the models, for example, the physical hydrological processes in the basin vary in the microscale. This has hindered the integration of climate and hydrological models. Currently, there are several projects around the globe that seek to better estimate the parameters to make these models more reliable.

Regional Climate Models: Method for Evaluation

Downscaling methods have been developed for assessing the regional consequences of climate change.

The term downscaling refers to techniques for the transfer of meteorological information derived from GCMs (usually with resolution greater than 100 km) for smaller or regional scales (5 to 50 km resolution). Some downscaling methodologies have been used (e.g., Fowler et al., 2007, Bardossy and Pegran, 2012 & Teutschbein et al., 2011).

Downscaling techniques can be framed in two conceptually distinct approaches: statistical and dynamic downscaling.

Statistical downscaling derives from the statistical or empirical relationship between the variables simulated by the GCMs, and local hydrological variables, (Mujumdar and Kumar, 2012). Different techniques of statistical downscaling have been developed over the last decade. Each method generally falls into one of three broad categories: regression methods or transfer functions, and stochastic climate generators.

Dynamic downscaling or Regional Climate Models (RCMs) are numerical models similar to GCMs, but with finer resolution, using as initial data and boundary conditions the result of GCM simulation. This technique consists of the nesting of an atmospheric regional model to a GCM, the regional one being forced unidirectionally throughout the domain by the information generated by the GCM.

The RCMs incorporate regional characteristics such as topography, vegetation, soil, etc., not present or simplified by global models.

The application of the regional climate model allows the evaluation of impacts in smaller areas, and the improvement in spatial resolution allows a more detailed analysis of the local impacts. Despite the uncertainties surrounding the process of atmospheric modeling and climate projections, several studies on the use of regional climate models suggest a better performance in the results obtained compared to GCMs (e.g., Kawazoe and Gutowski, 2013, Vautard et al., 2013).

Since AR4, the resolution of the RCMs has increased from approximately 50 km to 25 km. However, the RCMs are computationally expensive, require a lot of processing time, and cannot be easily transferred to new regions. The results of regional climate models are also sensitive to the choice of baseline conditions.

Novel computational tools may hold the potential to address these uncertainties. Schneider et al. (2017) outline a blueprint for a next-generation climate model that would employ advancements in data assimilation and machine learning techniques to learn continuously from real-world observations and high-resolution simulations.

This approach gets at the heart of the uncertainty that plagues today's models: a mathematical technique known as parameterization. Earth system models integrate atmospheric, oceanic, chemical, and biological processes, many of which are too complex or occur at scales too small to simulate directly (e.g., formation of individual clouds). Parameterization allows for simplified simulation (e.g., a statistical description of cloudiness) based on underlying conditions, or parameters.

Some Evidence of Studies, Control, and Adaptation to Climate Change

Evidence of climate change has become more and more numerous, such as the permafrost of the Arctic region that is disintegrating in Northwest Canada and flowing into rivers, dumping large quantities of sludge and carbon-rich sediments. This soil is composed of organic matter that is estimated to have almost twice as much carbon in permafrost as in the Earth's atmosphere. This phenomenon will lead to an increase in global warming (Steven et al., 2017).

This evidence connects with what is predicted in the studies of Peci et al., 2017, where, from analyzes of several regions of the world, it is revealed that climate changes are accelerating the Earth's ecosystems, it is estimated the greater number of epidemics, due to the movement of mosquitoes; a new design of agricultural geography, by the lack of pollinators; and migrations of entire traditional communities in search of food.

The Climate Observatory composed by the NOAA, NASA, and Metoffice concluded in 2017 that the warming identified in 2016 broke the previous record of 2015, which in turn had beaten the record for 2014. Given this, CRED stresses that achieving the objective to eliminate poverty, as defined in the Sustainable Development Objectives may be compromised, as the observed climate changes, due to their effects and disasters, threaten this quest.

COP21, which took place in December 2015 in Paris, brought together 196 countries, which committed to planning and implementing measures to curb greenhouse gas emissions and combat the impacts of global warming. As much as each country lives a different social, cultural and economic condition, the idea of signing an agreement is that everyone can act within their means to minimize the impacts of global warming by adapting to its effects. In the case of developing countries, the agreement provides financial and technological assistance to enable them to cope with the damage caused by the change in temperature.

The Intergovernmental Panel on Climate Change stresses that mitigation and adaptation measures are urgent and should be assessed according to the degree of vulnerability and adaptability of countries or regions. Due to its complexity, the implementation of these measures is not a simple task, especially for developing countries with less financial resources. In general, agricultural sectors and small coastal or hillside areas are more vulnerable to climatic events. Thus, each situation must be analyzed individually, and adaptation actions must be planned to minimize potential losses and optimize recovery capacity.

ADAPTATION, VULNERABILITY, RESILIENCE, AND CLIMATE CHANGE

We can conceptualize adaptation as being adjustments in ecological or socioeconomic systems in response to climate change, resulting from practices, processes, measures or changes. Adaptation to climate change must be guided by the climatic scenarios that are being presented to society by the scientific community. These scenarios are contained in the IPCC's Fifth Assessment Report (AR5) (2014), which is based on a more comprehensive forecasting system than the previous reports, considering emissions impacts. These scenarios serve to estimate the degree of impact these changes will have on various sectors, countries, and regions so that society can prepare for these changes (Carvalho and Furtado, 2014). In today's scenario, even if a global and intensive action is implemented to reduce emissions, the climate will continue to change due to emissions from previous decades. Thus, the inevitability of climate change forces all countries to seek individual adaptation strategies. (IPCC - AR5, 2014).

However, adaptive actions should occur from the identification of elements of an analysis system, considering sensitivity and vulnerability. That is the degree to which a system is negatively or positively affected by climate-related stimuli and the susceptibility of a system or its inability to cope with the adverse effects of climate change. The greater the sensitivity of livelihoods to climate effects, combined with the lesser the ability to cope with them, the greater the susceptibility to their impact (Wolf, 2011). The ability of a system to adapt to climate change will depend on a variety of factors such as wealth, access to technologies, information systems, among others, which minimize harm and provide and cope with its consequences (IPCC, 2014).

Refers to the propensity level of an individual, community or system to be affected by climate change. It is a reason given for exposure and sensitivity for the ability to adapt to climate change. Where, exposure defines the degree to which an individual is susceptible to a climate hazard, sensitivity to the degree of the impact of a climate hazard on an individual, and ability to adapt to an individual's ability to adjust to changes in the environment.

An adaptation strategy should promote the reduction of vulnerability to climate change and risk management associated with this phenomenon, involving the identification of exposure to current and future impacts based on climate projections. The identification and analysis of the vulnerability to these possible impacts and the definition of actions and guidelines that promote the adaptation to each sector can be realized considering the following sectors (National Climate Adaptation Plan - Brazil, 2016): Agriculture; Water resources; Food and nutrition security; Biodiversity; Cities; Disaster Risk Management; Industry and Mining; Infrastructure; Vulnerable Peoples and Populations; Health; Coastal Areas.

The concept of Resilience is linked to the ability of an element to return to its initial state after undergoing external influence. The Resilience Engineering approach represents a new way of thinking about safety, where your field of study admits that systems can return to their normal state after suffering some wobble/disturbance, ensuring safety and better risk management. Given the complexity of the system, it is necessary to characterize the elements and/or indicators that contribute to a resilient system and its adaptive capacity. The skills that configure resilient systems can be considered as: Reaction to adverse events; Monitoring of ongoing processes; Anticipating future threats and opportunities and Learning from experience (Hollnagel, 2006).

Through this concept, it is possible to perceive that resilience is more than adaptation to the climate, but in this case, we will make the specific approach to climate resilience. The Paris Agreement, aligned at COP21, addressed adaptation to climate change and resilience with great relevance. As populations grow and climate-related disasters impact cities, adaptation, and resilience need to be increasingly integrated into urban development. Some cities are already prioritizing this adaptation, such as New York, London, Paris and dozens of other cities, setting up special working groups and partnerships at the highest levels of their governments to tackle climate change.

Cities Resilient to Climate Changes

Resilience in ecosystems can be characterized by a dynamics of systems that may have a balance for some time and then undergo a rapid or unexpected change, called the "adaptive cycle" (Holling, 2002). Mintzberg et al. (2009) refers to communities, cities, and regions as processes of continuous and cyclical changes, relating the cycle of urban development to the adaptive cycle of resilience. This describes cities in a cycle which, when immersed in crisis, tend to strengthen their weaknesses to potentialize opportunities.

The urban system uses the resources and processes them according to their needs, generating the technical and socioeconomic processes that happen in a city. After the confusion (crisis) the adaptive cycle of these passes to the exploration (face of slope), emphasizing the creativity and remodeling to a new reality. This system will never be as before. However, three beacons will be fundamental to characterize them as resilient or vulnerable: level/speed/volume of growth; production and accumulation of energy; amount of waste. We can consider as examples of cities that present proposals to become resilient to the changes in the climate:

- Rotterdam, Netherlands

Rotterdam, with 80% of its land below sea level and one of the world's largest ports, aims to be 100% weatherproof by 2025. Rotterdam has accumulated centuries of experience regarding resilience, particularly in the areas of integrated water resources management and innovative climate adaptation. Rotterdam's efforts are still underway in the wake of the 1953 flood, which killed nearly 2,000 people and caused widespread property damage. It was an event that underscored the destructive power of the sea and stimulated the modern flood management industry in the country. In 2007, this Dutch city announced its ambition to become 100% weatherproof by 2025 - able to continue to function economically and socially with the least disturbance under any extreme weather situation (Rotterdam Resilience Strategy, 2016)

- Nova York, USA

New York City aims to be the world's largest sustainable city and global leader in the fight against climate change. Among the actions developed are the reduction of dangerous emissions of greenhouse gases and have cleaner air and water minimizing the environmental footprint. The city is developing its goal of reducing greenhouse gases by 80% by 2050 ("80 x 50"), expanding from an initial focus on buildings to include energy, transportation, and solid waste as part of an action plan to achieve this goal. As a resilient city, New York will be able to prepare and respond to adverse events, provide basic functions and services to all residents after the crash events by the year 2050. The city will enhance private and public buildings to be more efficient in energy use and resilient to the impacts of climate change; adapt infrastructures such as transport, telecommunications, water, and energy to withstand severe weather events; and strengthen our coastal defenses against flooding and rising sea levels (One New York, 2016).

- Porto Alegre, Brazil

The city of Porto Alegre is a world reference in participatory processes. The selection of Porto Alegre for the first city team of the 100 Resilient Cities Challenge promoted by the Rockefeller Foundation in December 2013 represents the city's commitment to building a better future for its citizens. To this end, the city faced the complex challenge of articulating, researching, building ideas, concepts and planning to create the City's first Resilience Strategy. The Porto Alegre Resilience Strategy proposes the vision to make the city more resilient and a reference for other cities in Latin America in this theme. Some of the actions proposed for this are: To have a diversified economy that fosters creative, collaborative economies and new technologies; develop degraded areas of the City; exploit the agricultural potential of the rural area and stimulate the production of organic and family farming. In order to achieve this strategic objective, the City commits itself to achieve substantial progress in the following goals until 2022 (Resilience Strategy of Porto Alegre, 2017).

Case Study: Climate Change and Hydrological Modeling

Relationships are currently sought to understand the processes of climate change in order to assess their impacts and mitigate them. In the context of the hydrographic basin, the hydrological modeling of the possible impacts of climate change has been the most applied methodology. Different methodologies have been used to estimate the impacts of climate change on water resources.

According to Mujumdar and Kumar (2012), the general procedure for estimating the impacts of climate change on hydrological behavior is as follows: (1) Determine parameters of a hydrological model for the river basin using as input data observed climate and flow variables, thus validating the use of the model; (2) Design, through the use of GCMs and RCMs, future climate variables for the hydrographic region under study; (3) Simulate the hydrological characteristics of the basin under the projections of climatic models; and (4) Compare the simulations of the climatic projections between the reference period and the future period.

In a study conducted by Adam (2016), the impacts of climate change on the flow regimes of the Uruguay River Basin, located in southern Brazil, were evaluated. The area bounded for study comprises 110000 km^2, with the Uruguay River being the main watercourse. This region has great importance for the country due to the agroindustrial activities developed and its hydroelectric potential.

Figure 8 shows the location of the study region as well as the points of analysis. These points, where the flow results were generated, correspond to the exudation of each of the seven sub-basins delimited in this study.

Table 2 shows the location of each point and characteristics of the respective sub-basins.

According to ANA report (2013), the main consumptive uses of water in the BHRU region are for human supply, industrial activities, livestock and agriculture (irrigation). As for non-consumptive uses, the hydroelectric power generation, with

Figure 8. Location Uruguay River basin and analysis points

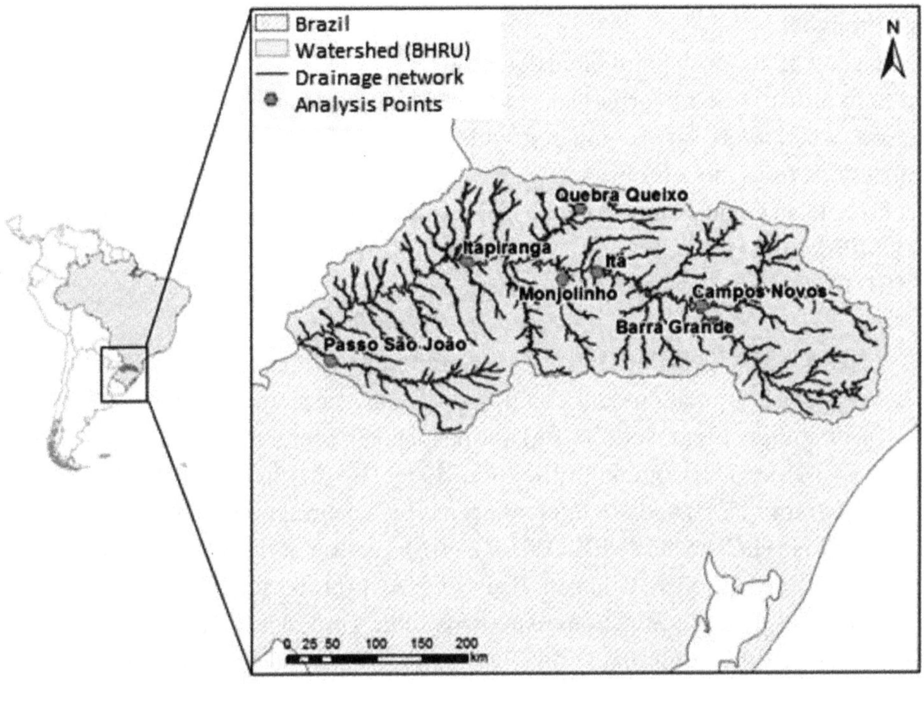

Table 2. Identification of the points analyzed in the BHRU

Cód/Sub-Basin	Lat	Long	River	Area Drainage (km^2)
215/Barra Grande	-27.776	-51.195	Pelotas	12016
216/Campos Novos	-27.604	-51.327	Canoas	14100
092/Itá	-27.277	-52.382	Uruguai	38440
220/Monjolinho	-27.345	-52.731	Passo Fundo	3754
286/Quebra Queixo	-26.656	-52.546	Chapecó	2670
292/Itapiranga	-27.163	-53.683	Uruguai	72146
103/Passo São João	-28.139	-55.049	Ijuí	9541

a potential of 13.6 GW, corresponds to 11.2% of the total installed capacity in the country.

The natural water availability of the BHRU is greatly influenced by important spatial and temporal variations of some of the climatic parameters, such as the pluviometric regime, which is reflected in the activities developed in the region.

Critical events are frequent in the region and mainly affect populations along the Uruguay River and some of its tributaries. According to ANA (2013), 242 municipalities of BHRU decreed Emergency Situation (SE) or State of Public Calamity (SPC) due to floods and 26 municipalities decreed SE or ECP for droughts and droughts.

In general, for each point of analysis, a series of observed flows were simulated through a distributed hydrological model, which corresponds to the current data period, which were later compared with series of simulated flows from climate projections from models climate change.

Five global models, summarized as follows, were selected from CMIP3: NCCCSM (CCSM3), MPEH5 (ECHAM5 / MPI-OM), GFCM21 (GFDL-CM2.1), MRCGCM (MRI- CGCM2.3.2) and HADCM3 HadCM3).

1. NCCCSM (CCSM3): The CCSM (Community Climate System Model) model is a global climate model composed of four geophysical models simulating Community Land Model (CLM), atmosphere (Community Atmosphere Model (CAM), sea ice (Community Sea-Ice Model - CSIM) and ocean (Parallel Ocean Program - POP) and are interconnected by a central coupling;
2. MPEH5 (ECHAM5 / MPI-OM): The fifth version of this model is the latest of the ECHAM models. Compared to its ECHAM4 processor, it is more flexible and because of the modifications made to its code, it produces significantly different weather simulations than previous versions. The model consists of a dynamic

spectral nucleus, an advective tracing scheme for the components of water (solid, liquid and vapor) and chemical substances, a physical parametrization system involving short and longwave radiation schemes, stratification of clouds, horizontal and vertical diffusion, processes on the surface of the Earth, among others (Roeckner et al., 2003);

3. GFCM21 (GFDL-CM2.1): This model was developed to simulate more realistically a range of phenomena of diurnal and temporal scale fluctuations (extreme events) in a synoptic scale for climate over centuries. Likewise, this model is composed by other models (atmosphere, surface, ocean and sea-ice) interconnected by a central coupling. The CM2.1 model incorporates several changes from the earlier CM2.0 version, which aim to reduce the bias of the simulations, such as in the land surface component, evaporation is suppressed when the soil is frozen to a depth greater than 30 cm (Delworth et al., 2006);
4. MRCGCM (MRI-CGCM2.3.2): This model was developed to be able to examine transient climate changes associated with anthropogenic forcing such as the emission of greenhouse gases and sulfate aerosols. The model has already presented good performances for simulations of the ENSO (El Niño and Southern Oscillation) phenomena, the Asian summer monsoons, among others. It is also an example of a coupled global model (Yukimoto et al., 2001);
5. HADCM3 (UKMO-HadCM3): This is one of the most used models in the forecasts and analyzes issued by the Third and Fourth Reports of the IPCC. It was the first model that did not need artificial adjustments of heat and water flows to obtain good simulations. It was developed to produce simulations for periods of a thousand years. It consists of two components: atmospheric model (HadCM3) and ocean model (which includes the sea-ice model). The simulations are done for 360 days, 30 days each month (Gordon et al., 2000). Modelo Climático Regional ETA/CPTEC.

This dynamic downscaling (RCM) is the result of a partnership between the National Institute for Space Research (INPE) and the Hadley Center (Met Office Hadley Center MOHC) which offered the global climate scenarios with sufficient time resolution to run the Brazilian regional model ETA-CPTEC. Through this cooperation between INPE and MOHC, data sets were developed and made available for global boundaries of the HadCM3 global model for the present climate (1961-1990) and the future (2010-2100).

The elaboration of the ensembles of this RCM was based on multiple changes in the GCM parameters HadCM3 and described in Chou et al. (2014), from the 17 members of the original set created by the Hadley Center team, INPE selected a subset of four HadCM3 members to serve as a boundary condition to the ETA-CPTEC regional model, and the control member and three other members.

From the ETA / CPTEC model, daily data of the following variables were generated: temperature at 2 meters above the Earth's surface (ºC); dewpoint temperature at 2 meters above the Earth's surface (ºC); atmospheric pressure to the Earth's surface (hPa); total precipitation (mm); wind at 10 m (m.s-1); and mean short-wave radiation incident to the Earth's surface (W.m-2).

These variables were made available in five sets: four members of the Eta 40 km model, named High, Low, Medium, and Control, according to the climate sensitivity in each member of the rounds of the global HadCM3 model and a Eta 20km round, obtained from the control member of the HadCM3 model.

In all simulations of global and regional climate models, the period from 1961 to 1990 was used to represent the current climate. The period from 2011 to 2100, which represents the climate of the future in the 21st century, was divided into three subintervals of 30 years each (Future 1: 2011-2040; Future 2: 2041-2070; Future 3: 2071-2100). All rounds of climate models were obtained considering the emission scenario of Greenhouse Gases (GHG) A1B.

The large-scale MGB-IPH hydrological model is a semi-distributed process-based model using physical and conceptually based equations to simulate continental hydrological cycles (Collischonn et al., 2007). The application ranges of modified versions also include assessments of climate change impacts, hydrological forecasting, land-use change analyses, input data for decision-support systems, water quality analysis and sediment transport modeling.

In the current version of the MGB-IPH model, the basin is divided into small unit-catchments (Fan and Collischonn, 2014), and then further into hydrological homogeneous regions termed hydrological response units (HRUs), generally defined from a combination of soil and vegetation type maps (Haghnegahdar et al., 2015). Each unit-catchment has a unique river reach, where the river routing processes is performed.

Figure 9. Variation of the climatic variables of CT40, LOW, MID, HIGH and CT20: solar radiation, temperature, relative humidity, wind speed and atmospheric pressure

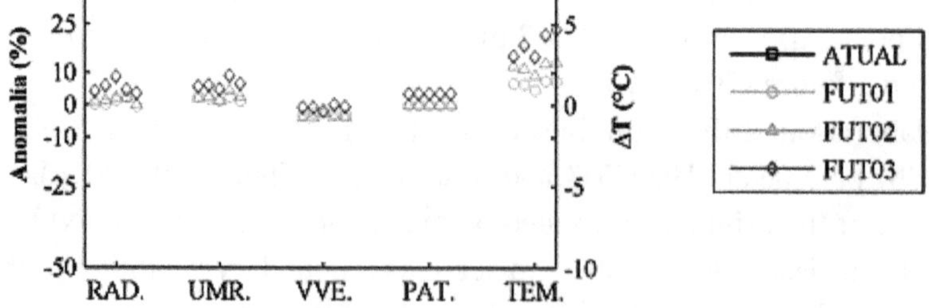

The hydrological modeling of the BHRU, both of the observed flow series and the simulated series (obtained from climatic projections), was performed using the MGB / IPH hydrological model. These series were simulated for the current period: 1960 to 1990 and three more future horizons Fut 1: 2011 to 2040; Fut 2: 2041 to 2070 and Fut 3: 2071 to 2100.

Observed variables (daily precipitation and flow as well as normal climatological data regarding temperature, relative air humidity, wind speed, atmospheric pressure and insolation) were extracted from the National Water Agency (ANA) database for the period from 1960 to 1990.

The analysis of the flow modifications is based on the comparison of hydrological modeling results in different periods. These differences in the series of simulated flows are used to characterize and quantify the possible impacts of climate change on the BHRU.

Due to the large amount of data resulting from this work item, the results shown here refer to the Itapiranga sub-basin, exúrio of the Uruguay River. Figure 9 represents the variation of the climatic variables, obtained by the regional model, considered in the simulations besides the precipitation: solar radiation, temperature, relative air humidity, wind speed and atmospheric pressure.

Figure 10 represents the variation of the climatic variables, obtained by the global models, considered in the simulations besides the precipitation: solar radiation, temperature, relative air humidity, wind speed and atmospheric pressure.

Figure 11 shows the changes in mean annual rainfall, in terms of anomalies in relation to the current period, projected by different climate models at different time periods. The chart on the left refers to the results of the RCMs and the one on the right shows the results of the GCMs.

Figure 10. Variation of the climatic variables of GFCM, MPEH, MRCG, HADCM, and NCCC: solar radiation, temperature, relative air humidity, wind speed and atmospheric pressure

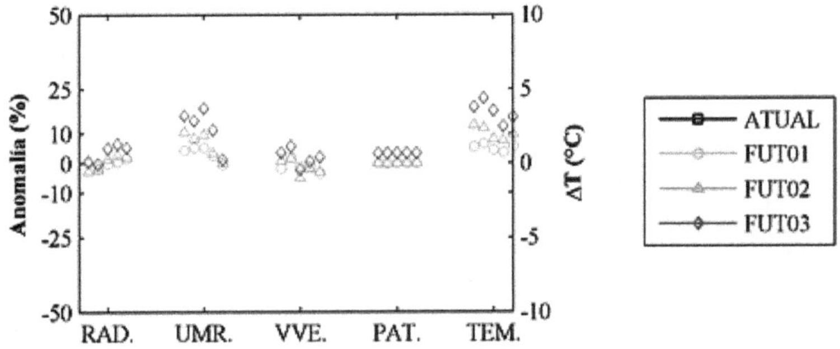

It is observed in Figure 11 that all versions of the regional model predict that precipitation should increase, reaching anomaly values greater than 40%, in general, growing over time, throughout the 21st century. As for the global models, the same agreement between the results is not observed, with the GFCM model being the most divergent. This model suggests that there will be a small increase in precipitation in the Itapiranga sub-basin in the future1 followed by a reduction of more than 10% in precipitation in both Future 2 and Future 3. The other global models suggest that precipitation in the BHRU should increase to in the 21st century, in relation to the current climate, however, none of the global models predicts a change as large as those foreseen by the regional model, reaching a maximum of 20% of anomaly.

The sensitivity of BHRU to rainfall increase is shown in Figure 12. This graph shows, for each climate model and for each period of the future, the relationship between the relative precipitation variation and the relative variation of the average flow.

Figure 11. Projected changes in rainfall in the Uruguay River basin upstream of the Itaipiranga sub-basin (left-hand side: regional models)

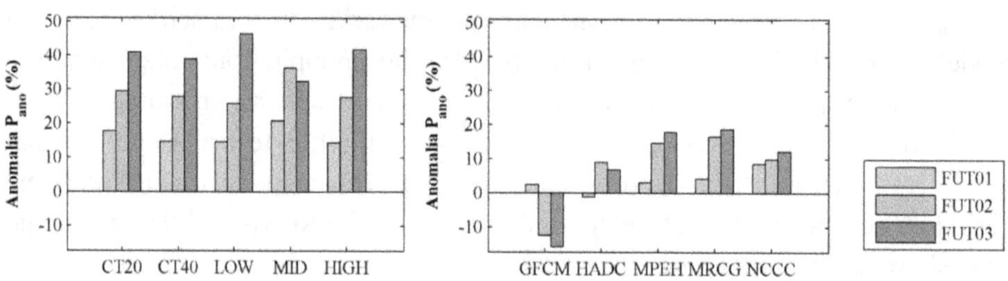

Figure 12. Relation between predicted changes in precipitation and the mean flow rate of the Uruguay River in the Itapiranga sub-basin considering different climate models and different periods of the future

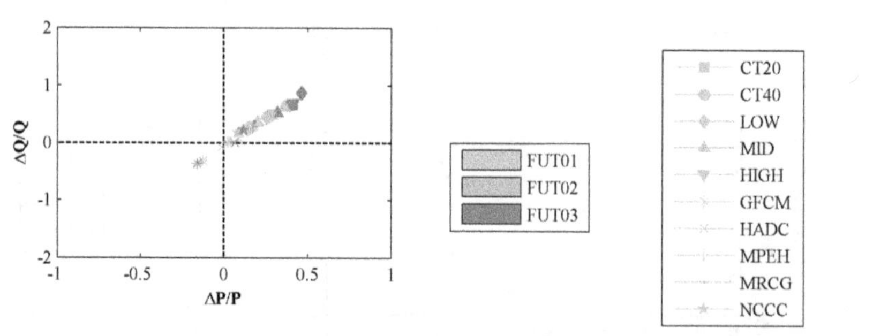

Figure 13 shows the results of projections of Uruguay River flow changes in the Itapiranga HPP in terms of average monthly flows obtained based on projections of climatic changes of the Eta model. It is observed that in all versions of the ETA model (CT20, CT40, MID, HIGH and LOW) a change in the hydrological regime is foreseen, with the appearance of a bi-modal regime, with a maximum in May and a maximum in November.

Figure 13. Average monthly flows in the Itapiranga subbasin under the conditions of the current period (black line) and in the three periods of the 21st century, considering the climate change projections of the different versions of the ETA / CPTEC regional models

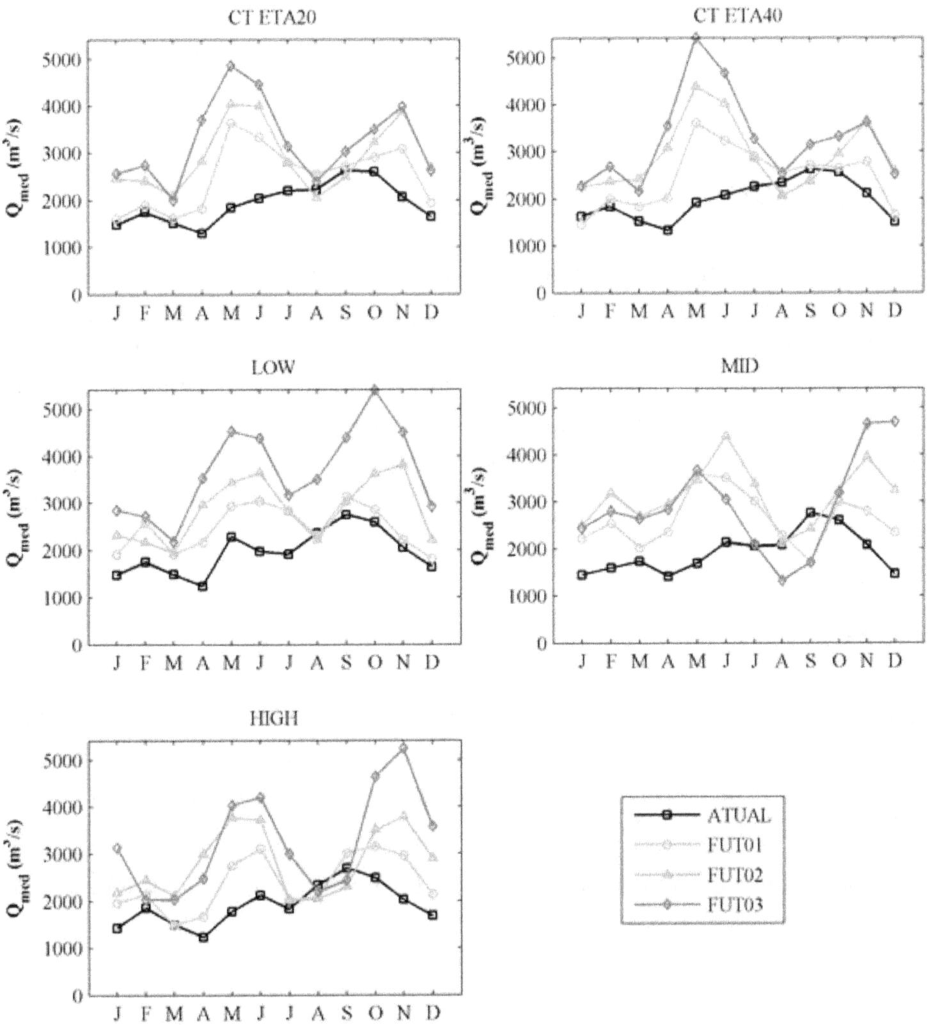

Based on the results presented in this item: (a) in general the results presented for the Itapiranga sub-basin are similar to those of the other points analyzed, with variations of magnitude, but not in relation to the variation signal; (b) The regional models present positive flow variations, generally increasing with the analyzed time horizon. This occurs at all points analyzed in the BHRU; (c) The global models present positive and negative variations, increasing the uncertainty range of mean flow predictions; (d) The differences found between the series CT40 and CT20 are small, mainly in relation to the average flows. In this case the increase of resolution of the model did not influence the final result; Despite the magnitude differences between the series of maximums and minimums obtained, the regional models agree with each other on the signal of the change, in this case positive. On the other hand, the global models are divergent (as for the variation signal), increasing the uncertainty range of the average flow series.These conclusions demonstrate the importance of an adequate analysis of uncertainty for the application of the results in this study in the management of water resources.

CONCLUSION

The importance of the use of environmental parameters is associated with its use as an instrument for planning and managing urban spaces, serving to make better use of natural resources and also as a preventive measure against environmental degradation. In the face of the degradation of atmospheric air quality and the impacts caused by climate change, an integrated assessment of this compartment is required. The data presented in this chapter will provide reliable bases and methodologies for environmental control, supporting the adoption of more restrictive environmental policies.

REFERENCES

Adam, K. N. (2016). Incertezas e impactos de mudanças climáticas sobre o regime de vazões na bacia hidrográfica do rio Uruguai (Doutorado em Recursos Hídricos e Saneamento Ambiental). Universidade Federal do Rio Grande do Sul - Instituto de Pesquisas Hidráulicas. Porto Alegre.

Agência Nacional de Águas (ANA). (2013). Conjuntura dos recursos hídricos no Brasil. Agência Nacional de Águas.

Alves, N. O., Loureiro, A. L. M., Santos, F. C., Nascimento, K. H., Dallacort, R., Vasconcellos, P. C., ... Medeiros, S. R. B. (2011). Genotoxicity and composition of particulate matter from biomass burning in the eastern Brazilian Amazon region. *Ecotoxicology and Environmental Safety*, *74*(5), 1427–1433. doi:10.1016/j.ecoenv.2011.04.007 PMID:21496924

Ares, A., Fernandes, J. A., Aboal, J. R., & Carballheira, A. (2011). Study of the air quality in industrial areas of Santa Cruz de Tenerife (Spain) by active biomonitoring with *Pseudoscleropodium purum*. *Ecotoxicology and Environmental Safety*, *74*(3), 533–541. doi:10.1016/j.ecoenv.2010.08.019 PMID:20828820

Bárdossy, A., & Pegram, G. (2012). Multiscale spatial recorrelation of RCM precipitation to produce unbiased climate change scenarios over large areas and small. *Water Resources Research*, *48*(9), W09502. doi:10.1029/2011WR011524

Barry, R. G., & Chorley, R. J. (2013). *Atmosfera, Tempo e Clima (9th ed.)*. Porto Alegre: Bookman.

Basha, S., Jhala, J., Thorat, R., Goel, S., Trivedi, R., Shah, K., ... Jha, B. (2010). Assessment of heavy metal content in suspended particulate matter of coastal industrial town, Mithapur, Gujarat, India. *Atmospheric Research*, *97*(1-2), 257–265. doi:10.1016/j.atmosres.2010.04.012

Beelen, R., Hoek, G., Van den Brandt, P. A., Goldbohm, R. A., Fischer, P., Schouten, L. J., ... Brunekreef, B. (2008). Long-term effects of traffic-related air pollution on mortality in a Dutch cohort (NLCS-AIR study). *Environmental Health Perspectives*, *116*(2), 196–202. doi:10.1289/ehp.10767 PMID:18288318

Bocchi, C., Bazzini, C., Fontana, F., Pinto, G., Martino, A., & Cassoni, F. (2016). Characterization of urban aerosol: Seasonal variation of mutagenicityand genotoxicity of PM2.5, PM1and semi-volatile organic compounds. *Mutation Research*, *809*, 16–23. doi:10.1016/j.mrgentox.2016.07.007 PMID:27692295

Bouwer, L. M., Aerts, J. C. J. H., Van De Coterlet, G. M., Giesen, N., Gieske, A., & Mannaerts, C. (2004). Evaluating downscaling methods for preparing global circulation model (GCM) data for hydrological impact modelling. In Climate change in contrasting river basins: adaptation strategies for water, food and environment. Wallingford: CABI.

Brown, S. G., Frankel, A., Raffuse, S. M., Roberts, P. T., Hafner, H. R., & Anderson, D. J. (2007). Source apportionment of fine particulate matter in Phoenix, AZ, using positive matrix factorization. *Journal of the Air & Waste Management Association*, *57*(6), 741–752. doi:10.3155/1047-3289.57.6.741 PMID:17608008

Carreras, H. A., Calderón-Segura, M. A., Gómez-Arroyo, S., Murillo-Tovar, M. A., & Amador-Muñoz, O. (2013). Composition and mutagenicity of PAHs associated with urban airborne particles in Córdoba, Argentina. *Environmental Pollution*, *178*, 403–410. doi:10.1016/j.envpol.2013.03.016 PMID:23624338

Carvalho, S. A. D., & Furtado, A. T. (2014). Os desafios da adaptação às mudanças climáticas globais. Climacom –. *Cultura Cientifica*, *2*, 1–11.

Chou, S. C., Lyra, A., Mourão, C., Dereczynski, C., Pilotto, I., Gomes, J., ... And Marengo, J. (2014). Assessment of Climate Change over South America under RCP 4.5 and 8.5 Downscaling Scenarios. *American Journal of Climate Change*, *3*(05), 512–527. doi:10.4236/ajcc.2014.35043

Cohen, D. D., Crawford, J., Stelcer, E., & Atanacio, A. J. (2012). Application of positive matrix factorization, multi-linear engine and back trajectory techniques to the quantification of coal-fired power station pollution in metropolitan Sydney. *Atmospheric Environment*, *61*, 204–211. doi:10.1016/j.atmosenv.2012.07.037

Collischonn, W., Allasia, D. G., Silva, B. C., & And Tucci, C. E. M. (2007). The MGB-IPH model for large scale rainfall-runoff modeling. *Hydrological Sciences Journal*, *52*(5), 878–895. doi:10.1623/hysj.52.5.878

Crispim, B. A., Vaini, J. O., Grisolia, A. B., Teixeira, T. Z., Mussury, R. M., & Seno, L. O. (2012). Biomonitoring the genotoxic effects of pollutants on *Tradescantia pallida* (Rose) D.R. Hunt in Dourados, Brazil. *Environmental Science and Pollution Research International*, *19*(3), 718–723. doi:10.100711356-011-0612-3 PMID:21912955

Crouse, D. L., Peters, P. A., Van Donkelaar, A., Goldberg, M. S., Villeneuve, P. J., Brion, O., ... Burnett, R. T. (2012). Risk of nonaccidental and cardiovascular mortality in relation to long-term exposure to low concentrations of fine particulate matter: A Canadian national-level cohort study. *Environmental Health Perspectives*, *120*(5), 708–714. doi:10.1289/ehp.1104049 PMID:22313724

Dallarosa, J., Teixeira, E. C., Meira, L., & Wiegand, F. (2008). Study of the chemical elements and polycyclic aromatic hydrocarbons in atmospheric particles of PM10 and PM2.5 in the urban and rural areas of South Brazil. *Atmospheric Research*, *89*(1-2), 76–92. doi:10.1016/j.atmosres.2007.12.004

Dibike, Y. B., & Coulibaly, P. (2005). Hydrologic impact of climate change in the Saguenay watershed: Comparison of downscaling methods and hydrologic models. *Journal of Hydrology (Amsterdam)*, *307*(1-4), 145–163. doi:10.1016/j.jhydrol.2004.10.012

Divan, A. M. Junior, Oliva, M. A., & Ferreira, F. A. (2008). Dispersal pattern of airbone emissions from aluminium smelter in Ouro Preto, Brazil, as expressed by foliar fluoride accumulation in eight plant species. *Ecological Indicators*, *8*(5), 454–461. doi:10.1016/j.ecolind.2007.04.008

Environmental Protection Agency (EPA). (2017). *Positive matrix factorization (PMF) 5.0: fundamentals and user guide. Research and Development, 2014*. Available at: http://www.epa.gov/heasd/documents/PMF_5.0_User_Guide.pdf

Estratégia de Resiliência de Porto Alegre. (2017). Disponível em: http://100resilientcities.org/wpcontent/uploads/2017/07/PortoAlegre_Strategy_Portuguese_PDF.pdf

European Environment Agency (EEA). (2015). *Air pollution due to ozone: health impacts and effects of climate change*. Available: https://www.eea.europa.eu/data-and-maps/indicators/air-pollution-by-ozone-2/assessment

Faloona, I., Conley, S. A., Blomquist, B., Clarke, A. D., Kapustin, V., Howell, S., ... Bandy, A. R. (2010). Sulfur dioxide in the tropical marine boundary layer: Dry deposition and heterogeneous oxidation observed during the Pacific Atmospheric Sulfur Experiment. *Journal of Atmospheric Chemistry*, *63*(1), 13–32. doi:10.100710874-010-9155-0

Fan, F. M., & Collischonn, W. (2014). Integraçao do Modelo MGB-IPH com Sistema de Informaçao Geografica. *RBRH - Rev. Bras. Recur. Hídricos*, *19*, 243–254.

Fioletov, V. E., McLinden, C. A., Krotkov, N., & Li, C. (2015). Lifetimes and emissions of SO2 from point sources estimated from OMI. *Geophysical Research Letters*, *42*(6), 1969–1976. doi:10.1002/2015GL063148

Foresti, A. J. (2015). *Um arranjo interdisciplinar para a gestão de riscos de desastres socionaturais com base na Engenharia da Resiliência, RS* (Tese apresentada, como requisito para obtenção do título de Doutora). ao Programa de Pós-graduação em Meio Ambiente, da Universidade do Estado do Rio Grande do Sul.

Forster, P. M., Andrew, T., Good, P., & Gregory, J. M. (2013). Evaluating adjusted forcing and model spread for historical and future scenarios in the CMIP5 generation of climate models. *Journal of Geophysical Research, D, Atmospheres*, *118*(3), 1139–1150. doi:10.1002/jgrd.50174

Fowler, H. J., Blenkinsop, S., & Tebaldi, C. (2007). Linking Climate Change Modelling to Impacts Studies: Recent Advances in Downscaling Techniques for Hydrological Modelling. *International Journal of Climatology*, *27*(12), 1547–1578. doi:10.1002/joc.1556

Gao, J., Tian, H., Cheng, K., Lu, L., Wang, Y., Wu, Y., ... Hao, J. (2014). Seasonal and spatial variation of trace elements in multi-size airborne particulate matters of Beijing, China: Mass concentration, enrichment characteristics, source apportionment, chemical speciation and bioavailability. *Atmospheric Environment*, *99*, 257–265. doi:10.1016/j.atmosenv.2014.08.081

Guha-Sapir, D., Hoyois, P., & Below, R. (2016). *Annual Disaster Statistical Review 2015: The Numbers and Trends*. Brussels: CRED.

Guimarães, E., Domingos, M., Alves, E. S., Caldini, N. Jr, Lobo, D. J. A., Lichtenfels, A. J. F. C., & Saldiva, P. H. N. (2000). Detection of the genotoxicity of air pollutants in and around the city of São Paulo (Brazil) with the *Tradescantia*-micronucleus (Trad-MCN) assay. *Environmental and Experimental Botany*, *44*(1), 1–8. doi:10.1016/S0098-8472(00)00050-2 PMID:10927123

Haghnegahdar, A., Tolson, B. A., Craig, J. R., & Paya, K. T. (2015). Assessing the performance of a semi-distributed hydrological model under various watershed discretization schemes. *Hydrological Processes*, *29*(18), 4018–4031. doi:10.1002/hyp.10550

Herring, S. C., Hoell, A., Hoerling, M. P., Kossin, J. P., Schreck Iii, C. J., & Stott, P. A. (2016). Explaining Extreme Events of 2015 from a Climate Perspective. *American Meteorological Society*, *9*(12), S1–S145.

Holling, C. S., & Gunderson, L. H. (2002). *Resilience and adaptive cycles. In Panarchy: understanding transformations in human and natural systems* (pp. 25–62). Washington, DC: Island Press.

Hollnagel, E., Woods, D. D., & Leveson, N. (2006). *Resilience engineering: Concepts and precepts*. Aldershot, UK: Ashgate.

Illi, J. C., Vancetta, T., Alves, D. D., Osório, D. M. M., Bianchin, L., Quevedo, D. M., & Juchem, F. (2017). Integrated assessment of air pollution by metals and source apportionment using ryegrass (*Lolium multiflorum Lam.*) in southern Brazil. *Environmental Science and Pollution Research International*, *24*(3), 2790–2803. doi:10.100711356-016-8041-y PMID:27837473

IPCC. (2013). Summary for Policymakers. In: Climate Change 2013: The Physical Science Basis. Contribution of Working Group I to the Fifth Assessment Report of the Intergovernmental Panel on Climate Change. Cambridge University Press.

IPCC. (2014). *Fifth Assessment Report – IPCC-AR5. Climate Change – Synthesis Report*. Disponível em: http://www.ipcc.ch/report/ar5/syr/

Jalota, S. K., Kaur, H., Kaur, S., & Vashisht, B. B. (2013). Impact of climate change scenarios on yield, water and nitrogen-balance and use efficiency of rice-wheat cropping system. *Agricultural Water Management, 116*, 29–38. doi:10.1016/j.agwat.2012.10.010

Kawazoe, S., & Gutowski, W. J. Jr. (2013). Regional, Extreme Daily Precipitation in CMIP5 Simulations. *Journal of Hydrometeorology, 14*(4), 1228–1242. doi:10.1175/JHM-D-12-0112.1

Kim, E., & Hopke, P. K. (2007). Source Identifications of airborne fine particles using positive matrix factorization and U.S. Environmental Protection Agency Positive Matrix Factorization. *Journal of the Air & Waste Management Association, 57*(7), 811–819. doi:10.3155/1047-3289.57.7.811 PMID:17687996

Klumpp, A., Ansel, W., & Klumpp, G. (2004). *Urban Air Pollution, bioindication and environmental awareness*. Göttingen: Cuvillier Verlang.

Knutti, R., & Sedlácek, J. (2012). Robustness and uncertainties in the new CMIP5 climate model projections. *Nature Climate Change*. doi:10.1038/nclimate1716

Kumar, P., Robins, A., Vardoulakis, S., & Britter, R. (2010). A review of the characteristics of nanoparticles in the urban atmosphere and the prospects for developing regulatory controls mode particles in an urban street canyon. *Atmospheric Environment, 44*(39), 5035–5052. doi:10.1016/j.atmosenv.2010.08.016

Lago, A. A. C. (2013). *Conferências de desenvolvimento sustentável*. Brasília: FUNAG.

Li, H., Hopke, P. K., Liu, X., Du, X., & Li, F. (2015). Application of positive matrix factorization to source apportionment of surface water quality of the Daliao River basin, northeast China. *Environmental Monitoring and Assessment, 187*(3), 80. doi:10.100710661-014-4154-2 PMID:25655122

Lipsett, M. J., Ostro, B. D., Reynolds, P., Goldberg, D., Hertz, A., Jerrett, M., ... Bernstein, L. (2011). Long-term exposure to air pollution and cardiorespiratory disease in the California teachers study cohort. *American Journal of Respiratory and Critical Care Medicine, 184*(7), 828–835. doi:10.1164/rccm.201012-2082OC PMID:21700913

Marcott, S. A., Shakun, J. D., Clark, P. U., & Mix, A. C. (2013). A Reconstruction of Regional and Global Temperature for the Past 11,300 Years. *Science, 339*(6124), 1198–1201. doi:10.1126cience.1228026 PMID:23471405

Marengo, J. A. (2011). *Riscos das mudanças climáticas no Brasil: Análise conjunta Brasil-Reino Unido sobre os impactos das mudanças climáticas e do desmatamento na Amazônia*. Projeto colaborativo realizado pelo Centro de Ciência do Sistema Terrestre (CCST) do Instituto Nacional de Pesquisas Espaciais (INPE) do Brasil e o Met Office Hadley Centre (MOHC) do Reino Unido.

McLinden, C. A., Fioletov, V., Boersma, K. F., Kharol, S. K., Krotkov, N., Lamsal, L., ... Yang, K. (2014). Improved satellite retrievals of NO2 and SO2 over the Canadian oil sands and comparisons with surface measurements. *Atmospheric Chemistry and Physics*, *14*(7), 3637–3656. doi:10.5194/acp-14-3637-2014

Meireles, J. R. C., & Cerqueira, E. M. M. (2011). *Use of the Micronucleus Test on Tradescantia (Trad-MCN) to Evaluate the Genotoxic Effects of Air Pollution*. Air Pollution – New Developments.

Miller, K. A., Siscovick, D. S., Sheppard, L., Shepherd, K., Sullivan, J. H., Anderson, G. L., & Kaufman, J. D. (2007). Long-term exposure to air pollution and incidence of cardiovascular events in women. *The New England Journal of Medicine*, *356*(5), 447–458. doi:10.1056/NEJMoa054409 PMID:17267905

Mintzberg, H., Ahlstrand, B., & Lampel, J. (2009). *Strategy Safari: Your complete guide through the wilds of strategic management*. London: Prentice Hall.

Morawska, L., Ristovski, Z., Jayaratne, E. R., Keogh, D. U., & Ling, X. (2008). Ambient nano and ultrafine particles from motor vehicle emissions: Characteristics, ambiente processing and implications on human exposure. *Atmospheric Environment*, *42*(35), 8113–8138. doi:10.1016/j.atmosenv.2008.07.050

Moreira, H. M., & Giometti, A. B. R. (2008). O Protocolo de Quioto e as Possibilidades de Inserção do Brasil no Mecanismo de Desenvolvimento Limpo por meio de Projetos em Energia Limpa. Contexto Internacional, 30, 9-47. doi:10.1590/S0102-85292008000100001

Moss, R. H., Richard, H. M., Jae, A. E., Kathy, A. H., Martin, R. M., Steven, K. R., ... Thomas, J. W. (2010). The next generation of scenarios for climate change research and assessment. *Nature*, *463*(7282), 747–756. doi:10.1038/nature08823 PMID:20148028

Mujumdar, P. P., & Kumar, N. D. (2012). *Floods in a Changing Climate: Hydrologic Modeling*. Cambridge University Press. doi:10.1017/CBO9781139088428

Murakami, M., Abe, M., Kakumoto, Y., Kawano, H., Fukasawa, H., Saha, M., & Takada, H. (2012). Evaluation of ginkgo as a biomonitor of airborne polycyclic aromatic. *Atmospheric Environment*, *54*, 9–17. doi:10.1016/j.atmosenv.2012.02.014

Nakicenovic, N., & Swart, R. (2000). *Special Report on Emissions Scenarios: A Special Report of Working Group III of the Intergovernmental Panel on Climate Change*. Cambridge, UK: Cambridge University Press.

National climate adaptation plan – 1° Nap Monitoring and Evaluation Report - 2016/2017 – Brazil. (n.d.). Disponível em: http://www.mma.gov.br/clima/adaptacao/plano-nacional-de-adaptacao?tmpl=component&print=1

Neeling, J. D. (2011). *Climate Change and Climate Modeling*. Cambridge university press.

Oliva, M. A., & Figueiredo, J. G. (2005). Gramíneas biondicadoras da presença de flúor em regiões tropicais. *Revista Brasileira de Botanica. Brazilian Journal of Botany*, *28*(2), 389–397. doi:10.1590/S0100-84042005000200017

Palacio, I. C., Oliveira, I. F., Franklin, R. L., Barros, S. B. M., & Roubicek, D. A. (2016). Evaluating the mutagenicity of the water-soluble fraction of air particulate matter: A comparison of two extraction strategies. *Chemosphere*, *158*, 124–130. doi:10.1016/j.chemosphere.2016.05.058 PMID:27258903

Pecl, G. T., Araujo, M. B., Bell, J. D., Blanchard, J., Bonebrake, T. C., Chen, I., ... Williams, S. E. (2017). Biodiversity redistribution under climate change: Impacts on ecosystems and human well-being. *Science*, *355*(6332), 1–10. doi:10.1126cience.aai9214 PMID:28360268

Perry, C. T. (2010). *Psidium guajava* as a bioaccumulator of nickel around an oil refinery, southern Brazil. *Ecotoxicology and Environmental Safety*, *73*(4), 647–654. doi:10.1016/j.ecoenv.2009.10.001 PMID:20036423

Pope, C. A. III, Ezzati, M., & Dockery, D. W. (2009). Fine-particulate air pollution and life expectancy in the United States. *The New England Journal of Medicine*, *360*(4), 376–386. doi:10.1056/NEJMsa0805646 PMID:19164188

Reff, A., Eberly, S. I., & Bhav, E. P. V. (2007). Receptor modeling of ambient particulate matter data using positive matrix factorization: Review of existing methods. *Air & Waste Management Association*, *57*(2), 146–154. doi:10.1080/10473289.2007.10465319 PMID:17355075

Rotterdam Resilience Strategy - Consultation Document – 2016. (n.d.). Disponível em: http://100resilientcities.org/wp-content/uploads/2017/06/strategy-resilient-rotterdam.pdf

Roy, A., Chatterjee, A., Tiwari, S., Sarkar, C., Das, S. K., Ghosh, S. K., & Raha, S. (2016). Precipitation chemistry over urban, rural and high altitude Himalayan stations in eastern India. *Atmospheric Research*, *181*, 44–53. doi:10.1016/j.atmosres.2016.06.005

Schneider, T., Lan, S., Stuart, A., & Teixeira, J. (2017). Earth System Modeling 2.0: A Blueprint for Models That Learn From Observations and Targeted High-Resolution Simulations. *Geophysical Research Letters*, *44*(24), 12,396–12,417. doi:10.1002/2017GL076101

Seinfeld, J. H., & Pandis, S. N. (2006). *Atmospheric chemistry and physics: from air pollution to climatechange* (2nd ed.). New York: John Wiley & Sons.

Silveira, C. S., Filho, F. A. S., Cabral, L. S., Lemos, W. E. D., & Porto, V. C. (2013). Avaliação de desempenho dos modelos de mudança climática do CMIP5 quanto à representação da precipitação sobre as bacias do setor elétrico brasileiro no século XX e análise das projeções para o cenário rcp8.5. Em: XX Simpósio de Brasileiro de Recursos Hídricos.

Stein, A. F., Draxler, R. R., Rolph, G. D., Stunder, B. J. B., & Cohen, M. D. (2015). NOAA's HYSPLIT Atmospheric Transport and Dispersion Modeling System. American Meteorological Society, 2059-2077.

Steven, V., Kokelj, T. C., Lantz, J. T., Rebecca, S., & Denis, L. (2017). Climate-driven thaw of permafrost preserved glacial landscapes, northwestern Canada. *Geology*.

Stone, D. A. S., & Knutti, R. (2010). Weather and Climate. In C. F. Fung, A. Lopez, & M. New (Eds.), *Modelling the Impact of Climate Change and Water Resources*. John Willey &Sons. doi:10.1002/9781444324921.ch2

Tasic, M., Mijic, Z., Rajsic, S., Zekic, A., Perisic, M., & Stojic, A. (2010). Characteristics and application of receptor models to the atmospheric aerosols research. In *Air Quality*. Rijeka: SciyoIntech. Disponível em: http://www.intechopen.com/books/airquality/characteristics-and-application-of-receptor-models-to-the-atmospheric-aerosols-research

Teixeira, E. C., Feltes, S., & Santana, E. R. R. (2008). Estudo das emissões de fontes móveis na Região Metropolitana de Porto Alegre, Rio Grande do Sul. *Quimica Nova*, *31*(2), 244–248. doi:10.1590/S0100-40422008000200010

Teixeira, E. C., & Mattiuz, I. (2012). Estimated atmospheric emissions from biodiesel and characterization of pollutants in the metropolitan area of Porto Alegre-RS. *Anais da Academia Brasileira de Ciências*, *84*(3), 655–667. doi:10.1590/S0001-37652012000300008 PMID:22886159

Teutschbein, C., Wetterhall, F., & Seibert, J. (2011). Evaluation of different downscaling techniques for hydrological climate-change impact studies at the catchment scale. *Climate Dynamics*, 1–19.

Trenberth, K. E., & Jones, P. D. (2007). Observations: surface and atmospheric climate change. In Climate Change 2007: The Physical Science Basis. Cambridge University Press.

United Nations Framework Convention on Climate Change. (2015). *Paris Agreement*. Availableat:http://unfccc.int/files/meetings/paris_nov_2015/application/pdf/paris_agreement_english_.pdf

Van, V. (2011). Representative Concentration Pathways: An overview. *Climatic Change*, *109*(1-2), 5–31. doi:10.100710584-011-0148-z

Vautard, R., Jacob, D., Gobiet, A., & Belda, M. (2013). The simulation of European heat waves from an ensemble of regional climate models within the EURO-CORDEX project. *Climate Dynamics*, *41*(9-10), 2555–2575. doi:10.100700382-013-1714-z

Wehner, M. (2013). Methods of Projecting Future changes in Extremes. In *Extremes in a Changing Climate: Detection, Analysis and Uncertainty*. London: Springer. doi:10.1007/978-94-007-4479-0_8

Wolf, J. (2011). Climate change adaptation as a social process. In J. D. Ford & L. Berrang-Ford (Eds.), *Climate change adaptation in developed nations – from theory to practice* (pp. 21–32). London: Springer. doi:10.1007/978-94-007-0567-8_2

World Health Organization (WHO). (2005). *Air Quality Guidelines Global Update 2005*. Available:http://www.euro.who.int/__data/assets/pdf_file/0005/78638/E90038.pdf?ua=1

Xiao, H. W., Xiao, H. Y., Long, A. M., Wang, Y. L., & Liu, C. Q. (2013). Chemical composition and source apportionment of rainwater at Guiyang, SW China. *Journal of Atmospheric Chemistry*, *70*(3), 269–281. doi:10.100710874-013-9268-3

Xiao, J. (2016). Chemical composition and source identification of rainwater constituents at an urban site in Xi'an. *Environmental Earth Sciences*, *75*(3), 209. doi:10.100712665-015-4997-z

York, O. N. (n.d.). *The Plan for a Strong and Just City – 2016.* Disponível em: http://100resilientcities.org/wp-content/uploads/2017/07/OneNYC-ilovepdf-compressed.pdf

Zhang, N., Cao, J., He, Y., & Xiao, S. (2014). Chemical composition of rainwater at Lijiang on the Southeast Tibetan Plateau: Influences from various air mass sources. *Journal of Atmospheric Chemistry, 71*(2), 157–174. doi:10.100710874-014-9288-7

KEY TERMS AND DEFINITIONS

Active Biomonitoring: Is a tool used to evaluate the air quality of environments with different magnitudes of environmental impacts and consists of the systematic analysis of the responses of organisms introduced in a standardized way in the natural environment for a specified period.

Air Pollution: Can be defined as the presence of harmful substances in the atmosphere in an amount sufficient to affect its composition or equilibrium, damaging the environment and the most varied life forms.

Anthropogenic Sources: Such as combustion processes, are responsible for the increasing emission of particles in the atmosphere, but the influence of natural sources (continental dust, marine spray, and biogenic particles).

Bioindicators: Are organisms that react to environmental changes by modifying their vital functions, and their chemical composition, providing information about the environmental scenario.

Climate Models: Are extremely sophisticated computer programs that simulate the climate system and its complex interactions between atmosphere, ocean, land surface, snow, ice, and the global ecosystem, as well as a variety of chemical and biological processes.

Dynamic Downscaling or Regional Climate Models (RCMs): Are numerical models similar to GCMs.

General Circulation Models (GCMs): Can provide a reasonable representation of a regional climatic characterization such as ENSO, Hadley cells, trade winds, and jets in the atmosphere.

Chapter 5
Sustainable Development in LED Lights With Air Pollution:
A Case Study of Shengottai Munciplaity, India

B. K. Ramesh
Governemnt of Tamilnadu, India

G. Venkatesan
Anna University, India

J. Thirumal
Anna University, India

ABSTRACT

Environmental pollution and remediation on a global scale have drawn attention to the vital need for new environmentally friendly, clean chemical technologies and processes. Ultraviolet LED (UVLEDs) are mainly employed for the photo catalytic degradation of organic pollutants present in air and water. LEDs are available as both chip and bulb types. As LEDs are energy efficient and mercury free, Shengottai municipality proposed to change all exiting streetlights to LED lights in order to reduce the electrical energy as well as the air pollution.

DOI: 10.4018/978-1-5225-7289-3.ch005

INTRODUCTION

United Nations adopted 17 sustainable development goals with 169 targets in 2015 to transform the planet by 2030 ("Advantages of LED Lighting", 2016). Environmental pollution and remediation on a global scale have drawn attention to the vital need for new environmentally friendly, technologies and process. Air pollution due to human activity mainly the various industrial, mining activities and motor vehicle emissions. The removal of toxic gases has been mainly performed through the use of adsorbents. Road transport is one of the major sources of particulate matter (PM) in urban area (BP, 2018)

Inhalation if urban respiration PM is detrimental to human health not only because of its size but also due to chemical composition (Bloomberg, 2016).

Photocatalytic degradation of pollutants is mainly by the application of conventional ultraviolent lamps. Recently solid state technology has resulted in the development of compact lower cost, and environmentally friendly light emitting diodes. LEDs can emit light of different wave length (infrared, visible or near-ultraviolet) based on the composition of the semiconducting materials.

Ultraviolet LED (UVLEDs) is mainly employed for the photo catalytic degradation of organic pollutants present in air and water. LEDs are available as both chip and bulb types. As LEDs are energy efficient and mercury free. Not all LED lights is optimal, when used as street lighting.Improper design of the lighting fixture can result in glare,creating a road hazars condition ('Council on Science', 2012).

The first application of UV-LED as an irradiation source for the purification of air was carried out by Johnson 2003 (Dahl et. al, 2006). The photo catalytic removal of pollutants under irradiation of LEDs has been mainly carried out using ultraviolet LEDs, However some visible LED have also been applied to the degradation of a number of organic compounds in water and volatile organic compounds in air.

The emission of odorous pollutants from wastewater treatment plants, landfills, livestock facilities, exhaust emissions from automobiles, and waste water from paper production plants presents serious environmental and health concerns. Many of the emitted compounds are highly toxic and harmful to the environment. Diethyl sulfide (DMS) and nitrogen monoxide are representative compounds. LEDs have recently have been employed for the removal of gasses from air (Municipal Corporation Jaipur and Bhubaneswar).

India's urban system is the second largest in the world with an urban population surging to 31.16% in 2011 from 27.81% in 2001 (Census of India, 2011) of the total population. As per the 18th Electric power survey of CEA, the estimated energy consumption in Indian public lighting sector with 2009-10 as the base year is about 8,478 million kWh in 2012-13. The sector will grow at a CAGR of 7% during the XII and XIII plan periods (Gentler, Gillies, and Pierson, 2000)

In Government of India, Ministry of Power (MoP) and Bureau of Energy Efficiency (BEE) are entrusted with the task of preparing the implementation plan for the National Mission on Enhanced Energy Efficiency (NMEEE) under National Action Plan on Climate Change (NAPCC) ("Council on Science",2012) This mission has a component which deals with the market based Measures to improve the energy efficiency in energy intensive area with the certification of energy savings which could be transferred. In this context, the Government of Tamil Nadu has initiated energy efficiency in street lighting by converting the exciting fluorescent tube lamps, Sodium vapour lamb and mercury lamp in to light emitting diode (LED) lamps and achieved about 50% saving on its annual energy consumption (Huang et. Al, 2009).

DESCRIPTION OF SENGOTTAI MUNICIPALITY

The Municipality (ULB) is responsible for development, operation and maintenance of Public Street Lighting system. The Public Street Lighting System has grown significantly over the past years with growth in population, infrastructure (roads,streets and extenstion areas) in the Municipality and focus on improving lighting levels. The Sengottai Municipality in Tirunelveli district with the area of 2.68 sq.km and the population of 26,823, as per 2011 census.

Existing Public Street Lighting System

Sengottai Municipality consists of around 1023 street lighting fixtures. Out of them, 67% are Fluorescent tube lights, 15% are Sodium Vapour Lamps (SVL), 18% are Compact Fluorescent Tube Lights. The total connected load from all the lighting fixtures is 62 kW.(Table. 1)(IPCC, 2005), with an annual savings of 0.13 million units(Kwh)electrical energy and the amount of rupees 6.00lakhs .

PROJECT DETAILS

Project Objective

The objective is to reduce electricity consumption in street lighting, all lights to be converted to LED lights as seen in Figure 1. In order to take this concept forward to measure the actual savings in power consumption, departmentaly survey ("Municipal Corporation Jaipur and Bhubaneswar") would be carried out to establish actual energy consumption, which would act as a reference point for potential reductions in energy consumption.(Table. 2 and Table 3). This will lead to improvement in the service

Table 1. Energy saving potential in various conventional lamps

Present Lighting System	Equivalent LED Upgrade	Potential Saving in Energy Consumption
Fluorescent Lamp 40W	LED 16W	61%
Sodium Vapour Lamp 70 W	LED 30W	61%
Sodium Vapour Lamp 150 W	LED 60W	63%
Sodium/Metal Halide 250W	LED 120W	56%
Sodium Vapour lamp 400W	LED 240W	46%

Table 2. Energy saving potential in the Shengottai Municipality

Exsisting Lighting Fixtures	Quantity	Total Power (kw)	Retrofitting LED Lights (W)	LED Power (kw)	Power Saving Potential (kw)
FTL 40W	685	27.4	20	13.70	13.70
SVL 250W	68	17.00	120	8.16	8.84
SVL 150W	87	13.05	60	5.22	7.83
CFL 24W	183	4.39	18	3.29	1.098
Total	1023	61.84		27.08	30.37

delivery as well as reduction in greenhouse gas emissions. Already the Municipal Corporation of Akola in Maharastra, Jaipur in Rajesthan and Bhuvaneshwer in Odisha completed like this project and achieved their energy savings as well control the air pollution (IPCC, 2014).

Energy Savings Verifications

Savings are determined for entire street lighting system by Measurement of actual power drawn at all switching points of street lighting system in an ULB. Since lighting load is constant throughout its operation, Aggregation of all the switching loads multiplied with operational hours during that season will provide energy consumption ("Municipal Corporation Jaipur and Bhubaneswar")

Savings = (Baseline Energy – Reporting Period Energy)

± Routine Adjustments ± Non-Routine Adjustments

- Routine Adjustments include Regular Power cuts
- Non-Routine Adjustments include unscheduled power cuts

Table 3. Lighting details of Shengottai Munciplaity

Particulars	Qty	
Total Lighting Fixtures	1023	Nos.
Aggregate Power (kW)	61.8	kW
Defunct Light/Fixtures	41	Nos.
Percentage of Defunct Fixtures	4.0%	
Aggregate Power at Switching Points	51	kW
Aggregate Power of Defunct Fixtures	1.64	kW
Existing LEDs	0.00	kW
Aggregate Power excluding LED	51	kW
Distribution/Cable loss, etc.	5	kW
Distribution /Cable loss, etc.	10%	
Operational Hours during Winter (Nov-Feb)	12	per day
Operational Hours during Mar-Oct)	11	per day
Hours of Power cut, etc.	0.67	40 Min. per day
Average St. Lights operational Hours	10.5	per day
Annual Electricity Consumption	196759	kWh
Annual Electricity Consumption as per ULB Records (FY 2016-17)	300000	kWh
Proposed LED Fixtures (Retrofitting)	1023	Nos.
Aggregate proposed LED Power	27.1	kW
Distribution loss expected to reduce by 50%	2.6	kW
Power Savings Potential	51.2	kW
Annual Energy saving Potential	196215	kWh
Energy saving potential	65.4%	

Environmental and Social Impact

Energy Efficiency through street lighting is an initiative with certain objectives such as; reduction in electricity consumption, reduction in carbon dioxide emissions and reduction of expenses in urban local bodies on annual electricity bills and maintanence cost. With this initiative certain impacts are observed (Johnson, 2003). The long-term and short- term environmental and social impacts with its mitigation measures are discussed as,

- **Reduction in Green House Gases** – As the project reduces the consumption of energy for street lighting, the electricity required for municipal infrastructure

Figure 1. Global primary energy consumption
(Source: www.ourworldindata.org)

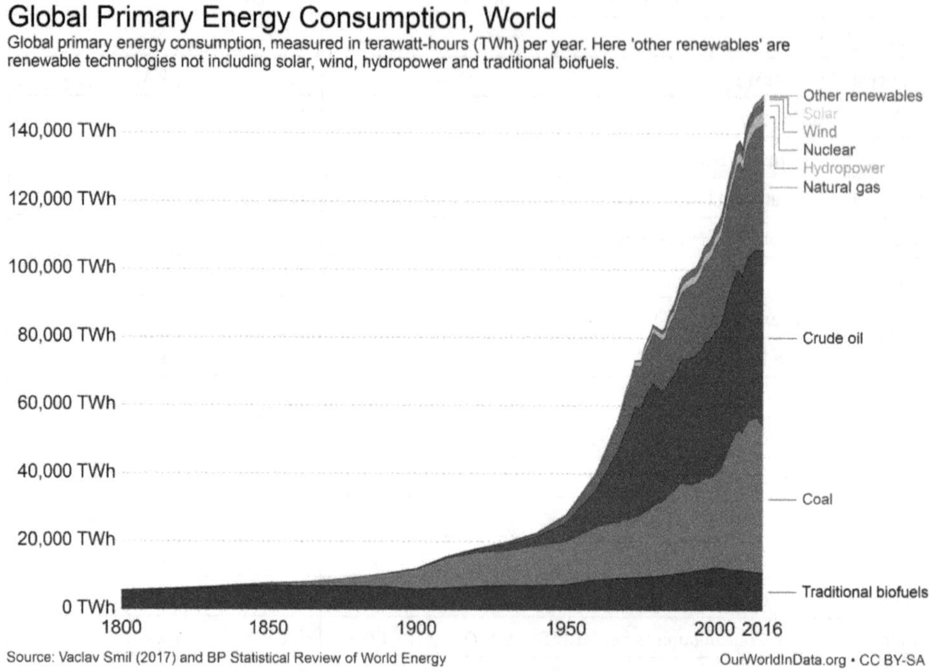

reduces and thus the green house gas emission at generation point will also reduces.

- **Particulates in Air** –This is also controlled, if the fixtures and bulbs are broken on site such as breakage in lead or mercury lamps, during replacement.
- **Noise** - Site work activities such as replacement of energy efficient lamps with old lamps create noises in the area and this can be disturbing for local residents.
- **Reduction in energy consumption and costs:** Modern LED based energy efficient lighting systems consume about 45%- 60% less energy than older and inefficient systems that were installed in this Municipality (ULB).
- **Reduction in GHG Emissions**: Reduction in Green House Gas (GHG) emissions that are equivalent to 231 t CO2 annually from the saved energy as seen in Figure 2 (United Nations General Assembly, 2015).
- **Social benefits:** LEDs not only improves safety conditions for vehicular traffic and pedestrians but also improve the safety and security of every town citizens.

Figure 2. Annually saved energy countries
(Source: www.ourworldindata.org)

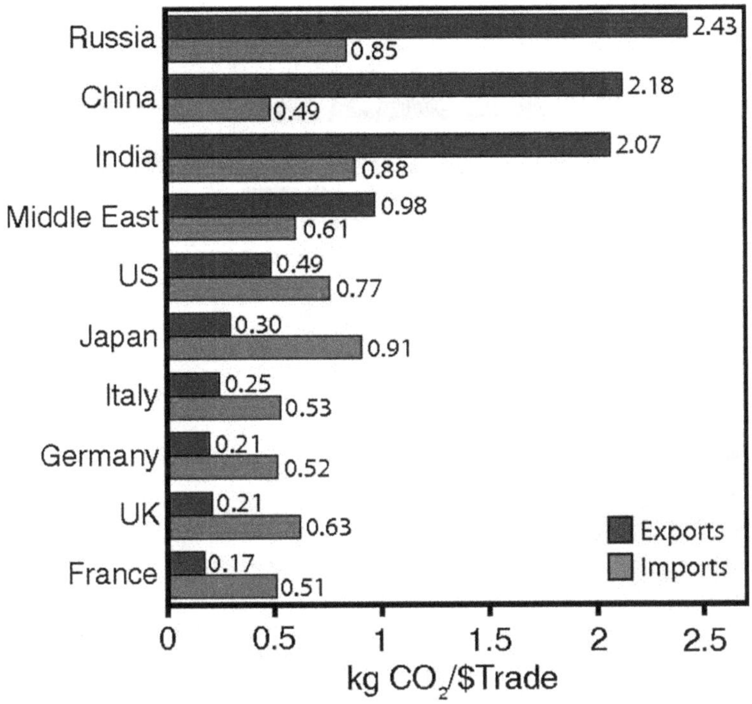

CONCLUSION

Access to energy is a key pillar for human wellbeing, economic development and poverty alleviation. If we want to reduce our global greenhouse gas emissions, the world has to transition from an energy system dominated by fossil fuels to a low-carbon one this is what most countries have set long-term targets to achieve within the Paris climate agreement as seen in Figure 3 (Vaclav Smil Energy, 2017) . With the exception of carbon capture and storage (CCS) technology, we have two options to achieve this: renewable technologies (including bioenergy, hydropower, solar, wind, geothermal, and marine energy) and nuclear energy. Both of these options produce very low CO_2 emissions per unit of energy compared with fossil fuels. We call this process of transitioning from fossil fuels to low-carbon energy sources 'decarburization' (Lin et. al, 2014).

Intergovernmental Panel on Climate Change (IPCC) and International Energy Agency (IEA) still regard carbon capture and storage to be a crucial element in meeting our global carbon targets (Order of Commissioner of Municipal Administration, Government of TamilNadu).

Figure 3. Renewable energy investment
(Source: www.ourworldindata.org)

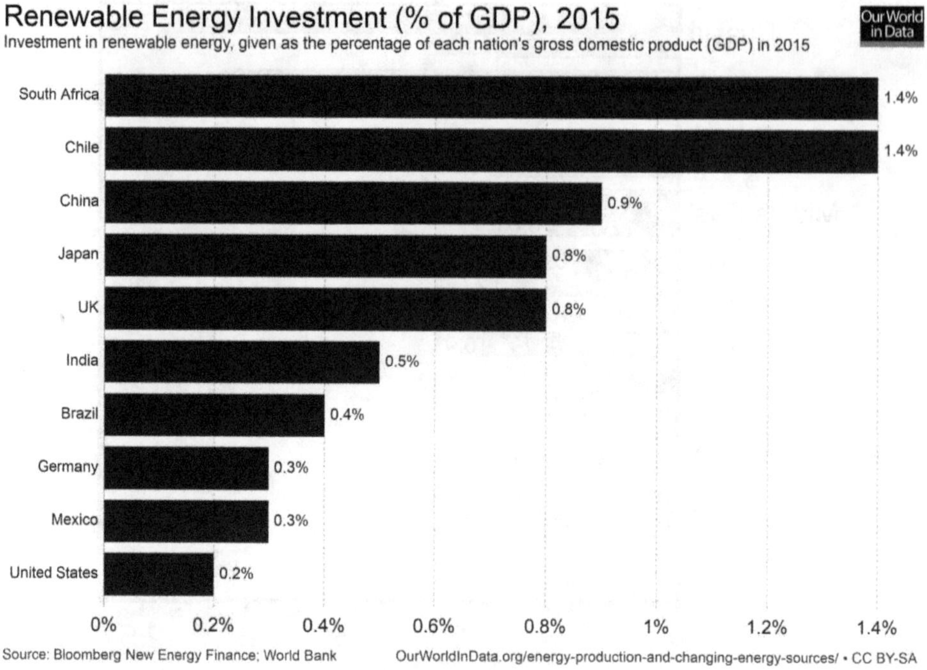

LED lighting has no mercury or lead, and does not release any toxic substance .The light output is very consistent across cold or warm temperature gradients (IPCC, 2005). Considering conversion to energy efficient LED street lighting use lower CCT lights that will minimize potential health and environmental effects (The World Bank, 2016).

REFERENCES

Advantages of LED Lighting. (2016). Retrieved from http://energy.ltgovernors.com/in-depth-advantages-of-led-lighting.html

American Medical Association. (2012). *Council on Science and Public Health Report 4, light pollution. Adverse effects of night time lighting*. American Medical Association.

Bloomberg New Energy Finance. (2016). *Global trends in renewable energy finance 2016*. Author.

Dahl, A., Gharibi, A., Swietlicki, E., Gudmundsson, A., Bohgard, M., Ljungman, A., ... Gustafsson, M. (2006). Traffic – generated emission of ultrafine particles from pavement – tire interface. *Atmospheric Environment*, *40*(7), 1314–1323. doi:10.1016/j.atmosenv.2005.10.029

Gentler, A. W., Gillies, J. A., & Pierson, W. R. (2000). an assessment of the mobile source contribution in U.S. *Water, Air, and Soil Pollution*, *123*(1/4), 203–212. doi:10.1023/A:1005263220659

Huang, X., Wang, H., Yin, S., Chen, X., Chen, W., & Yang, H. (2009). sterlizatioin system for air purifier by combining ultraviolet light emitted diodes. *Journal of Chemical Technology and Biotechnology (Oxford, Oxfordshire)*, *84*(10), 1437–1440. doi:10.1002/jctb.2180

IPCC. (2005). *2005: Carbon Dioxide Capture and Storage Summary for Policymakers.* Intergovernmental Panel on Climate Change. Available.

IPCC. (2014). Summary for Policymakers. In Climate Change 2014: Mitigation of Climate Change. Contribution of Working Group III to the Fifth Assessment. IPCC.

Johnson, B. D. (2003). High –power short –wave LED purifies air. *Photonics Spectra*, *37*, 111.

Lin, Y., Liu, Y., Sun, Y., Zhu, X., Lai, J., & Heynderickz, I. (2014). Model predicting discomfort glare caused by LED road light. *Optics Express*, *22*(15), 18056–18071. doi:10.1364/OE.22.018056 PMID:25089425

The World Bank. (2016). *World Development Indicators*. Author.

United Nations General Assembly. (2015). *Transforming Our World: The 2030 Agenda for Sustainable Development.* Draft resolution referred to the United Nations summit for the adoption of the post-2015 development agenda by the General Assembly at its sixty-ninth session. UN Doc. A/70/L.1 of 18 September (2015).

Vaclav Smil Energy Transitions. (2017). *Global and National Perspectives*. BP Statistical Review of World Energy.

Chapter 6
Effects of Atmospheric Pollutants on Biodiversity

Akash
Gurukula Kangri University, India

Navneet
Gurukula Kangri University, India

B. S. Bhandari
HNB Garhwal University, India

Kamal Bijlwan
SGRR University, India

ABSTRACT

Biodiversity is severely affected by the various anthropogenic activities. Humans depend on natural resources for survival. The resources include food, clean drinking water, timber for construction, fibers for clothing, natural gas, and coal for industries. The biggest threat to the atmosphere is pollutants caused by different human activities. The earth atmosphere is a layer of gases that surrounds the planet and allows us to breathe and live. The atmosphere is made up of Nitrogen, Oxygen, and others types of gases. This mixture of gases is known as air. Air pollution results from a variety of pollutants like NO_2, SO_2, CO, CFCs, and Ozone. Carbon dioxide, which is a greenhouse gas, warms the earth. This chapter will explore the effects of atmospheric pollutants on biodiversity due to various types of adulterants in the atmosphere and its effects on the survival of earth. Additionally, it will sketch various tools that can reduce these harmful activities.

DOI: 10.4018/978-1-5225-7289-3.ch006

INTRODUCTION

Climate change is expected to worsen the quality of air in densely populated regions of world by changing atmospheric ventilation and dilution, precipitation band resulting into the losses of biodiversity (Fiore *et al.,* 20015). Every organism gets affected due to the changes in climate. Climate is an important factor that influences the quality of air. Reduction in the quality of air will directly affect ecosystem in ways that also could affect human health (Haase *et al.,* 2014; Kinney, 2008). and Who, 2013). As the climate continues to change in future these impacts would be more drastic in future. Various studies reveal that climate change has already affected the quality of air in atmosphere. Further the changes in climate from pre-industrial resulted in up to 111,000 and 21,400 additional premature fine particle- and ozone-related deaths due to climate change. (Fang *et al.,* 2013; Silva et al., 2013).Human activity has made adverse impact on the atmosphere and its components (Uysal *et al.,* 2003 and Health and Environment Alliance, 2009).The study of Intergovernmental Panel on Climate Change 2007 document "most of the observed rise in globally averaged temperatures of earth since the mid-20th century is very likely due to the recorded increase in anthropogenic greenhouse gas concentrations" (Trenberth *et al.,*2007). Global greenhouse gas and their further emissions due to the various human activities have been growing day by day with an increase of 70% between 1970 and 2004. One of the most important green house gas is carbon dioxide which increases due to the anthropogenic greenhouse gas (Trenberth *et al.,* 2007). The negative effect of Atmospheric pollutants on health has a quantifiable impact, not only in terms of morbidity but also on the mortality of respiratory diseases. The recent increase in morbidity associated with respiratory diseases is the continuous degradation of the quality of air as a result of increasing levels of outdoor Atmospheric pollutants such as vehicle emissions. The major Atmospheric pollutants that are toxic, causing serious disease in the human and also effects the flora and fauna which are nitrogen dioxide, sulfur dioxide, and particulate matter. Some other aeroallergens which are derived from pollen grains, lead to bronchial obstruction in predisposed individuals. The pollen allergy is one of the models that study the relationship between respiratory allergic disease and atmospheric pollutants.

CLIMATIC CHANGES AND LOSSES IN BIODIVERSITY

The main climatic changes in atmospheric composition is mainly due to the combustion of fossil fuels, excessive use for the generation of energy as well s the transportation. Different types of Atmospheric pollutants have been reported which differ in their reaction properties, chemical composition, emission as well as the persistence in environment and their ability to be transported in long and short distances as well as their eventual impacts on the survivals. These different pollutants share some similarities which can be grouped to four classes. The major effects of climatic change on the quality of air area degrading the removal process like dispersion, precipitation and amplifying the atmospheric chemistry (Fiore *et al.*,1995), resulting into the severe effects on the organism of earth. Some are gaseous pollutants like ozone, $SO_{2,}$ CO, NOx, Volatile Organic Compounds, some are persistent organic pollutants (for e.g. dioxins) and some are heavy metals (for e.g. lead, mercury and particulate Matter (PM). Different anthropogenic emissions, future climate will change the emissions of biogenic volatile organic compounds due to higher temperatures of earth and changed plant metabolism which can alter secondary organic aerosols resulting in changes in secondary particle levels (Carslaw *et al.,* 2010). The hydroxyl radical formation is associated with methane gas and another greenhouse gas (Fiore *et al.,* 2014). These all interaction into the atmosphere causes different types of diseases and health problems among the organisms. Plants, animals and microbial biomass also get affected due to all these interaction into the environment. Combustible particles can affect cardiopulmonary mortality, hospitalization as well as the respiratory disease like bronchitis, rhinitis, asthma and chronic (WHO, 2013). Recent evidence from different worker revels that the unfair quality of air enhance rheumatic diseases (Sun *et al.,* 2016), diabetes (Thiering *et al.,* 2015), neurodegenerative diseases, cognitive functioning (Clifford *et al.,*2016) and other harmful gases such as the secondary pollutant are related to all circulatory and respiratory mortalities (Who, 2013 and Turner *et al.,*2016) as well as chronic respiratory diseases such as asthma (Who, 2013). Studies showed that higher ozone concentrations causes preterm birth (Olsson and Morgen, 2013), reproductive health (Hansen *et al.,*2010) as well as cognitive decline (Xu *et al.,*2016). The other natural factors are dust particles and storms which also increases in future due to climatic changes (Fuzzi *et al.,*2015). Jacob and Winner (2015), observed that due to climatic change especially due to circulation as well as the decreasing frequency of mid-latitude cyclones, the annual mean particulate matter 2.5(PM) concentrations will change by $\pm$ 1 µg/m3 in the united states of America and other European countries. These all react with hydroxyl radical resulting into the formation of ground-level ozone. The rising temperature due to the climatic changes lead to the increasment of natural VOC emission which affect the ozone concentration (Sillman *et al.,* 1995). Further the primary and secondary pollutants can gear up the climatic change that in turn of public health. The secondary pollutants such as ozone can severely affect the different crop yields which in combination with climate, can affect food security and public health (Patz *et al.,*2014 and McGrath *et al.,*2015). In the last few century, the greatest increase in the emissions of Atmospheric pollutants due to the industrial and economic growth has made poor quality of air. This is a major problem especially in industrialized countries and an emerging problem for the rest of the world in future. The increasment in global temperature is sometime also due to the Soot (black carbon) which can absorb heat (De-Angelo *et al.,* 2013). Sometime at the surface of ozone, the various secondary pollutants formed from the interaction of precursor compounds with the light from sun including UV radiation (Jacob and Winner, 2015). The precursor are VOCs, CO, CH4. Higher concentrations of PM2.5 due to the anthropogenic sources can result from changes in precursor emissions or in meteorology and in the physical as well as the chemical behaviour of various particles into the atmosphere (Liu *et al.,*2010). The other thing is wildfires which is associated with climatic changes, could severely reduce air quality in future (Liu *et al.,*2010).The Increased concentrations of many greenhouse gases like CO_2, methane, the earth's atmosphere have already warmed the planet substantially, causing more severe and prolonged heat waves, droughts, and floods, Atmospheric pollutants, temperature variability, forest fires. All of the effects due to climatic changes and pollution cause severe respiratory health risk. (Intergovernmental Panel on Climate Change, 2007 and Solomon *et al.,* 2007). The changes in air quality and climate have a quantifiable impact, not only on the morbidity but also on the mortality for different types of respiratory diseases (Peters *et al.,*2004 and Dockery *et al.,*2007). Global earth temperature has markedly risen up over the last five decades due to the greenhouse gas emissions.The inhalable particles are those particles whose aerodynamic diameter is less than 10 μm (PM10), while fine inhalable particles are less than 2.5 μm (PM2.5) and coarse inhalable particles are 2.5 -10 μm.(WHO, 2013). The size of these particles surface as well as the chemical composition of the particulate material determine the category of risk that exposure to this agent represents for human health. The fine particles, owing to their smaller size, penetrate deeply into the respiratory system of the living being and severely affect the alveoli. Atmospheric pollutants show these effects by causing different types of direct cellular injury or by enhancing the intracellular signalling pathways and transcription factors that are very sensitive to the oxidative stress.(WHO, 2014 and Cecchi *et al.,*2013)

Ozone is a powerful oxidant which has been associated with persistent structural airway and lung tissue damage as well. In many European country it has been observed that it is contributing to more severe symptoms of asthma as well as in increase in respiratory hospital admissions (Filippidou, 2011). It has been estimated for U.K, in 2020 the annual ozone level >1,500 ppm will associates with deaths of many people due to various respiratory disease. The Positive associations also have been observed between urban Atmospheric pollutants and respiratory symptoms in children. Different study reveals the relations between motor vehicle exhausts and the level of acute or chronic respiratory symptoms in children living near traffic area (Jacob *et al.,* 2009; Gryparis *et al.,* 2004 and-Jerrett *et al.,*2009) The bad quality of Atmospheric pollutants can negatively influence the development of lungs in children and in adolescents (Gryparis *et al.*, 2004; Epton *et al.*,, 2008 and Gauderman *et al.*,2007). Different study of different authors have suggested that the Children adversely gets effected due to Atmospheric pollutants . The deficits in lung function was directly correlated with set of pollutants, including acid vapour, nitrogen dioxide, fine particulate matter (PM2.5), as well as the elemental carbon. The dysfunction in lung during childhood may enhance the risk of respiratory conditions, such as episodic wheezing which happens during a viral infection. However, the larger effect of pollution-related deficits can occur during later in life, since dysfunction in lung is a strong risk factor for complications and death during adulthood. (Li S *et al.,* 2012). The prediction about the effects of climate change due to Atmospheric pollutants and on health-related Atmospheric pollutants is hampered by several limits which are, future emissions which depend on numerous factors like economic development, energy use, population growth, economic development and production. The indoor pollution is another approach and its level of Atmospheric pollutants other than environmental tobacco smoke which have also been also related to asthma prevalence or symptoms by sparse studies (Hulin *et al.,* 2012; Bentayeb etal.,2013 and Annesi-Maesano *et al.*,2013). Environmental tobacco smoke is one of the most important categories of risks for respiratory symptoms and diseases in the world. The evidence is also reliable for indoor particulate matter and nitrogen dioxide which have been associated with respiratory diseases like asthma. Atmospheric pollutants is the changes in the levels of quality as well as in the purity of the air due to various natural or anthropogenic emissions of biological chemical and biological substances. Different pollutants affecting the air quality include inhalable particles.The vehicle exhaust, particle pollution as well as the ground-level ozone are the most important types of harmful pollutants.

CLIMATE CHANGE AND ATMOSPHERIC POLLUTANTS

There are the different types of environmental pollutants in the atmosphere which affect the biodiversity in different way. Ozone is one of the most important known green house gas. It is an important greenhouse gas and plays significant role in the radiative budget in the stratosphere through its interactions with both long wave and short wave radiation (Gauss *et al.,* 2003). It is a precursor in oxidizing reactions because it forcefully influences the affect of greenhouse gases as methane and hydrofluorocarbons (Gauss *et al.,* 2006). It also affect indirectly climate due to the limiting sequestration of CO_2 by vegetationn (Sitch *et al.,* 2007). The direct radiative affects of ozone provide the positive radiative forcing of the atmospheric gases especially green house gases (Gauss *et al.,* 2003).The ground-level ozone is an important air pollutant that severely affects human and the health of other organism

Table 1. Common environmental pollutants and their source

Gropus	Pollutants' Sources	Emitted Pollutants
Agricultural pollutants	Due to open burning	Carbon monoxide, SPM and VOC
Pollutants from mining	Coal mining	Sulphur dioxide, SPM and NOx,
	Ammonium sulphate	VOC
	Plastic, Syntheticresins, material, fibres	SPM, CS_2, VOC and H_2S,
	Varnishes, Paints, lacquers	VOC and SPM
	Soap	SPM
	Petroleum and natural gas production	Sulphur dioxide
	Non ferrous ore mining	Lead and SPM
	Stone quarrying	SPM
	Pesticide production, Fertilizers	Ammonia and SPM
	Ammonium nitrate	SPM, Ammonia and nitric acid
	Beverges and tobacco	VOC, Sulphur dioxide, carbon mono oxide and H_2S
	Textiles and leather industries	VOC and SPM,
	terephtalic acid, Maleic anhydride	VOC and carbon mono oxide
	Paper products, printing	SPM,VOC, hydrogen sulphide and Sulphur dioxide
	Ammonia	Oxides of nitrogen, Sulphur dioxide, VOC, SPM, and ammonia
		Trinitrotoluene
	Hydrochloric acid	Hydrochloric acid
	Hydrofluoric acid	SiF_4 and Hydrogen fluoride
	Sulphuric acid	SO_3 and Sulphur dioxide
	Nitric acid	Oxides of nitrogen
	Wood products	VOC, SPM,
	Phosphoric acid	F_2 and SPM
	Lead oxide and pigments	Lead and SPM
	Alkyl lead	Lead
	Phallic anhydride	SPM, carbon mono oxide and Sulphur dioxide
	Carbon black and printing ink	NOx, CO, SO_2,VOC, hydrogen sulphide and SPM,
Fuel refineries pollutants	Products of petroleum and coal	VOC, F2, carbon dioxide, SPM and sulphur dioxide
	clay products	
	Cement lime and plaster	Sulphur dioxide
Wholesale pollutants	Gas, Steam Electricity	Oxides of nitrogen, carbon monoxide,SO_3, lead and sulphur dioxide
Transport pollutants	filling operation	VOC
Community services pollutants	Municipal incinerators	lead, SPM, sulphur dioxide, Oxides of nitrogen carbon monoxide and VOC,
Metal industries pollutants	Non -ferrous industries	Sulphur dioxide, F, lead and SPM,
	Iron and steel	NOx,,SO_2, carbon monoxide VOC and lead

SPM: Suspended particulate matter, VOC: Volatile organic compounds
Source: Economopoulos, 1993

as well as the growth of forests worldwide (Ainsworth *et al.,* 2012). The increasing level of the emission of the precursor of ozone like NOx, CO have affected different lives on earth (Brasseur *et al.,* 2001).

The long-range of trans-Pacific transport of contaminant air masses from Asia is a substantial part of the background ozone in western North America (Vingarzan, 2004). It has been seen that there is significance increasment in the concentration of ozone in the free troposphere during 1984 (0.63-0.34 ppb/ year) in spring season the intercontinental transport across the Pacific is most efficient (Cooper *et al.,* 2010; Law, 2010). The emission from the Asian continent influence, emissions from Canada and Mexico which contribute to background level of ozone concentrations in the eastern region of the United States and in California as well (Wang *et al.,* 2009). In U.S.A, It has also been observed that the highest ozone pollution occurs in the south-western region of United States, especially in California where the fourth highest maximum averages value ranges from 71 to 106 ppb. On the other hand Moderately elevated O3 levels (51 ppb in Montana to 78 ppb in New Jersey) occurs in the U.S.A (Bytnerowicz *et al.,* 2008). The other contaminant of atmosphere is reactive nitrogen which is generally results from different types human activities in the environment which severely causes harm to human health as well as the ecological problems. The components of nitrogen like NOx results mostly from different combustion processes, while ammonia and N_2O mostly results from agricultural activities. The compounds of nitrogen can affect climate in different ways. The oxide of nitrogen and ammonia contribute to climatic changes indirectly by the altering the production and loss of atmospheric climatic forcers, both aerosols scattering incoming solar radiation and greenhouse gases. These different compounds severely affects climate in different way. The oxide of nitrogen especially N_2O can affect the environment by increasing the ozone formation and also by increasing the removal of CH4 (in cooling effect). So NH3 and NOx can enhance light-scattering aerosols (Pinder *et al.,* 2013). In some areas of U.S.A, the concentration of NOx and nitric acid and atmospheric NOx and nitric acid have decreased as per the report of Chestut, 2005) due to Environmental Protection Agency of the government. Deposition have remained stable, mainly due to a lack of regulatory control of agricultural emission and use of nitrogen as a fertilizer (Davidson *et al.,* 2012). Some of the phytotoxic effects of the reactive nitrogen has been seen by Bytnerowicz *et al.,* 1998. The application of synthetic nitrogen fertilizers as well as the cultivation of N-fixing crops are major parts of the U.S agricultural economy (Smil, 2002 and Houlton *et al.,* 2013). The nitrogen fixation from anthropogenic activity creates essentials industrial products like plastics, nylon and explosive (Domene and Ayers, 2001) but various human problems and environmental issues has developed from the use of unintentional leakage of nitrogen. High rate of mortality, morbidity results from climatic changes and pollution. The contamination of water sources by nitrite, severity of toxic algal blooms as well as the hypoxia in freshwater and coastal marine areas and global climatic change results from the emission of the potent greenhouse and ozone-depleting gas oxides of nitrogen (Davidson *et al.,* 2012).

The level of SO_2 still contributes to acidic deposition and its direct phytotoxic effects are limited to small areas in local geographic areas (Legge *et al.,* 1998; US EPA, 2008). Study revel that wet deposition of sulphur has continuously decreased in recent decades (Greaver *et al.,* 2012; NADP, 2010).

The small scale changes in the concentration of the sulphur dioxide can cause large changes in atmospheric chemistry by several reasons. Sulphur dioxide when oxidized it forms H_2SO_4 which has low vapour pressure for allowing it to adsorb on tiny dry particles and after that to increase the size of those particles by molecular condensation of more acid (Ehhalt,1999). This resulting molecules increases the concentration of aerosol. Sulphur dioxide reacts quickly with OH leaving little OH to oxidize many gases like methane and carbon monoxide. So increasment in the concentrations of methane and carbon monoxide indicate decreasing oxidizing capacity. The sufficient concentration of sulphur dioxide condenses to a liquid below−10 °C. The temperature in the troposphere layer decreases with increasing altitude at −6.4 °C/km to −55 °C at the tropopause and after that increases with altitude in the overlying stratosphere. In most of the cases, troposphere layer and lower stratosphere layer sulphur dioxide is potentially a liquid and this liquid form has good soluble ability and can be oxidized readily by hydrogen per oxide in cloud droplets and also in precipitation (Ehhalt,1999). Study from Ehhalt (1999) revels that 70% of the available sulphur dioxide, oxidized in this way. This type of oxidation process does not lead to nucleation of new particles (Alexander *et al.,* 2005). Laj et al.1990, observed that the depletion of hydrogen peroxide in the ice of Greenland revels that sulphur dioxide reached in Greenland as a gas and after that it was oxidized in precipitation. Another study from Barrie and Hoff (Barrie *et al.,*1984) observed that sulphur dioxide in the arctic region originating from fossil fuel burning in Eurasia.

Biodiversity Losses Due to Atmospheric Pollutants

Atmospheric pollutants like sulphur dioxide, ozone, carbon monoxide, volatile organic compounds (VOCs), particulate matter (PM10 andPM2.5) increasing day by day and resulting in the serious disease in the biotic components. Atmospheric pollutants affects survivals on earth by many ways. The hazardous pollutants escape to the environment by different types of natural or anthropogenic activities and may severely effects on human, plants as well as the health of the animals. Different types of harmful chemical or pollutants Increases into the atmosphere which differ in their chemical composition, emission, time of disintegration, reaction properties as well as the ability to diffuse in long distances.The different types of Atmospheric pollutants severely affects human health in different ways. The composition of Atmospheric pollutants and their dose as well as the time of exposure resulting into the diverse impacts on human health. The polluted air has bad health effects on human which can range from nausea and difficulty in breathing as well as skin cancer. Atmospheric pollutants can include different types of defects like birth defects, reduced activity of the immune system, developmental delays in children, and other diseases. Epidemiological and animal model data revels that primarily affected systems from the pollutants level in air are the cardiovascular and the respiratory system. But the function of several other organs can be also influenced (Cohen *et al.,* 2005; Huang and Ghio, 2006; Kunzli and Tager, 2005; Sharma and Agrawal, 2005). Humans comes direct contact with different level of Atmospheric pollutants in their surrounding primarily via ingestion and inhalation but the dermal contact represents a minor route of exposure. Atmospheric pollutants contributes to the contamination of and water and food which makes ingestion in different cases the major route of pollutant intake (Thron, 1996). With the help of respiratory tract and gastrointestinal path, absorption of pollutants may occur in human body, on the other hand a number of toxic substances can be found in the general circulation and some are deposited to different tissues in the body. The elimination occurs only to a certain degree by excretion (Madden and Fowler, 2000).Different studies from the workers revels that different types of Atmospheric pollutants affects human in different way in terms of their concentration. Different effects are also observed with the rate of long-term exposure to lower pollutant concentrations. The different symptoms on human such as throat and nose irritation, broncho constriction as well as the dyspnoea, especially in asthmatic individuals which are usually experienced when exposure to increased levels of sulphur dioxide (Balmes *et al.,* 1987), nitrogen oxides (Kagawa, 1985). On the other way, particulate matter which penetrates the alveolar epithelium (Ghio and Huang, 2004), ozone level initiate lung inflammation (Uysal and Schapira, 2003). The patients infected with lung lesions or diseases related to the lungs, pollutant-initiated inflammation will worsen their condition. Some Atmospheric pollutants like nitrogen oxides increase the susceptibility to respiratory infections (Chauhan *et al.,* 1998) and chronic exposure to ozone and certain heavy metals reduces functional level of lungs (Rastogi *et al.,* 1991; Tager *et al.,* 2005). The high exposure of ozone is also responsible for emphysema, asthma, lung cancer (Kuo *et al.,* 2006; Nawrot *et al.,* 2006). Micro expose of nitrogen dioxide causes Emphysema-like lesions (Wegmann *et al.,* 2005). Another harmful pollutant is dioxins, which induce liver cell damage (Kimbrough *et al.,* 1977), as it indicated by an increase levels of different enzymes in the blood (Mandal, 2005). The Green house gas like carbons monoxide binds to haemoglobin modifying its conformation and reduces its capability to transfer oxygen in blood (Badman and

Jaffe, 1996). Oxygen reduction level can severely affect the function of different organs, resulting in the slow reflexes, impaired concentration as well as the confusion. Particulate matter also induces systemic inflammatory changes affecting equally blood coagulation (Riediker *et al.,* 2004). Different level of Atmospheric pollutants causes changes in blood clotting and induces lung irritation resulting to angina or even to myocardial infarction (Vermylen *et al.,* 2005). Tachycardia, increased level of blood pressure as well as anaemia in human occurs due to an inhibitory effect on haematopoiesis which have been observed as a results of heavy metal pollution like arsenic, mercury, nickel (Huang and Ghio, 2006)).The heavy metals can also induces damages in kidney like tubular dysfunction, excretion of low molecular weight proteins. The can also risk of stone formation or nephrocalcinosis (Damek-Poprawa and Sawicka-Kapusta, 2003; Vamvakas *et al.,* 1993; Jarup, 2003; Loghman-Adham, 1997) and renal cancer (Boffetta *et al.,* 1993).

The different effects from may be poor quality of air which they inhale or ingest. In case of Atmospheric pollutants, mostly inhalation triggers the health problems in animals but occasionally deposition of particles from different industrial source exhaust on pasture land which may affect health of animals directly resulting in toxic residue in their eggs, meat, milk or in other products obtained from animals. High level of dioxin in milk of dairy cows and zinc-induced arthritis in growing foals are the few examples of pasture grass pollutants by deposits of smoke from

Table 2. Effects of environmental pollutants on human beings

Environmental Pollutants	Damages	References
Co	Systemic inflammatory changes, Lung inflammation	Riediker *et al.,* 2004.
	Myocardial infraction	Vermylen *et al.,* 2005.
O3	Reduces lung function	Rastogi *et al.,* 1991 and Tager *et al.,* 2005
	Emphysema and asthma,	Kuo *et al.,* 2006 and Nawrot *et al.,* 2006.
So_2	Dyspnoea	Balmes *et al.,* 1987
Other types of atmospheric pollutants	harm developing foetus	Schell *et al.,* 2006
	Congenital malformations	Bellinger, 2005
Oxides of Nitrogen (No, No_2 etc)	Increases tract respiratory disease	Birch *et al.,* 2011 and van Grinsven *et al.,* 2013
	Respiratory tract system problems and cardiovascular	Cohen *et al.,* 2005;
	Increased colon cancer risk	Grinsven *et al* 2013
	Declining visibility—loss of aesthetics	Birch *et al.,* 2011
	Increased effects of airborne disease	Grinsven *et al* 2013

various types nearby industrial activities. Cattle, poultry, Pigs, goats and sheep are kept in indoor areas for a variable part of their life, often for all of their life. These restriction and closed areas affects the health of these animals in different ways. It has been observed that air quality of this area is sometime much better in compare to that of the closed facilities for swine and poultry (Wathes *et al.,* 1998). Dunham (1991), observed that maximal concentrations of gases or pollutants in piggeries are: 2.4 mg dust /m^3; 0.08 mg endotoxin/m^3, 7 ppm ammonia, 105 colony-forming units (CFUs) of total microbes/m^3; as well as 1,540 ppm. carbon dioxide. The different concentrations of bacteria up to 1.1 x106 CFU/m^3, inhalable dust content of 0.26 mg/m3 and the concentration of ammonia was 27 ppm. These observations have also been reported by Scherer & Unshelm, 1995 in facilities during winter, while at summer lower concentrations but less difference have been found in in- and outdoor temperature in summer and winter.The aerosolization of different microorganism and their secreted endotoxins, odours, as well as the dust particles are inevitable consequences of the generation and the handling of different types of wastes from the food production chain and from animals. Animals faces different level of the Atmospheric pollutants from the environment. Although they kept in huge facilities but often diseased due to self-made indoor Atmospheric pollutants. Different types of pollutants are found in manure which containing enteric bacteria and endotoxin resulting in the negative effects on the health of animals (Pickrell, 1991). The different concentration of the airborne bacteria and the endotoxin produced by them is related to the level of pen cleanliness. Generation of toxic gasses like ammonia and its higher concentrations in the air are not affected severely to the health of animals, but also by volume of the building, pig flow management and pig density and (Scherer & Unshelm, 1995). Similar factors on the variants levels of ammonia are known to play a role in farrowing units and in poultry houses (Harry, 1978). The pollutants ammonia is considered as one of the most important inhaled toxicant which has negative effect on the health of animals. According to Dodd & Gross (1980) the 1000 ppm level of ammonia for less than 24 hour causes impaired ciliary activity, caused mucosal damage as well as secondary infections in laboratory animals. The respiratory disease in swine is mostly due to the different combination of primary and opportunistic infectious agents, whereby adverse atmosphere and management conditions are the triggers. The primary respiratory infectious agents have capability to cause serious disease on their own, however, often uncomplicated infections have also been observed. If the primary infection becomes complicated serious respiratory disease may occurs in animals by opportunistic bacteria. Some of the pathogens responsible for the various diseases in animals are respiratory syndrome virus, porcine reproductive, and other caused by *Bordetella bronchiseptica, Haemophilus parasuis, Actinobacillus suis, Streptococcus suis, Mycoplasma hyopneumoniae.* It has been reported that, pig or poultry facilities are exposed to the same increased levels of ammonia, carbon monoxide, hydrogen sulphide, as well as the dust particles from feed and manure of the animals (Pickrell, 1991; Catcott, 1961; Manzo *et al.*, 2010). That is why, workers involved in swine production tend to have greater rates of asthma and types of respiratory infections. The lesions caused due to Atmospheric pollutants in production animals mainly include inflammatory processes.

There is a strong bonding between stomatal conductance and demage of plants by ozone (Reich and Amundson, 1985). Deposition of the ozone in terrestrial ecosystems is mostly controlled by stomata (Ainsworth *et al.,* 2012). These the epidermal cell situated in the plants leaves. The non-stomatal processes are also essentials and sometimes dominate the total ozone deposition to forests (Kurpius and Goldstein, 2003). The High level of this green house gas may cause visible injury in foliar parts of plants and affect gas exchange as well as the photosynthesis, resulting in shortened longevity of foliage, reduced plant height and bad health (Grulke *et al.,* 2009). The long-term chronic exposures to ozone can also lower the rate of stomatal conductance and decrease the level of photosynthsis in plants (Reich and Amundson, 1985). The exposure to chronic levels of ozone may limit the ability in closing and opening of the stomata(Grulke *et al.,* 2006), and increasing the water use (Sun *et al.,* 2012 . The long term and short term reduction of stomatal activity caused by the high level of ozone reduce the capability of plants to sequester carbon dioxide and is the basis for the indirect ozone greenhouse effect (Sitch *et al.,* 2007). It has been observed from various study that increased level of the ozone gas since the industrial revolution has decreased the rate of photosynthetic and uptake of carbon dioxide as well as the stomatal conductance of trees by 11% - 13% (Wittig *et al.,* 2007). Study from Felzer *et al.* (2004) revels a 2.6–6.8% reduction rate of the annual net primary productivity in United States in response to the levels of ozone since 1950 with the largest decreases in the late 1980s and early 1990s. The adverse effects of the ozone on loss of carbon sequestration are similar in magnitude to land

Table 3. Affected animals due to atmospheric pollutants

Scientific Name	Effects Due to Poor Quality of Air
Parnassius apollo	Decline in polluted areas due to the bad air quality(Holm, 1985).
Lutra lutra	Decline in otter populations due to different level of pollutants which has been suggested for Galloway, Scotland (Fry and Cook, 1984), and also for UK (Mason and Macdonald, 1987)
Lepidoptera spp.	Decline in polluted atmosphere Barbour,1986)
Collembola spp.	The poor quality of air (Koposzki,1992).
Aphantopus hyperantus	High SO_2 levels was found (Heath *et al.,* 1984).
Anas rubripes	Duckling mortality due to the pollutants level (Haramis,1984)

use change as well as in the offset increases in carbon sequestration due to global warming, N fertilization, CO2. The different levels of nitrogen deposition may have many adverse effects like changes in biodiversity, contamination of water and soils (Greaver *et al.,* 2012; Pardo *et al.,* 2011; US EPA, 2008). The long-and short term reduction of stomatal activity caused by elevated level of ozone, carbon dioxide may also reduce the ability of plants to sequester CO2 which is the basis for the indirect ozone greenhouse effect (Sitch *et al.,* 2007). The Simulation of ozone effects on hardwood of trees in the north-eastern region of United States using 1987–1992 ozone, data showed the declines of annual net primary productivity ranging from 3% to 16% with most pronounced effects at the highest ozone levels as well as on the soils with high water-holding capacity (Ollinger *et al.,* 1997).

Suggestion for Atmospheric Management

For the general regulations of the quality of air effective actions should mainly focus on traffic access control and technological innovation to reduce emission level of low-carbon-motor and energy related emission reduction by improving energy efficiency, or energy switching, ban of coal sales as well as energy conservation. Different types of strategies can be helpful in reducing the levels of the Atmospheric pollutants .Climate change strategies on the emission of greenhouse reductions also reduced the concentrations of contaminants and provided related health co-benefit. In energy related strategies, health benefits were obtained mainly from the reduction in the concentrations of sulphur dioxide and the oxide of nitrogen whereas the traffic control strategies was more geographically biased because of the higher health benefit within the immediate catchment of the interventions in a region. Different study of the world has focused with the strategies aimed to mitigate climatic changes or to reach climate change targets. Green houses gases like carbon dioxide, methane, carbon monoxide. There is a clear need to develop different models that stimulates the multiple biotic and abiotic factors while designing future ecosystem changes in order to understand the potential risks factors to forest resources. These models could provide science-based outcomes and future scenarios as well as improve the human understanding or thinking of what adaptive or mitigation measures could be recommended to air resource and land managers. There is ample evidence from models and observations that the twenty first century climate change will worsen the ozone pollution. The emissions of pollutant are expected to respond to climate change. Day by day of the rising of the temperature increase the demand for air conditioning in summer when ozone and particulate matter concentrations are highest. The emission from the evaporation of anthropogenic NMVOCs increases but the effect determined for mobile sources are generally weak in range of 1.3–5% K_1 (Cardelino and Chameides, 1990; Rubin *et al.,* 2006). The level of ozone and

Table 4. Affected Phyto diversity due to atmospheric pollutants

Name	Scientific Name	Notes and Source
Bryophytes	*Bryum, Hypnum cupressiforme, Grimmia pulvinata*, and *Orthotrichium* .	Susceptible to damage by sulphur dioxide and noitrogen do oxide
Fungi	Many mycorrhizal fungi including *Cantharellus cinabrius, Russula* spp, *Lactarius* spp, *Hygrphorus* spp and *Hygrocybe* spp	These fungi are affected by nitrogen deposition in acidified forests.
Alage	*Nostoc, Scytonema* etc	Greatly affected due to air high level of air pollutants
Lichens	Many foliar species, eg *Usnea, Ramalina.*	Declined due to SO_2 pollution.
Fringe moss	*Anuginosum and Racomitrium*	Decline due to Nitrogen deposition.
Bog mosses (Bryophytes)	*Sphagnum* spp.	*Sphagnum* species are damaged by sulphur dioxide and perhaps also by nitrogen dioxide
Zymnopserms	*Picea, Larix europeaus* and *abies,*	Affected by ozone and other air other types of pollutants.
Herbaceous pants	Many species, including *Primula veris, Vicia sepium,Trifolium medium, Melica nutans, Hepatica nutans*, etc	Declined due to soil acidification
Broadleaved trees	*Quercus robur, Quercus alba, Acer saccarina, Populus tremulens* and Others	Pollutants indirect effects by soil acidification and by increased nitrogen deposition.
Aquatic plants	*Echinospora, Littorella, Lobelia dortmanna, uniflora*, and *Isoetes*	Affected due to acidification

its background become an important topic for the management of the air quality. This background also increases significantly in future because of the global rises in methane and oxide of nitrogen like NOx (Fiore *et al.,* 2002). In country like U.S.A, Climate has provide some relief and comfortable in summer. Climate change may provide some relief, at least in summer. Wu *et al.,* 2008b Observed that the U.S. policy-relevant background defined by the U.S. Environmental Protection Agency (EPA) as the surface ozone concentration due to the absence of North - American anthropogenic emissions, will reduces by up to

2 ppb in summer as a result of 2000–2050 climate change. Similar observation was given by Lin *et al.,* 2008b. Further Wu et al. 2008b project that climatic changes will fully offset the impacts of increasing the emission from anthropogenic activities in eastern U.S in summer, as there will still be a 2–5 ppb rises in the PRB in the west. Finally, as the world moves forward to develop energy and transportation policies as well as the other policy directed at mitigating climate change, it will be

significant factor into these different policies which co- or dis-benefits for regional Atmospheric pollutants. The energy policy offers an opportunity to dramatically improve the quality of air through transition to non-polluting sources of energy. In contrast, a switch to biofuels would not necessarily benefit the quality of air and could possibly be detrimental (Jacobson, 2007).

Role of Plants in Cleaning of Atmosphere

Different studies reveals that plants contribute in atmospheric pollutants reduction (Boonen *et al.,* 2015; Angelo *et al.,* 2013). The different plants species show substantial effect on Atmospheric pollutants control (Escobedo *et al.,* 2011). Number of different studies have identified and quantified various ecosystem services in context of Atmospheric pollutants reduction due to dry deposition function (Nowak, 2000; Nowak *et al.,* 2006). The most important process in removal of pollutants is dry deposition, in which plants removes air pollutant from the troposphere in non-precipitation period (Cavanagh, 2006; Schlesinger, 1979). Dry deposition is mainly depends on nature as well as the property of air pollutant such as gaseous/particles form etc.). Sometime property of surfaces such as size, chemical nature, roughness, and weather variables like wind direction and speed as well as the temperature is also play significant role in removal of pollutants (Roupsard, 2013). Once the Atmospheric pollutants are accumulated on the surface of plant leaf, they are removed primarily in two different ways. Firstly through stomata (epidermal cell) uptake of gaseous pollutants and leaf interception of particulate matter (Nowak *et al.,* 2006). The first process leads to the diffusion of air pollutant into the inner part of plant leaves. The gases form of pollutants may also be absorbed or react with the surface of plant. Secondly removal may be by re-suspension of intercepted particles from the leaf surfaces through the action of wind. Now it is well known that plant remediate the environmental pollutants by different mechanism. Different plant species like *Penninsetum pedicellatum, Zinnia anguistifolia, Triticum aestivum, Hordeum vulgare, Populus deltoids*, *Azolla pinnata,* Quercus robur, Fraxinus excelsior, Aesculus hippocastanum, *Zea mays, Astragalus sinicus* removes the environmental pollutants in different way. Below the table is given for various plants species involving removal of atmospheric pollutants.

Research and Management Needs

Long-term monitoring of ozone and other green house gases and development of physiologically relevant ozone indices are needed to verify model outputs and understanding of future risk. The study delve with the empirical evidences provided to support that the different Atmospheric pollutants have direct implications and

Table 5. Plants as remediator of pollutants

Plant Species	Pollutants	References
Triticum aestivum	Benthiocarb	Sato, 1989
Beta vulgaris	Polychlorinated biphenyls(PCBS)	Brazil *et al.*, 1995
Zea mays	Atrazine	Seibert *et al.*,1981
Saccharum spontaneum	2,4-D	Sandmann and Loose,1984
Penninsetum pedicellatum	Chlorpyrifos and Cypermethrie, Fenvalerate	Dubey and Fulekar., 2011
Fragmites australies	Fixed Nitrogen	Haleem *et al.*, 2000
Zinnia anguistifolia	Mefenoxam	Pai *et al.*, 2001
Alfalfa	Hexadecane and PAHs	Nichols *et al.*,1997
Triticum aestivum	2,4-D	Kingley *et al.*, 1994
Hordeum vulgare	2,4-D	Jacobsen, 1997
Populus deltoids nigra	1,4-dioxane	Schnoor *et al.*, 1998
Astragalus sinicus	Cd	Sriprang *et al.*,2002
Poplus spp.	1,4-dioxane	Kelley *et al.*, 2001
Zea mays	3-methylbenzoate	Ronchel *et al.*, 2001
Azolla pinnata	Diesel fuel	Cohen *et al.*,2002
Kolchia	Atrazine,metolachlor and trifluralin	Anderson *et al.*,1994
Rye Grass	Chlorpyrifos	Korade and Fulekar., 2009

Source: Modified from Kala, 2014

addition pressure on the human as well as the ecosystem services. Different improved techniques are needed to estimate the level of air pollutant exposure. Such techniques would greatly expand our ability to evaluate the future risks caused by various toxic gases and atmospheric deposition of pollutants . Better observation of spatial as well as the temporal distribution of particulate matter and other Atmospheric pollutants are needed for identification and their sources to develop control strategies. So the study demands more robust studies in the field of Atmospheric pollutants .

CONCLUSION

This chapter presents the adverse effects of different types of atmospheric adulterants on human health. As mention above major impairments of different organs can be observed due to the harmful atmospherics pollutants. The observation also focussed that, in view of increased exposure of different pollutants plants are still providing a protective covering to human as well as the animals from different types of Atmospheric pollutants . Humans burning fossil fuels are currently emitting high level of sulphur dioxide and various gases like carbon mono oxide, oxides of nitrogen every day. Such high level of these gases causing not only global warming but widespread extinction of many species of plants and animals, the different atmospheric pollutants also causing several harmful diseases in human beings. Both carbon and sulphur dioxide emissions must be reduced significantly to reverse global warming, but reducing of sulphur dioxide should be the highest priority in future.

REFERENCES

Ainsworth, E. E., Yendrek, C. R., Sitch, S., Collins, W. J., & Emberson, L. D. (2012). The effects of tropospheric ozone on net primary productivity and implications for climate change. *Annual Review of Plant Biology*, *63*(1), 637–661. doi:10.1146/annurev-arplant-042110-103829 PMID:22404461

Alexander, B., Park, R. J., Jacob, D. J., Li, Q. B., Yantosca, R. M., Savarino, J., ... Thiemens, M. H. (2005). Sulfate formation in sea-salt aerosols: Constraints from oxygen isotopes. *Journal of Geophysical Research, D, Atmospheres*, *110*(D10).

Anderson, T. A., Kruger, E. L., & Coats, J. R. (1994). Enhanced degradation of a mixture of three herbicides in the rhizosphere of a herbicide-tolerant plant. *Chemosphere*, *28*(8), 1551–1557. doi:10.1016/0045-6535(94)90248-8

Angelo, J., Andrade, L., Madeira, L. M., & Mendes, A. (2013). An overview of photocatalysis phenomena applied to NOx abatement. *Journal of Environmental Management*, *129*, 522–539. doi:10.1016/j.jenvman.2013.08.006 PMID:24018117

Annesi-Maesano, I., Baiz, N., Banerjee, S., Rudnai, P., & Rive, S. (2013). Sinphonie Group. Indoor air quality and sources in schools and related health effects. *Journal of Toxicology and Environmental Health. Part B, Critical Reviews*, *16*(8), 491–550. doi:10.1080/10937404.2013.853609 PMID:24298914

Avery, M. A. (2010). Increasings pringtime ozone mixing ratios in the free troposphere over western North. *American Naturalist*, *463*, 344–348.

Badman, D. G., & Jaffe, E. R. (1996). Blood and Atmospheric pollutants: State of knowledge and research needs. *Otolaryngology - Head and Neck Surgery*, *114*(2), 205–208. doi:10.1016/S0194-5998(96)70166-3 PMID:8637733

Balmes, J. R., Fine, J. M., & Sheppard, D. (1987). Symptomatic bronchoconstriction after short-term inhalation of sulfur dioxide. *The American Review of Respiratory Disease*, *136*(5), 1117–1121. doi:10.1164/ajrccm/136.5.1117 PMID:3674573

Barbour, D. A. (1986). *Why are there so few butterflies in Liverpool? An answer.* London: Royal Entomological Society.

Barrie, L. A., & Hoff, R. M. (1984). The oxidation rate and residence time of sulphur dioxide in the Arctic atmosphere. *Atmospheric Environment, 18*(12), 2711-2722.

Bellinger, D. C. (2005). Teratogen update: Lead and pregnancy. *Birth Defects Research. Part A, Clinical and Molecular Teratology, 73*(6), 409–420. doi:10.1002/bdra.20127 PMID:15880700

Bentayeb, M., Simoni, M., Norback, D., Baldacci, S., Maio, S., & Viegi, G. (2013). Indoor Atmospheric pollutants and respiratory health in the elderly. *J Environ Sci Health A Tox Hazard Subst Environ Eng, 48*, 1783-9.

Birch, M. B. L., Gramig, B. M., Moomaw, W. R., Doering, O. C. III, & Reeling, C. J. (2011). Why metrics matter: Evaluating policy choices for reactive nitrogen in the chesapeake bay watershed. *Environmental Science & Technology, 45*(1), 168–174. doi:10.1021/es101472z PMID:20853823

Boffetta, P., Merler, E., & Vainio, H. (1993). Carcinogenicity of mercury and mercury compounds. *Scandinavian Journal of Work, Environment & Health, 19*(1), 1–7. doi:10.5271jweh.1510 PMID:8465166

Boonen, E., Akylas, V., Barmpas, F., Boréave, A., Bottalico, L., Cazaunau, M., ... Beeldens, A. (2015). Construction of a photocatalytic De-polluting field site in the Leopold II tunnel in Brussels. *Journal of Environmental Management, 155*, 136–144. doi:10.1016/j.jenvman.2015.03.001 PMID:25863437

Brasseur, G. P., Muller, J.-F., Tie, X., & Horowitz, L. (2001). Tropospheric ozone and climate: past, present and future. In T. Matsuno & H. Kida (Eds.), *Present and Future of Modeling Global Environmental Change: Toward Integrated Modeling* (pp. 63–75). Tokyo: TERRAPUB.

Brazil, G. M., Kenefick, L., Callanan, M., Haro, A., De Lorenzo, V., Dowling, D. N., & O'gara, F. (1995). Construction of a rhizosphere pseudomonad with potential to degrade polychlorinated biphenyls and detection of bph gene expression in the rhizosphere. *Applied and Environmental Microbiology, 61*(5), 1946–1952. PMID:7646029

Bytnerowicz, A., Arbaugh, M., Schilling, S., Fraczek, W., & Alexander, D. (2008). Ozone distribution and phytotoxic potential in mixed conifer forests of the San Bernardino Mountains, southern California. *Environ. Pollut, 155*(3), 398–408. doi:10.1016/j.envpol.2008.01.046 PMID:18359541

Cardelino, C. A., & Chameides, W. L. (1990). Natural hydrocarbons, urbanization, and urban ozone. *Journal of Geophysical Research, 95*(D9), 13,971–13,979. doi:10.1029/JD095iD09p13971

Carslaw, K. S., Boucher, O., Spracklen, D. V., Mann, G. W., Rae, J. G. L., Woodward, S., & Kulmala, M. (2010). A review of natural aerosol interactions and feedbacks within the Earth system. *Atmospheric Chemistry and Physics, 10*(4), 1701–1737. doi:10.5194/acp-10-1701-2010

Catcott, E. J. (1961). Effects of Atmospheric pollutants on animals. In *World Health Organization Monograph series, no.41* (pp. 221–231). Geneva: WHO.

Cavanagh, J. E. (2006). *Potential of Vegetation to Mitigate Road-generated Atmospheric pollutants*. Retrieved April 21st, 2016 from https://www.niwa.co.nz/sites/niwa. co.nz/files/import/attachments/co1x0405 p05.pdf

Cecchi, L. D., Amato, G., & Annesi-Maesano, I. (2013). Climate, urban Atmospheric pollutants, and respiratory allergy. In R. A. Pielke (Ed.), *Climate vulnerability: understanding and addressing threats to essential resources* (pp. 105–113). Waltham, MA: Academic Press. doi:10.1016/B978-0-12-384703-4.00116-7

Chandra, S. V., & Garg, N. (1991). A cross-sectional study of pulmonary function among workers exposed to multimetals in the glass bangle industry. *American Journal of Industrial Medicine, 20*(3), 391–399. doi:10.1002/ajim.4700200311 PMID:1928114

Chauhan, A. J., Krishna, M. T., Frew, A. J., & Holgate, S. T. (1998). Exposure to nitrogen dioxide (NO2) and respiratory disease risk. *Reviews on Environmental Health, 13*, 73. PMID:9718623

Chestnut, L., & Mills, D. M. (2005). Afresh look at the benefits and costs of the US acid rain program. *Journal of Environmental Management, 77*(3), 252–266. doi:10.1016/j.jenvman.2005.05.014 PMID:16171931

Clifford, A., Lang, L., Chen, R., Anstey, K. J., & Seaton, A. (2016). Exposure to Atmospheric pollutants and cognitive functioning across the life course–A systematic literature review. *Environmental Research, 147*, 383–398. doi:10.1016/j.envres.2016.01.018 PMID:26945620

Cohen, A. J., Ross Anderson, H., Ostro, B., Pandey, K. D., Krzyzanowski, M., Kunzli, N., ... Smith, K. (2005). The global burden of disease due to outdoor Atmospheric pollutants. *Journal of Toxicology and Environmental Health. Part A., 68*(13-14), 1301–1307. doi:10.1080/15287390590936166 PMID:16024504

Cohen, M. F., Williams, J., & Yamasaki, H. (2002). Biodegradation of diesel fuel by an *Azolla*-derived bacterial consortium. *Journal of Environmental Science and Health. Part A*, *37*(9), 1593–1606.

Cooper, O. R., Parrish, D. D., Stohl, A., Trainer, M., Nédélec, P., Thouret, V., ... Leblanc, T. (2010). Increasing springtime ozone mixing ratios in the free troposphere over western North America. *Nature*, *463*(7279), 344–348. doi:10.1038/nature08708 PMID:20090751

Damek-Poprawa, M., & Sawicka-Kapusta, K. (2003). Damage to the liver, kidney, and testis with reference to burden of heavy metals in yellow-necked mice from areas around steelworks and zinc smelters in Poland. *Toxicology*, *186*(1-2), 1–10. doi:10.1016/S0300-483X(02)00595-4 PMID:12604166

Davidson, E. A. (2009). The contribution of manure and fertilizer nitrogen to atmospheric nitrous oxide since 1860. *Nature Geoscience*, *2*(9), 659–662. doi:10.1038/ngeo608

Davidson, E. A. (2012). Excess nitrogen in the US environment: Trends, risks, and solutions. *ESA Issues Ecology*, *15*, 1–16.

DeAngelo, B. J. (2013). Bounding the role of black carbon in the climate system: A scientific assessment. *Journal of Geophysical Research, D, Atmospheres*, *118*(11), 5380–5552. doi:10.1002/jgrd.50171

Dockery, D. W., & Stone, P. H. (2007). Cardiovascular risks from fine particulate Atmospheric pollutants. *The New England Journal of Medicine*, *356*(5), 511–513. doi:10.1056/NEJMe068274 PMID:17267912

Dodd, K. T., & Gross, D. R. (1980). Ammonia inhalation toxicity in cats: A study of acute and chronic respiratory dysfunction. *Archives of Environmental Health*, *35*(1), 6–14. doi:10.1080/00039896.1980.10667455 PMID:7362273

Domene, L. A. F., & Ayres, R. U. (2001). Nitrogen's role in industrial *systems. Journal of Industrial Ecology*, *5*(1), 77–103. doi:10.1162/108819801753358517

Donham, K. J. (1991). Association of environmental air contaminants with disease and productivity in swine. *American Journal of Veterinary Research*, *52*, 1723–1730. PMID:1767997

Dubey, K. K., & Fulekar, M. H. (2011). Effect of pesticides on the seed germination of *Cenchrus setigerus* and *Pennisetum pedicellatum* as monocropping and co-cropping system: Implications for rhizospheric bioremediation. *Romanian Biotechnological Letters*, *16*(1), 5908–5918.

Economopoulos, A. P. (1993). *Assessment of Sources of Air Water and Land Pollution. A Guide to Rapid Source Inventory Techniques and Their Use in Formulating Environmental Control Strategies. Part One: Rapid Inventory Techniques in Environmental Pollution. Part Two: Approaches for Consideration in Formulating Environmental Control Strategies (Unpublished document WHO/YEP/93.1.)*. Geneva: WHO.

Epton, M. J., Dawson, R. D., Brooks, W. M., Kingham, S., Aberkane, T., Cavanagh, J. A., ... Brown, L. (2008). The effect of ambient Atmospheric pollutants on respiratory health of school children: A panel study. *Environmental Health*, *7*(1), 16. doi:10.1186/1476-069X-7-16 PMID:18479529

Escobedo, F., Kroeger, T., & Wagner, J. (2011). Urban forests and pollution mitigation: Analyzing ecosystem services and disservices. *Environ. Pollut*, *159*(8-9), 2078–2087. doi:10.1016/j.envpol.2011.01.010 PMID:21316130

Fang, Y., Naik, V., Horowitz, L. W., & Mauzerall, D. L. (2013). Atmospheric pollutants and associated human mortality: The role of air pollutant emissions, climate change and methane concentration increases from the preindustrial period to present. *Atmospheric Chemistry and Physics*, *13*(3), 1377–1394. doi:10.5194/acp-13-1377-2013

Felzer, B., Kicklighter, D., Melillo, J., Wang, C., Zhuang, Q., & Prinn, R. (2004). Effects of ozone on net primary production and carbon sequestration in the conterminous United States using a biogeochemistry model. *Tellus*, *56B*(3), 230–248. doi:10.1111/j.1600-0889.2004.00097.x

Fiore, A. M. (2014). Atmospheric chemistry: No equatorial divide for a cleansing radical. *Nature*, *513*(7517), 176–178. doi:10.1038/513176a PMID:25209794

Fiore, A. M., Jacob, D. J., Field, B. D., Streets, D. G., Fernandes, S. D., & Jang, C. (2002). Linking ozone pollution and climate change: The case for controlling methane. *Geophysical Research Letters*, *29*(19), 1919. doi:10.1029/2002GL015601

Fiore, A. M., Naik, V., & Leibensperger, E. M. (2015). Air quality and climate connections. *Journal of the Air & Waste Management Association*, *65*(6), 645–685. doi:10.1080/10962247.2015.1040526 PMID:25976481

Fry, G. (1984). Australia and the South Pacific. *Diplomacy in the marketplace: Australia in world affairs, 90*, 175-77.Filippidou EC, Koukouliata A. Ozone effects on the respiratory system. *Progress in Health Sciences*, *2011*(1), 144–155.

Fuzzi, S., Baltensperger, U., Carslaw, K., Decesari, S., Denier Van Der Gon, H., Facchini, M. C., ... Nemitz, E. (2015). Particulate matter, air quality and climate: Lessons learned and future needs. *Atmospheric Chemistry and Physics*, *15*(14), 8217–8299. doi:10.5194/acp-15-8217-2015

Gauderman, W. J., Vora, H., McConnell, R., Berhane, K., Gilliland, F., Thomas, D., ... Peters, J. (2007). Effect of exposure to traffic on lung development from 10 to 18 years of age: A cohort study. *Lancet*, *369*(9561), 571–577. doi:10.1016/S0140-6736(07)60037-3 PMID:17307103

Gauss, M., Myhre, G., Isaksen, I. S. A., Grewe, V., Pitari, G., Wild, O., ... Sanderson, M. G. (2003). Radiative forcing in the 21st century due to ozone changes in the troposphere and the lower stratosphere. *Journal of Geophysical Research*, *108*(D9), 4292. doi:10.1029/2002JD002624

Ghio, A. J., & Huang, Y. C. (2004). Exposure to concentrated ambient particles (CAPs): A review. *Inhalation Toxicology*, *16*, 53.

Ghio, A. J., & Samet, J. (1999). *Atmospheric pollutants and health: metals and Atmospheric pollutants particles*. Academic Press.

Greaver, T. L., Sullivan, T. J., Herrick, J. D., Barber, M. C., Baron, J. S., Cosby, B. J., ... Novak, K. J. (2012). Ecological effects of nitrogen and sulfur Atmospheric pollutants in the US: What do we know? *Front. Ecol*, *10*(7), 365–372. doi:10.1890/110049

Grulke, N. E., Minnich, R. A., Paine, T. D., Seybold, S. J., Chavez, D. J., Fenn, M. E., . . . Dunn, A. (2009). Atmospheric pollutants increases forest susceptibility to wildfires: A case study in the San Bernardino Mountains in Southern California. Academic Press.

Grulke, N. E., Paoletti, E., & Heath, R. L. (2006). Direct measurements of foliar ozone uptake in crop and tree species. *Environ. Pollut.*, *146*, 640–647. doi:10.1016/j.envpol.2006.04.014 PMID:16777298

Gryparis, A., Forsberg, B., Katsouyanni, K., Analitis, A., Touloumi, G., & Schwartz, J. (2004). al. Acute effects of ozone on mortality from the "Atmospheric pollutants and health: A European approach" project. *American Journal of Respiratory and Critical Care Medicine*, *170*, 1080–1087. doi:10.1164/rccm.200403-333OC PMID:15282198

Haase, D., Larondelle, N., Andersson, E., Artmann, M., Borgström, S., Breuste, J., ... Kabisch, N. (2014). A quantitative review of urban ecosystem service assessments: Concepts, models, and implementation. *Ambio*, *43*(4), 413–433. doi:10.100713280-014-0504-0 PMID:24740614

Haleem, D., Von Wintzingerode, F., Moter, A., Moawad, H., & Göbel, U. B. (2000). Phylogenetic analysis of rhizosphere-associated β-subclass proteobacterial ammonia oxidizers in a municipal wastewater treatment plant based on rhizoremediation technology. *Letters in Applied Microbiology*, *31*(1), 34–38. doi:10.1046/j.1472-765x.2000.00760.x PMID:10886611

Hansen, C., Luben, T. J., Sacks, J. D., Olshan, A., Jeffay, S., Strader, L., & Perreault, S. D. (2010). The effect of ambient Atmospheric pollutants on sperm quality. *Environmental Health Perspectives*, *118*(2), 203–209. doi:10.1289/ehp.0901022 PMID:20123611

Haramis, G. M. (1984). Acid alteration of experimental wetlands: Effects on diversity andabundance of macroinvertebrates and growth, behaviours and survival of juvenile black ducks. Draft research summary, US Food and Wildlife Service, Eastern Energy and Land Use Team, Kearneyville, WV.

Harry, E. G. (1978). Atmospheric pollutants in farm buildings and methods of control: A review. *Avian Pathology*, *7*(4), 441–454. doi:10.1080/03079457808418301 PMID:18770401

Health and Environment Alliance. Climate Change. (2009). Retrieved from www.env-health.org/r/93

Heath, Pollard, & Thomas. (1984). *Atlas of Butterflies of Britain and Ireland.* Viking Press.

Huang, Y. C., & Ghio, A. J. (2006). Vascular effects of ambient pollutant particles and metals. *Current Vascular Pharmacology*, *4*(3), 199–203. doi:10.2174/157016106777698351 PMID:16842137

Huang, Y. C., & Ghio, A. J. (2006). Vascular effects of ambient pollutant particles and metals. *Current Vascular Pharmacology*, *4*, 199.

Hulin, M., Simoni, M., Viegi, G., & Annesi-Maesano, I. (2012). Respiratory health and indoor Atmospheric pollutants based on quantitative exposure assessments. *The European Respiratory Journal*, *40*(4), 1033–1045. doi:10.1183/09031936.00159011 PMID:22790916

Intergovernmental Panel on Climate Change (CH). Climate change. (2007). *The physical science basis. Geneva: Intergovernmental Panel on Climate Change; 2007 (cited 2007 Jan 01)*. Available from: http://www.ipcc.ch/pdf/assessment-report/ar4/wg1/ar4_wg1_full_report.pdf

Jacob, D. J., & Winner, D. A. (2009). Effect of climate change on air quality. *Atmospheric Environment, 43*(1), 51–63. doi:10.1016/j.atmosenv.2008.09.051

Jacobsen, C. S. (1997). Plant protection and rhizosphere colonization of barley by seed inoculated herbicide degrading Burkholderia (*Pseudomonas*) cepacia DBO1 (pRO101) in 2, 4-D contaminated soil. *Plant and Soil, 189*(1), 139–144. doi:10.1023/A:1004296615446

Jacobson, M. Z. (2007). Effects of ethanol (E85) versus gasoline vehicles on cancer and mortality in the United States. *Environmental Science & Technology, 41*(11), 4150–4157. doi:10.1021/es062085v PMID:17612204

Jerrett, M., Burnett, R. T., Pope, C. A., Thurston, G., & Krewski, D. (2009). Long-term ozone exposure and mortality. *The New England Journal of Medicine, 360*(11), 1085–1095. doi:10.1056/NEJMoa0803894 PMID:19279340

Kagawa, J. (1985). Evaluation of biological significance of nitrogen oxides exposure. *The Tokai Journal of Experimental and Clinical Medicine, 10*, 348.

Kala. (2014). Rhizoremediation: A Promising Rhizosphere Technology. *IOSR-JESTFT, 8*(8), 23-27.

Kelley, S. L., Aitchison, E. W., Deshpande, M., Schnoor, J. L., & Alvarez, P. J. (2001). Biodegradation of 1, 4-dioxane in planted and unplanted soil: Effect of bioaugmentation with *Amycolata* sp. CB1190. *Water Research, 35*(16), 3791–3800. doi:10.1016/S0043-1354(01)00129-4 PMID:12230161

Kimbrough, R. D., Carter, C. D., Liddle, J. A., Cline, R. E., & Phillips, P. E. (1977). Epidemiology and pathology of a tetrachlorodibenzodioxin poisoning episode. *Archives of Environmental Health, 32*(2), 77–86. doi:10.1080/00039896.1977.10667259 PMID:557961

Kingsley, M. T., Fredrickson, J. K., Metting, F. B., & Seidler, R. J. (1994). *Environmental restoration using plant-microbe bioaugmentation. In Bioremediation of chlorinated and polyaromatic hydrocarbon compounds* (pp. 287–292). Boca Raton, Fla: Lewis Publishers.

Kinney, P. L. (2008). Climate change, air quality, and human health. *American Journal of Preventive Medicine, 35*(5), 459–467. doi:10.1016/j.amepre.2008.08.025 PMID:18929972

Koposzki, H. (1992). Effects of acid and nitrogen deposition on the mesofauna, especially the collembolan. In *Proceedings of Atmospheric pollutants and Interactions between Organisms in Forest Ecosystems, 15th IUFRO International Meeting of Specialists on Atmospheric pollutants Effects on Forest Ecosystems* (p. 46). Academic Press.

Korade, D. L., & Fulekar, M. H. (2009). Rhizosphere remediation of chlorpyrifos in mycorrhizospheric soil using ryegrass. *Journal of Hazardous Materials*, *172*(2), 1344–1350. doi:10.1016/j.jhazmat.2009.08.002 PMID:19720454

Kunzli, N., Gutschmidt, K., Pope, A., Romieu, I., Samet, J. M., & Smith, K. (2005). The global burden of disease due to outdoor Atmospheric pollutants. *Journal of Toxicology and Environmental Health*, *68*(13-14), 1301–1307. doi:10.1080/15287390590936166 PMID:16024504

Kuo, C. Y., Wong, R. H., Lin, J. Y., Lai, J. C., & Lee, H. (2006). Accumulation ofchromium and nickel metals in lung tumors from lung cancer patients in Taiwan. *Journal of Toxicology and Environmental Health*, *69*(14), 1337–1344. doi:10.1080/15287390500360398 PMID:16760140

Kurpius, M. R., & Goldstein, A. H. (2003). Gas-phase chemistry dominates O3 loss to a forest, implying a source of aerosols and hydroxyl radicals to the atmosphere. *Geophysical Research Letters*, *30*(7), 1371. doi:10.1029/2002GL016785

Law, K. (2010). More ozone over North America. *Nature*, *463*, 307–308.

Li, S., Williams, G., Jalaludin, B., & Baker, P. (2012). Panel studies of Atmospheric pollutants on children's lung function and respiratory symptoms: A literature review. *The Journal of Asthma*, *49*(9), 895–910. doi:10.3109/02770903.2012.724129 PMID:23016510

Lilleskov, E. A., Liu, L., Lynch, J., Nadelhoffer, K., Perakis, S. S., Robin-Abbott, M. J., ... Dennis, R. L. (2011). Effects of nitrogen deposition and empirical nitrogen critical loads for ecoregions of the United States. *Ecological Applications*, *21*(8), 3049–3082. doi:10.1890/10-2341.1

Lin, J.-T., Patten, K. O., Liang, X.-Z., & Wuebbles, D. J. (2008). Effects of intercontinental transport on surface ozone over the United States at the present and future. *Geophysical Research Letters*, *35*(2), L02805. doi:10.1029/2007GL031415

Liu, Y., Stanturf, J., & Goodrick, S. (2010). Trends in global wildfire potential in a changing climate. *Forest Ecology and Management*, *259*(4), 685–697. doi:10.1016/j.foreco.2009.09.002

Loghman-Adham, M. (1997). Renal effects of environmental and occupational lead exposure. Environ. Health Perspect. 105, 928.Jarup, L., 2003. Hazards of heavy metal contamination. *British Medical Bulletin*, *68*, 167.

Madden, E. F., & Fowler, B. A. (2000). Mechanisms of nephrotoxicity from metal combinations: A review. *Drug and Chemical Toxicology*, *23*(1), 1–12. doi:10.1081/DCT-100100098 PMID:10711385

Mandal, P. K. (2005). Dioxin: A review of its environmental effects and its aryl hydrocarbon receptor biology. *Journal of Comparative Physiology*, *175*(4), 221–230. doi:10.100700360-005-0483-3 PMID:15900503

Manzo, N. D., Slade, R., & Dye, J. A. (2010). Interrelationship of lung inflammation and air pollutant exposure on cellular oxidative stress and epithelial injury. *Proceedings 27th Symposium of the Veterinary Comparative Respiratory Society.*

Mason, C. F., & MacDonald, S. M. (1987). Acidification and otter (Lutra lutra) distribution on a British river. *Mammalia*, *51*(1), 81–87.

McGrath, J. M., Betzelberger, A. M., Wang, S., Shook, E., Zhu, X. G., Long, S. P., & Ainsworth, E. A. (2015). An analysis of ozone damage to historical maize and soybean yields in the United States. *Proceedings of the National Academy of Sciences of the United States of America*, *112*(46), 14390–14395. doi:10.1073/pnas.1509777112 PMID:26578785

Nawrot, T., Plusquin, M., Hogervorst, J., Roels, H. A., Celis, H., Thijs, L., ... Staessen, J. A. (2006). Environmental exposure to cadmium and risk of cancer: A prospective population-based study. *The Lancet Oncology*, *7*(2), 119–126. doi:10.1016/S1470-2045(06)70545-9 PMID:16455475

Neas, L., Williams, R. W., & Devlin, R. B. (2004). Particulate matter exposure in cars is associated with cardiovascular effects in healthy young men. *American Journal of Respiratory and Critical Care Medicine*, *169*(8), 934–940. doi:10.1164/rccm.200310-1463OC PMID:14962820

Nichols, T. D., Wolf, D. C., Rogers, H. B., Beyrouty, C. A., & Reynolds, C. M. (1997). Rhizosphere microbial populations in contaminated soils. *Water, Air, and Soil Pollution*, *95*(1-4), 165–178. doi:10.1007/BF02406163

Nowak, D. J. (2000). Tree species selection, design, and management to improve air quality. *Construction Technology: Annual Meeting Proceedings of the American Society of Landscape Architects.*

Nowak, D. J., Crane, D. E., & Stevens, J. C. (2006). Atmospheric pollutants removal by urban trees and shrubs in the United States. *Urban Forestry & Urban Greening*, *4*(3-4), 115–123. doi:10.1016/j.ufug.2006.01.007

Ollinger, S. V., Aber, J. D., & Reich, P. B. (1997). Simulating ozone effects on forest productivity: Interactions among leaf-, canopy-, and stand-level processes. *Ecological Applications*, *7*(4), 1237–1251. doi:10.1890/1051-0761(1997)007[1237:SOEOFP]2.0.CO;2

Olsson, D., Mogren, I., & Forsberg, B. (2013). Atmospheric pollutants exposure in early pregnancy and adverse pregnancy outcomes: A register-based cohort study. *BMJ Open*, *3*(2), e001955. doi:10.1136/bmjopen-2012-001955 PMID:23386578

Pai, S. G., Riley, M. B., & Camper, N. D. (2001). Microbial degradation of mefenoxam in rhizosphere of Zinnia angustifolia. *Chemosphere*, *44*(4), 577–582. doi:10.1016/S0045-6535(00)00368-4 PMID:11482644

Pangtey, B. S., Chandra, S. V., & Garg, N. (1991). A cross-sectional study of pulmonary function among workers exposed to multimetals in the glass bangle industry. *American Journal of Industrial Medicine*, *20*(3), 391–399. doi:10.1002/ajim.4700200311 PMID:1928114

Pardo, L. H., Fenn, M. E., Goodale, C. L., Geiser, L. H., Driscoll, C. T., Allen, E. B., ... Haines, A. (2014). Climate change: Challenges and opportunities for global health. *Journal of the American Medical Association*, *312*(15), 1565–1580. doi:10.1001/jama.2014.13186 PMID:25244362

Peters, A., von Klot, S., Heier, M., Trentinaglia, I., Hörmann, A., Wichmann, H. E., & Löwel, H. (2004). Exposure to traffic and the onset of myocardial infarction. *The New England Journal of Medicine*, *351*(17), 1721–1730. doi:10.1056/NEJMoa040203 PMID:15496621

Pickrell, J. (1991). Hazards in confinement housing - gases and dusts in confined animal houses for swine, poultry, horses and humans. *Veterinary and Human Toxicology*, *33*, 32–39. PMID:2017863

Pinder, R.W., Bettez, N.D., Bonan, G.B., Greaver, T. L., Widier, W.R., Schlesinger, W.H., & Davidson, E. A. (2013). Impacts of human alteration of the nitrogen cycle in the US on radiative forcing. *Biogeochemistry, 114*, 25–40.

Rastogi, S. K., Gupta, B. N., Husain, T., Chandra, H., Mathur, N., Pangtey, B. S., ... Ravenscroft, J. (2006). Effects of pollution on human growth and development: An introduction. *Journal of Physiological Anthropology*, *25*(1), 103–112. doi:10.2114/jpa2.25.103 PMID:16617215

Reich, P.B., & Amundson, R.G. (n.d.). Ambient levels of ozone reduce net photosynthesis in tree and crop species. *Science, 230*, 566–570.

Riebau, A. R., & Andersen, C. (1985). Wildland Fires and Atmospheric pollutants. In *Developments in Environmental Science* (Vol. 8, pp. 365–403). Elsevier.

Riediker, M., Cascio, W. E., Griggs, T. R., Herbst, M. C., Bromberg, P. A., Neas, L., ... Devlin, R. B. (2004). Particulate matter exposure in cars is associated with cardiovascular effects in healthy young men. *American Journal of Respiratory and Critical Care Medicine*, *169*(8), 934–940. doi:10.1164/rccm.200310-1463OC PMID:14962820

Ronchel, M. C., & Ramos, J. L. (2001). Dual system to reinforce biological containment of recombinant bacteria designed for rhizoremediation. *Applied and Environmental Microbiology*, *67*(6), 2649–2656. doi:10.1128/AEM.67.6.2649-2656.2001 PMID:11375176

Roupsard, P. (2013). *Etude Phénoménologique Du dépôt Sec d'aérosols En Milieu Urbain* (Ph.D Thesis). University of Rouen.

Rubin, J. I., Kean, A. J., Harley, R. A., Millet, D. B., & Goldstein, A. H. (2006). Temperature dependence of volatile organic compound evaporative emissions from motor vehicles. *Journal of Geophysical Research*, *111*(D3), D03305. doi:10.1029/2005JD006458

Sandmann, E. R. I. C., & Loos, M. A. (1984). Enumeration of 2, 4-D-degrading microorganisms in soils and crop plant rhizospheres using indicator media; high populations associated with sugarcane (*Saccharum officinarum). Chemosphere*, *13*(9), 1073–1084. doi:10.1016/0045-6535(84)90066-3

Scherer, M., & Unshelm, J. (1995). Untersuchungen zur Belastung von Mastschweinen sowie der Umgebung von Mastschweinstallen durch atembaren Feinstaub, stallspezifische Bakterien und Ammoniak. *Zentralblatt fur Hygiene und Umweltmedizin*, *196*, 399–415. PMID:7727021

Schlesinger, R. B. (1979). Natural removal mechanisms for chemical pollutants in the environment. *Bioscience*, *29*(2), 95–101. doi:10.2307/1307745

Schnoor, J. L., Aitchison, E. W., Kelley, S. L., & Alvarez, P. J. J. (1998). Phytoremediation and bioaugmentation of 1, 4-dioxane. In *Abstr. Adv. Biol. Syst. Rem. Conf* (pp. 87-91). Academic Press.

Schroder, M., Linkert, K. H., Pedersen, S., Takai, H., Johnsen, J. O., Groot Koerkamp, P. W. G., ... Linke, S. (1998). Emissions of Aerial Pollutants in Livestock Buildings in Northern Europe: Overview of a Multinational Project. *Journal of Agricultural Engineering Research*, *70*(1), 3–9. doi:10.1006/jaer.1998.0278

Sharma, R. K., & Agrawal, M. (2005). Biological effects of heavy metals: An overview. *Journal of Environmental Biology*, *26*, 301. PMID:16334259

Shine, K. P., Stevenson, D. S., Sudo, K., Szopa, S., & Zeng, G. (2006). Radiative forcing since preindustrial times due to ozone change in the troposphere and the lower stratosphere. *Atmospheric Chemistry and Physics*, *6*(3), 575–599. doi:10.5194/acp-6-575-2006

Sillman, S., & Samson, P. J. (1995). Impact of temperature on oxidant photochemistry in urban, polluted rural and remote environments. *Journal of Geophysical Research, D, Atmospheres*, *100*(D6), 11497–11508. doi:10.1029/94JD02146

Silva, R. A., West, J. J., Zhang, Y., Anenberg, S. C., Lamarque, J. F., Shindell, D. T., & Horowitz, L. W. (2013). Global premature mortality due to anthropogenic outdoor Atmospheric pollutants and the contribution of past climate change. *Environmental Research Letters*, *8*(3), 034005. doi:10.1088/1748-9326/8/3/034005

Sitch, S., Cox, P. M., Collins, W. J., & Huntingford, C. (2007). Indirect radiative forcing of climatechange through ozone effects on the land-carbon sink. *Nature*, *448*(7155), 791–794. doi:10.1038/nature06059 PMID:17653194

Smil, V. (2002). Nitrogen and food production: Proteins for human diets. *Ambio*, *31*(2), 126–131. doi:10.1579/0044-7447-31.2.126 PMID:12078001

Solomon, S., Qin, D., Manning, M., Marquis, M., Averyt, K., & Tignor, M. I. (2007). Technical summary. In *Climate change. The physical science basis* (pp. 19–940). New York, NY: Cambridge University Press.

Sriprang, R., Hayashi, M., Yamashita, M., Ono, H., Saeki, K., & Murooka, Y. (2002). A novel bioremediation system for heavy metals using the symbiosis between leguminous plant and genetically engineered rhizobia. *Journal of Biotechnology*, *99*(3), 279–293. doi:10.1016/S0168-1656(02)00219-5 PMID:12385715

Sun, G., McLaughlin, S. B., Porter, J. H., Uddling, J., Mulholland, P. J., Adams, M. B., & Pederson, N. (2012). Interactive influences of ozone and climate on streamflow of forested watersheds. *Global Change Biology*, *18*(11), 3395–3409. doi:10.1111/j.1365-2486.2012.02787.x

Tager, I. B., Balmes, J., Lurmann, F., Ngo, L., Alcorn, S., & Kunzli, N. (2005). Chronic exposure to ambient ozone and lung function in young adults. *Epidemiology (Cambridge, Mass.)*, *16*, 751. doi:10.1097/01.ede.0000183166.68809.b0 PMID:16222164

Thiemens, M. H., & Geophys, M. H. (1999). D10307.D.H. Ehhalt. *Physical Chemistry Chemical Physics*, *1*(24), 5401.

Thiering, E., & Heinrich, J. (2015). Epidemiology of Atmospheric pollutants and diabetes. *Trends in Endocrinology and Metabolism*, *26*(7), 384–394. doi:10.1016/j.tem.2015.05.002 PMID:26068457

Thron, R. W. (1996). Direct and indirect exposure to Atmospheric pollutants. *Otolaryngology - Head and Neck Surgery*, *114*(2), 281–285. doi:10.1016/S0194-5998(96)70184-5 PMID:8637751

Trenberth, K. E., Jones, P. D., Ambenje, P., Bojariu, R. E., Parker, D., Rahimzadeh, F., ... Zhai, P. (2007). Observations: surface and atmospheric climate change. In Climate Change 2007: The Physical Science Basis. Contribution of Working Group I to the Fourth Assessment Report of the Intergovernmental Panel on Climate Change. Cambridge, UK: Cambridge University Press.

Turner, M. C., Jerrett, M., Pope, C. A. III, Krewski, D., Gapstur, S. M., Diver, W. R., ... Burnett, R. T. (2016). Long-term ozone exposure and mortality in a large prospective study. *American Journal of Respiratory and Critical Care Medicine*, *193*(10), 1134–1142. doi:10.1164/rccm.201508-1633OC PMID:26680605

US E.P.A. (2008). Integrated Science Assessment (ISA) for Oxides of Nitrogen and Sulfur— Environmental Criteria. Final report. Washington, DC: Author.

Uysal, N., & Schapira, R. M. (2003). Effects of ozone on lung function and lung diseases. *Current Opinion in Pulmonary Medicine*, *9*, 144.

Vamvakas, S., Bittner, D., & Koster, U. (1993, April). Enhanced expression of the protooncogenes c-myc and c-fos in normal and malignant renal growth. *Toxicology Letters*, *67*(1-3), 161–172. doi:10.1016/0378-4274(93)90053-Z PMID:8451757

Van Grinsven, H. J. M., Holland, M., Jacobsen, B. H., Klimont, Z., Sutton, M. A., & Willems, W. J. (2013). Costs and benefits of nitrogen for Europe and implications for mitigation. *Environmental Science & Technology*, *47*(8), 3571–3579. doi:10.1021/es303804g PMID:23473305

Vermylen, J., Nemmar, A., Nemery, B., & Hoylaerts, M. F. (2005). Ambient Atmospheric pollutants and acute myocardial infarction. *Journal of Thrombosis and Haemostasis*, *3*(9), 1955–1961. doi:10.1111/j.1538-7836.2005.01471.x PMID:16102102

Vingarzan, R. (2004). A review of surface ozone background levels and trends. *Atmospheric Environment*, *38*, 3431–3442.

Wang, H., Jacob, D. J., Le Sager, P., Streets, D. G., Park, R. J., Gilliland, A. B., & van Donkelaar, A. (2009). Surface ozone background in the United States: Canadian and Mexican influences. *Atmospheric Environment*, *43*(6), 1310–1319. doi:10.1016/j.atmosenv.2008.11.036

Wathes, C. M., Phillips, V. R., Holden, M. R., Sneath, R. W., Short, J. L., White, R. P. P., & Pedersen, S. (1998). Emissions of aerial pollutants in livestock buildings in Northern Europe: Overview of a multinational project. *Journal of Agricultural Engineering Research*, *70*(1), 3–9. doi:10.1006/jaer.1998.0278

Wegmann, M., Fehrenbach, A., Heimann, S., Fehrenbach, H., Renz, H., Garn, H., & Herz, U. (2005). NO2-induced airway inflammation is associated with progressive airflow limitation and development of emphysema-like lesions in C57bl/6 mice. *Experimental and Toxicologic Pathology*, *56*(6), 341–350. doi:10.1016/j.etp.2004.12.004 PMID:15945273

WHO Regional Office for Europe (DK). (2013). *Review of evidence on health aspects of Atmospheric pollutants: REVIHAAP Project*. Copenhagen: WHO Regional Office for Europe.

WHO Regional Office for Europe (DK). (2014). *WHO Expert Meeting: methods and tools for asessing the health risks of Atmospheric pollutants at local, national and international level*. Meeting report. Bonn, Germany, 12-13 May 2014. Copenhagen: WHO Regional Office for Europe.

Wilson, R., & Spengler, J. (1996). *Particles in our air*. Harvard University Press.

Wittig, V. E., Ainsworth, E. A., & Long, S. P. (2007). To what extent do current and projected increases in surface ozone affects photosynthesis and stomatal conductance in trees? A metaanalytic review of the last 3 decades of experiments. *Plant, Cell & Environment, 30*(9), 1150–1162. doi:10.1111/j.1365-3040.2007.01717.x PMID:17661752

Wu, S., Mickley, L. J., Jacob, D. J., Rind, D., & Streets, D. G. (2008). Effects of 2000–2050 changes in climate and emissions on global tropospheric ozone and the policy-relevant background ozone in the United States. *Journal of Geophysical Research, 113*(D18), D18312. doi:10.1029/2007JD009639

Xu, X., Ha, S. U., & Basnet, R. (2016). A review of epidemiological research on adverse neurological effects of exposure to ambient Atmospheric pollutants. *Frontiers in Public Health, 4*, 157. doi:10.3389/fpubh.2016.00157 PMID:27547751

Chapter 7
Air Pollution Meteorology

Ravichandran C.
Bishop Heber College, India

ABSTRACT

Air pollution is a major environmental problem. It is divided into indoor air pollution and outdoor air pollution. The pollutants released in the outdoor atmosphere are under the control of meteorological conditions prevailing at the time of emission and after. Thus, the subject, air pollution meteorology, has emerged. It explains the significance of meteorological aspects and their influences on air pollutants emitted in the outdoor atmosphere. Prevailing winds and atmospheric stability are the two major factors that determine the dispersion of the pollutants. In this chapter, a brief outline is presented on composition and structure of atmosphere, and its processes such as radiation through the atmosphere, winds, global circulation, and atmospheric stability. Mixing height and ventilation coefficients are explained. The prediction of air quality using Gaussian dispersion model with the input of data of emission and meteorology is also explained.

INTRODUCTION

Ever since humans learned to make fire, they started polluting the atmosphere. However, the realization of air pollution came in the later part of human history, especially after the Industrial Revolution and the Engine-Driven Transportation. The discovery of fossil fuels (coal, oil and natural gas) led to the invention of external and internal combustion engines which brought forth myriad benefits to the society. Nevertheless, they also caused various unanticipated environmental challenges. One among them is air pollution.

DOI: 10.4018/978-1-5225-7289-3.ch007

Air pollution caused many deaths, immediately or slowly, either directly or indirectly. It is estimated that ambient air pollution causes about 4.2 million deaths every year through stroke, heart disease, lung cancer and chronic respiratory diseases. Sadly enough, about 91% of the world's population lives in places of air pollution levels exceeding WHO limits (WHO, 2018). The adverse effects of air pollution on humans, plants, animals, materials and on environment are well documented. Recent details on air pollution exposure and its burden are available in "State of Global Air/2018" (HEI & IHME, 2018). The anthropogenic release of Green House Gases, especially CO_2, caused a major environmental problem – global warming which led to climate change.

As the problems caused by air pollution became worse in multifarious and multifaceted ways, the research was directed towards controlling air pollution through tall stacks and engineering measures. A new field, "Air Pollution Meteorology" was born to find better ways to dilute the pollutants in the atmosphere. The complete history of the field of "air pollution meteorology" till the 1970s, has been presented by Heidorn (1978).

Air pollution meteorology is a combination of two fields *viz.*, "air pollution" and "meteorology". It may be considered either one as a branch of other. The study of air pollution is incomplete without a study of air pollution meteorology. Thus, air pollution meteorology is an essential branch of air pollution. Meteorology is a "scientific study of atmospheric phenomena, particularly in the troposphere and lower stratosphere" (Encyclopedia Britannica, 2018). Meteorology deals with weather and its causes and serves in weather forecasting. As meteorology has many applications in almost every aspect of human life and environment, it has several branches: agrometeorology, aviation meteorology, biometeorology, and marine meteorology etc. One of them is air pollution meteorology.

Many atmospheric phenomena such as solar radiation, temperature variation (horizontal and vertical), winds, humidity, pressure etc. determine and influence the fate of the air pollutants in the atmosphere either directly or indirectly or both. They may favor the dilution of air pollutants to such a level that they can cause no harm or buildup of a concentration which will cause deleterious effects. Hence, the study of air pollution meteorology gains in importance. It helps us in understanding the of the atmosphere to dilute the pollutants under various atmospheric conditions, that differ, temporally and spatially. Air pollution meteorology also aids in the prediction of air pollution levels from point sources, area sources and line sources, through air pollution dispersion models (Zannetti, 1990; Visscher, 2014). As the basic knowledge of meteorology is necessary to understand the Air Pollution Meteorology, certain basic concepts of Meteorology are outlined in the following sections.

COMPOSITION OF THE ATMOSPHERE

The earth's atmosphere is a thin layer of gases around the planet. This thin layer of gases is commonly called, air. It is retained by the earth's gravity. It extends from a few meters below the surface of the earth to a height of about 60,000 kms. Most of the atmospheric mass is located near the surface, up to a height of 80 to 100 kms. As a result, the air exerts pressure vertically below it. The highest pressure is felt near the surface of the earth and the pressure decreases with increasing altitude. On average, 1013 mb of pressure is observed at the earth's surface. The density of air at the Earth's surface is about 1.2 kg/m^3 (Padmanabha Murty, 2004). The density too decreases with increasing height.

The atmosphere consists of constant gases, variable gases, water vapor and solid particles. The gases whose concentrations remain same and do not change over time are known as constant gases. Constant gases include, nitrogen (78.09%), oxygen (20.9%) and nitrous oxide and inert gases (0.97% together). The gases whose concentrations change from time to time and from place to place are called, variable gases. Carbon dioxide, ozone and water vapor are variable gases in the atmosphere. Solid particles of crustal origin and maritime origin play a vital role in cloud formation, in scattering and absorbing radiation, and affect visibility.

STRUCTURE OF THE ATMOSPHERE

The atmosphere may be divided vertically into homosphere and heterosphere. Homosphere is the layer extending up to a height of about 90 km from the ground, in which the atmospheric constituents are more or less uniform throughout. But, in heterosphere, they vary with height. The atmosphere may also be divided into four distinct layers, based on temperature changes with the altitude, namely: Troposphere, Stratosphere, Mesosphere and Thermosphere. These vertical layers are depicted in Figure 1.

- *Troposphere*: it is the bottom layer of the atmosphere, lying just above the earth's surface. Almost all the life forms including humans live in this layer. Most of the human activities take place within this layer. It extends to an average height of 12 km from the surface of the earth. The thickness of the troposphere, however, varies with the latitude. In the tropics, it is about 16 km; and in the polar regions, about 8 km. Generally, in this layer, the atmospheric temperature decreases with the height. The rate of decrease of the temperature with the altitude is called the lapse rate. The average lapse rate

Figure 1.

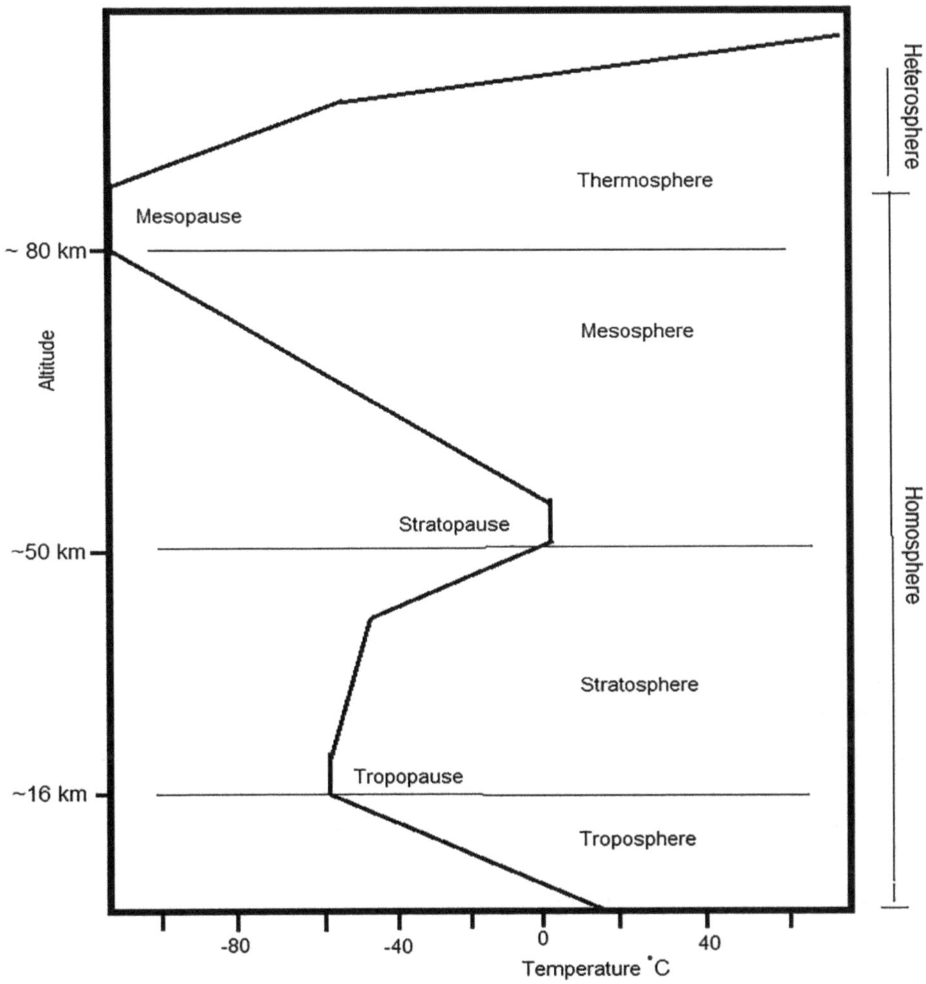

in the troposphere is about -6.4 °C/km. The negative sign is given to denote that the temperature decreases with increasing height. The troposphere ends at "Tropopause" in which the temperature neither increases nor decreases. It acts as a lid over the troposphere. The field, air pollution meteorology, is almost confined to the troposphere as the pollutants are released in this layer; and mixed and diluted in this layer. Many of the weather conditions are the result of the atmospheric phenomena occurring in the troposphere.

- *Stratosphere*: it lies above the tropopause and extends to a height of about 50 km above the earth's surface. Its thickness is about 35 km. The ozone layer exists within the stratosphere. The ozone layer, also called the ozonosphere, lies roughly between 15 km and 35 km above the earth's surface. The ozone layer protects the planet from harmful ultraviolet radiation. The temperature increases with the height in the stratosphere. The stratosphere ends in stratopause in which the temperature neither increases nor decreases.

- *Mesosphere*: The mesosphere lies above the stratopause; at the height of between about 52 kms and 80 km from the earth's surface. In this layer, the temperature decreases with the height. Above the mesosphere lies the mesopause in which the temperature remains constant.
- *Thermosphere:* The thermosphere lies approximately above 80 km from the earth's surface and extends to a height of 60,000 km. Above this height, it blends with space. The temperature increases with the height in this layer.

MAJOR PROCESSES IN THE ATMOSPHERE

The atmosphere is dynamic and active. Many processes occur in the atmosphere and these processes cause the weather and the climate of any geographical location. These processes influence and are influenced by the processes that take place in the other domains of the earth, namely, Hydrosphere, Biosphere and Lithosphere. Consequently, complex interactions exist among these spheres. Manahan (2011), in his book, "Environmental Science and Technology" included one more sphere in the earth's system called, "Anthrosphere". Anthrosphere is a domain of the earth dominated by human activities such as, agriculture, urbanization, industrialization etc. Evidently, anthrosphere exerts its influence in several ways on the processes of the other natural spheres, thus, adding more complexity. All these processes determine the weather and the climate at any geographical location.

In the atmosphere, winds are generated, and they flow from place to place with different speeds and in different directions. Water vapor enters the atmosphere through evaporation and transpiration (collectively called, "evapotranspiration"). It condenses after reaching the high altitudes to form clouds. The clouds thus formed, are transported by the prevailing winds to various places to bring precipitation over them. Over the deserts, the atmosphere is extremely dry and near the coastal regions, it is highly humid. At sometimes and in some places, there are no winds while at other times and in other places, they are extremely strong. Sometimes, winds become gusty, violent and stormy. Pressure variations over a large area create depressions which may develop into cyclones (typhoons, hurricanes). or anticyclones. There are extremely cold places and extremely hot places on the earth. All these variations exist due to various atmospheric phenomena and their interactions with the other spheres.

Radiation in the Atmosphere

All the above-mentioned processes do not happen by themselves. They require an input of energy. The incoming solar radiation (abbreviated as INSOLATION) is the ultimate source of energy for all of them. The incoming radiant energy is composed primarily of visible, and infrared, and to a lesser extent, of ultra-violet portions of the electromagnetic spectrum. Some of the energy is absorbed in the atmosphere by its constituents. The absorption of radiant energy takes place due to its interaction with the atmospheric constituents such as ozone, oxygen, water vapor and aerosols. Most of the UV radiation is absorbed in the ozone layer through the Oxygen-Ozone cycle. Only about 10% of the UV-B (280-320 nm) and most of the UV-A (320-400 nm) components reach the troposphere. Water vapor absorbs the solar radiation in the infrared region. Of the total incident energy striking the top of the atmosphere, about 18% on an average, is absorbed during its passage.

Of the remaining radiation, about 20% by clouds and particles and about 4 to 6% by the various earth's surfaces are reflected to the space. The reflected radiation is called, Albedo. About 6% of the incoming radiation is scattered by gases and aerosols. A part of the scattered radiation reaches the earth's surface while the remaining is left in space. Some radiation passes through the clouds and aerosols to reach the earth's surface. It is called, diffuse radiation. Diffuse radiation accounts for roughly 22%. Only about 30% of incident radiation reaches the earth's surface as direct radiation. Both direct radiation and diffuse radiation together make up about 52% of the radiation which strikes the earth's surface. The radiant energy reaching the earth's surface heats the surface and raises its temperatures. The extent of heating depends on the nature of the surface. If the radiation strikes a water body, some of it may be reflected and some will penetrate the deep layers. Reflection by water depends on the angle of incidence (i.e., the time of the day) and the maximum penetration occurs when the sun is perpendicular to the water. Water gets evaporated as the light energy falls on it. However, the heating of water takes place slowly, as its latent heat of vaporization is greater.

As the land surface and the water surface differ in their physical properties, so also the fate of the radiant energy striking them. The properties of the land surface such as the color and its texture etc. also affect the radiation. The land over the earth is not uniform; there are mountains, valleys, grass lands, forests, urban areas, etc. Due to these variations, the different land surfaces have different physical properties. For example, the dark surfaces absorb more radiant energy than the light surfaces; snow and ice reflect most of the radiation. It is estimated that fresh snow reflects up to 95% of radiation while dark colored soil surfaces reflect 10 to 20% of the radiation only (Pidwirny & Jones, 2009). Darker colored rocky and soil surfaces may absorb about 90% of the radiation striking them.

The radiation heats up the surface which it strikes. Both the land and the water surfaces are heated up, but the rate of heating differs. It also differs in different land surfaces. This heating of the land continues from dawn till dusk. After sunset, the heated land surfaces emit their energy in the form of infrared radiation. It is called, outgoing radiation. It is absorbed in the atmosphere by certain gases, called, Green House Gases. They are carbon dioxide, water vapor, methane, ozone, nitrous oxide and CFCs. Each of these gases absorbs infrared radiation at certain wavelengths. For others, they act as a "window", through which the outgoing radiation escapes into space. For example, CO_2 absorbs IR in the range between 14000 nm to 19000 nm mostly; it also absorbs in the range between 4000 nm and 4500 nm. For IR in other wavelengths, CO_2 acts as window. Due to the absorption of the outgoing radiation by the greenhouse gases, the atmosphere gets heated up. Consequently, the atmosphere is heated from below it. The atmosphere, in turn, emits heat energy in the form of infrared radiation, called, counter radiation. Without this counter radiation, the earth would become extremely cold – the earth's mean temperature would be, -20 °C instead of the current average temperature of 15 °C. This process acts like a blanket to keep the atmosphere warm during night. This process is called Green House Effect. Due to this heating, the atmosphere becomes warmer and the temperature keeps increasing with height. When temperature increases with height, it is called, inversion.

The incoming solar radiation and outgoing longwave radiation through the atmosphere is illustrated in Figure 2.

The concentration of carbon dioxide as well as of the other greenhouse gases keeps increasing in the atmosphere due to the large scale burning of fossil fuels and other human activities. It causes the enhancement of the greenhouse effect. This Enhanced Greenhouse Effect leads to Global Warming.

Figure 2.

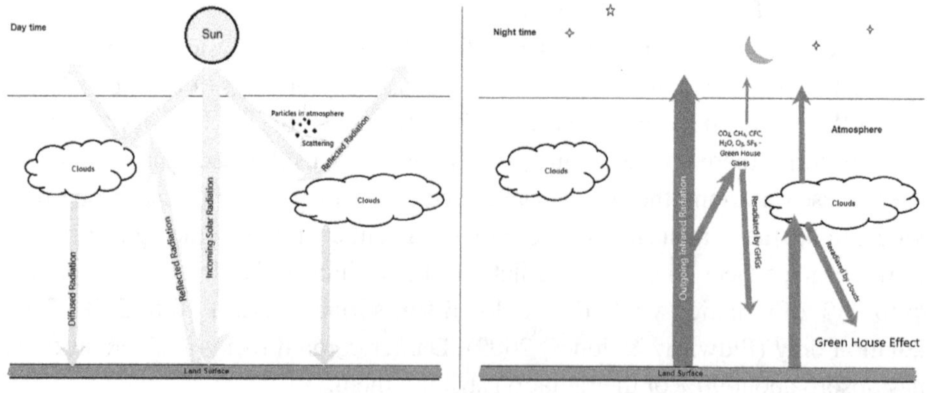

Winds in the Atmosphere

The earth's surface heated by INSOLATION will in turn, heat the overlying layer of air molecules through conduction. Consequently, the heated layer of air just above the ground becomes warmer than the air layer above it. As the warm layer is less dense, it starts moving upward, and this process is called, convection. Due to the upward movement of warm air, the pressure of that region becomes low. Hence, air from a relatively high-pressure region starts moving horizontally towards a low-pressure region. This horizontal movement of air is called, wind. As the winds are driven by pressure variation, *i.e.,* pressure gradient, the driving force is called pressure gradient force.

On a map of a large area, the pressure variation over the places may be depicted as Isobars – the lines connecting equal pressure values. The pressure gradient force initiates the air to move from the isobar of the highest pressure towards the isobar of the lowest pressure. If isobars are parallel to one another, the resulting wind would start flowing perpendicular to the isobars. But, it does not happen due to another force acting on the moving objects over the earth. This force is called, Coriolis force or Coriolis effect which is caused due to rotation of the earth. It is a deflective force, deflecting the wind to its right in the northern hemisphere and to its left in the southern hemisphere. The strength of Coriolis force is expressed as:

$$2v\omega \sin\varphi$$

Where,

v = velocity of the object = velocity of the wind
ω = angular velocity of the earth
φ = Latitude

The above equation explains that the Coriolis force is greater at high latitudes and that it keeps decreasing at low latitudes till it becomes zero at the equator. Due to the Coriolis effect, the wind which is supposed to move perpendicular to the isobars takes a right turn in the northern hemisphere (left turn in the southern hemisphere) and flows parallel to the isobars. This resultant wind is called, the Geostrophic wind. Figure 3 depicts the Geostrophic wind. The magnitude of wind speed increases as the pressure gradient force increases. The relationship between the wind speed and the pressure gradient force is linear and positive (Pidwirny & Jones, 2009). Accordingly, for the four-time increase of pressure gradient, there is

Figure 3.

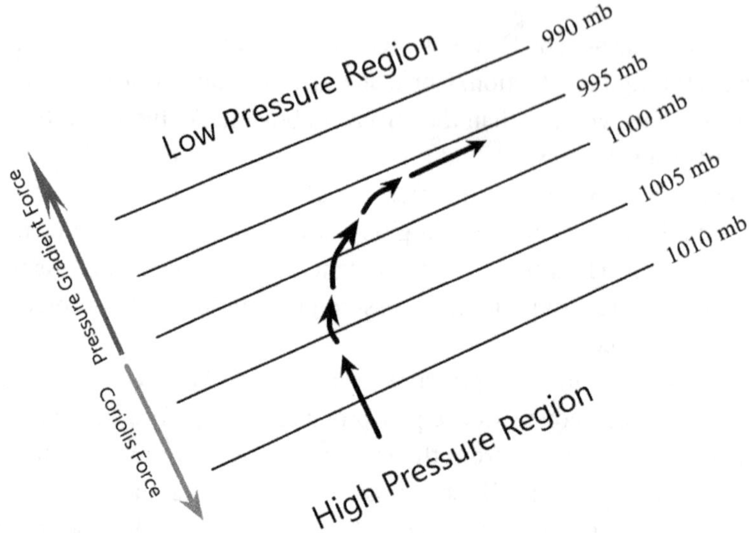

a four-time increase of wind speed. For example, if the pressure gradient increased from 0.1 mb/km to 0.4 mb/km, the resultant wind speed would be 4 times greater.

Winds that are close to the earth's surface may experience friction as an additional force, which will affect the flow of the wind. When winds flow over a water surface, they face less or no friction as the water surface is smooth. But, rough terrains like, rocks, buildings of varying heights in urban areas, tall trees and other vegetation, exert a frictional force of varying magnitudes over the winds. This is called Surface Roughness. It will reduce the speed and alter the direction of the wind. This force is called, the Frictional Force. Due to the frictional force, the resultant wind will flow at cross angles to the isobars. Due to this friction, the winds are diverted horizontally, laterally and / or vertically, thus, creating turbulence. Turbulence increases with the increase in wind speed. Nevertheless, the frictional effect will cease approximately at 1 km above the earth's surface. Thus, the winds above 1 km from the earth's surface will be the geostrophic wind – flowing parallel to the isobars. Moreover, the wind speed above 1 km from the earth's surface will be greater than that near the earth's surface. The speed of the wind increases with increasing altitude due to decrease in frictional force which is illustrated in Figure 4. The extent of frictional force is different over different surface features of the earth. For example, the water surface exerts extremely low or no friction to the moving air while, the land with tall buildings, trees etc. exert high friction.

Figure 4.

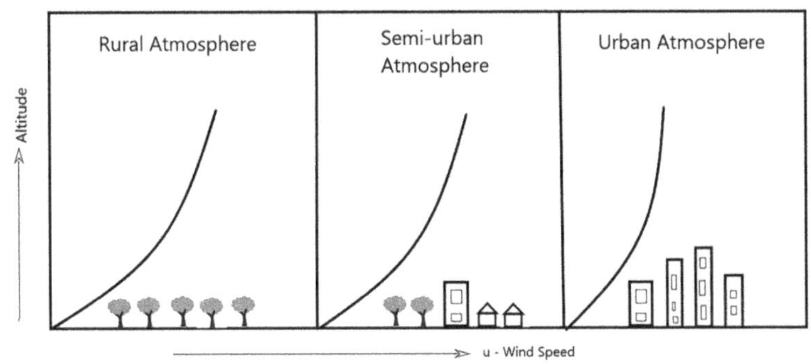

Global Circulation

The winds thus generated flow long distances over the continents and the oceans. There is a general pattern of the wind flow across the globe, which is called, Atmospheric Global Circulation. It is also called, Primary Circulation. However, there are variations regionally and locally at many places due to the differences in land features and land – marine contrasts. These differences result in Secondary Circulation Systems. Monsoon winds are one such system over South Asia, especially over the Indian subcontinent. The details of primary circulation and secondary circulation are available in many text books of meteorology.

Apart from these systems, there are many local winds such as Sea-Land Breeze in coastal regions and Mountain-Valley Breeze in the mountain-valley environment.

In coastal environments, during daytime, the land is heated up more rapidly than of the sea. Hence, the air over the land is warmer and rises upward. But, the air over the water is relatively colder. As a result, the air over water starts moving towards the land which is called, "Sea-Breeze". During night time, the opposite takes place. The land is cooled down more rapidly than of the sea. Hence, the air over the land is colder to move towards the sea which is called, "Land-Breeze". Sea-Land Breeze is illustrated in Figure 5.

In Mountain-valley system, the winds are channelized when valley is oriented at an acute angle to the wind direction. Part of the wind is peeled off by the valley and constrained to follow the valley floor.

During daytime, valley floor and the sloping mountain floor gets heated up making the air warmer to rise upward. When the warm air reaches the altitude of mountain top, it is cooled down and sinks downward. The sinking air returns to the valley. During night time, the radiative cooling of the sloping mountain floor

Figure 5.

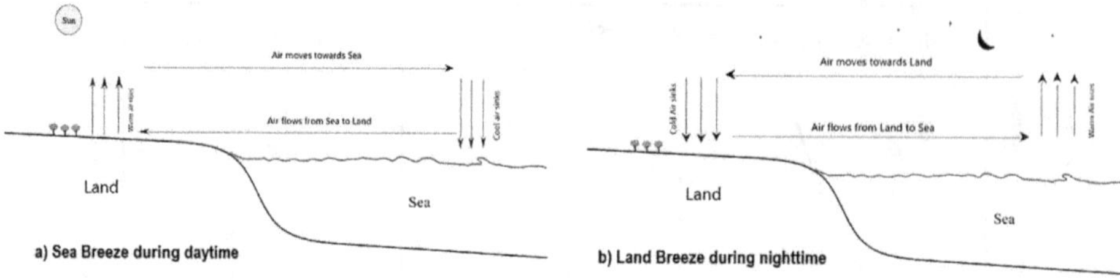

causes the air colder. As a result, the cold air from the sloping mountain floor slides downward towards the valley. Thus, the air is getting circulated within valley itself. This phenomenon is depicted in Figure 6.

METEOROLOGICAL ASPECTS OF AIR POLLUTION

Wind and Atmospheric stability are the two important factors that determine the fate of air pollutants in the atmosphere.

Transportation of Pollutants by Wind

The wind – horizontal movement of air – is a very important factor in transporting the air pollutants and thereby in their dilution. Winds carry the pollutants to long distances. As the pollutants travel along with the wind, they also diffuse laterally as well as vertically. Both the transportation and the diffusion together dilute the pollutants. Transportation and diffusion are collectively known as Dispersion. The

Figure 6.

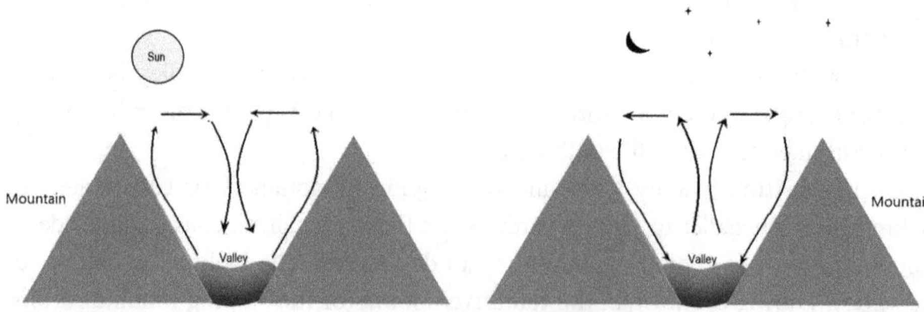

turbulent winds favor the dispersion effectively. The vertical diffusion is determined by the stability of the atmosphere.

The prevailing winds carry and transport the pollutants away from their sources, horizontally. The extent of the transportation depends on the speed and the turbulence of the wind. The direction of the wind, the direction from which it flows, determines the zone which will be affected by the pollutants. Areas located exactly opposite to the wind direction, which is the downwind direction and the adjoining areas will get affected. However, by the time they reach the locations in downwind direction, pollutants may get diluted by the dispersion to the extent not to cause harmful effects. Or, they may be less diluted so to cause adverse health effects on population living in the downwind directions. Wind direction and speed always vary over time, on an hourly basis of a day, on a daily basis of a month, and on a seasonal basis. Winds may also vary from year to year, from decade to decade and so on. The wind patterns over a period of time may be depicted in a diagram called, Wind Rose. Wind rose is a circular diagram depicting the frequencies (relative frequencies) of the wind from 16 different directions with different speed classes. An example of wind rose diagram is presented in Figure 7.

Wind roses are helpful in locating a proposed industrial site or residential site so that the residents will be least affected. From wind roses, the predominant wind direction in most of the days of a year can be deduced to help locating the industrial site / residential site while planning. From the wind rose diagrams average wind speed is determined which will be used in Air Quality Modelling. In preparation of Environmental Impact Assessment Reports, wind roses are presented as part of the air quality impact assessment.

Figure 7.

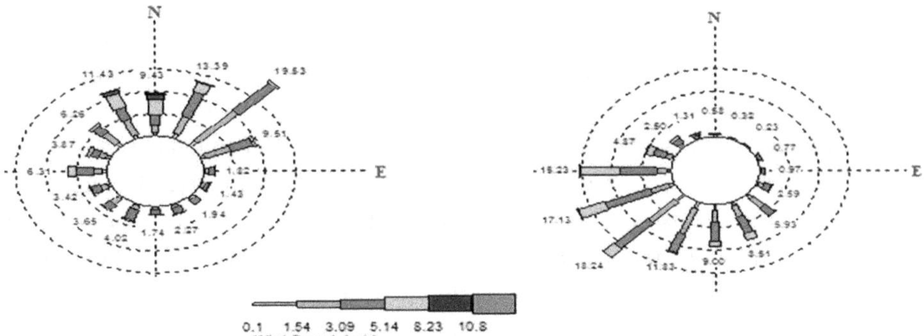

Stability of the Atmosphere

When a parcel of air receives an upward force and if it keeps moving upward, then the atmosphere is said to be Unstable; if it returns to its original position from where it was released, then the atmosphere is said to be Stable. The state of the atmosphere regarding the moving parcel of air is called, the Stability of the atmosphere. In other words, the atmospheric stability is the state of the atmosphere either to resist or to facilitate the vertical movement of the rising parcel of air. The existing atmospheric conditions at any given time either allow or resist the parcel of air to move upward. The stability of the atmosphere is determined by the lapse rate of the atmosphere. For determining the atmospheric stability, two different lapse rates are used: Environmental Lapse Rate (ELR) and Dry Adiabatic Lapse Rate (DALR).

Lapse rate, as already mentioned, is the rate of decrease of the temperature with height in the atmosphere. Environmental Lapse Rate (ELR) is the rate of decrease of ambient temperature with increasing height at any given time. It may vary hourly during the day and the night, daily, and seasonally. During the night time, the lapse rate in general, is positive, i.e., the temperature increases with increasing height. It is due to the greenhouse effect. On the other hand, Dry Adiabatic Lapse Rate (DALR) is the rate of decrease of temperature of the rising parcel of dry air. As the pressure decreases with height, a rising parcel of dry air will expand at a higher altitude. This expansion cools the air parcel adiabatically and, thereby, its temperature decreases. This rate of decrease of temperature is called dry adiabatic lapse rate. It is theoretically determined to be -1 °C/100 m approximately.

DALR deals only with the dry parcel of air. If the air parcel contains moisture in it, the lapse rate will be different. If the air parcel is saturated with water vapor, the lapse rate of such a rising air parcel is Wet Adiabatic Lapse Rate or Saturated Adiabatic Lapse Rate (SALR). When the air parcel with the water vapor in it, rises above, it expands and cools, and as a result, the water vapor condenses. The condensation of the water vapor releases the latent heat which will warm the air parcel. This will reduce the lapse rate and consequently SALR is different from DALR. Theoretically, the SALR is determined as -0.44 °C/100 m.

In air pollution meteorology, atmospheric stability is determined based on ELR in comparison with DALR. When ELR is greater than DALR, the rate of decrease of the temperature of the rising air parcel is less than the rate of decrease of the ambient temperature vertically, then the atmosphere is said to be Unstable. As the rising air parcel, reaches a certain height, it will be warmer than its surrounding and hence, it keeps rising. It is illustrated in Figure 8 a. During unstable conditions, the winds will be turbulent with moderate speeds. The pollutants released from any source will get diffused rapidly and effectively. Unstable conditions favor the dispersion more.

If ELR is less than DALR, the rate of decrease of the ambient temperature is less, but of the rising air parcel is more. The rising air parcel will be cooler than its surroundings and hence, it will sink and reach its original position. Thus, the atmosphere is resistant to the vertical diffusion. The atmosphere is said to be Stable which is depicted in Figure 8 b.

If the DALR and ELR are equal, i.e., the rate of decrease of the ambient atmospheric temperature and that of the rising parcel of air are the same, then the atmosphere is said to be Neutral. Figure 8 c depicts the neutral atmosphere. During neutral conditions, the winds may be stronger than those during unstable conditions.

When the ELR is positive – the ambient temperature increases with height. This condition is called, Inversion which is presented in Figure 8 d. This condition is also called Highly Stable. During inversion, the rising air parcel will cool down due to the adiabatic expansion but will be cooler than the surroundings. As a result, the air parcel will come down to its original position. It inhibits vertical dispersion of pollutants.

Atmospheric Stability of a place at a given time can be determined indirectly following Paquill's (1961) method. Pasquill (1961) classified atmospheric stability into six categories - A, B, C, D, E and F to represent Extremely Unstable, Moderately Unstable, Slightly Unstable, Neutral, Slightly Stable, and Moderately Stable respectively. They are determined using observed data of wind speed, INSOLATION conditions and Cloud Cover which is presented in Table 1.

Plume behavior under different stability conditions

The pollutants released from a source, due to its transportation and diffusion assume a shape resembling a feather as it spreads. Hence, the traveling pollutants from a source is called, Plume.

During unstable conditions, due to turbulence, the plume will travel at moderate speeds (as the winds are moderate) but with greater diffusion. As a result, the plume will travel as of a loop, and the plume is called the Looping Plume (Figure 9 a).

During neutral conditions, plume travels downwind assuming cone shape, called the Coning Plume (Figure 9 b). The pollutants travel faster but with moderate diffusion.

During stable conditions, the plume will travel with less vertical diffusion. Under inversion (highly stable), the plume will not diffuse vertically but will travel as a thick one with lateral oscillation. This type of plume is called the Fanning Plume (Figure 9 c).

At times, especially during the early morning hours, after sunrise, the inversion will start breaking from the ground, resulting in either a neutral or unstable condition near the ground and inversion above. This will make the plume diffuse vertically downward, resulting in, Fumigation (Figure 9 d). This will result in the high ground level concentration of pollutants.

Figure 8.

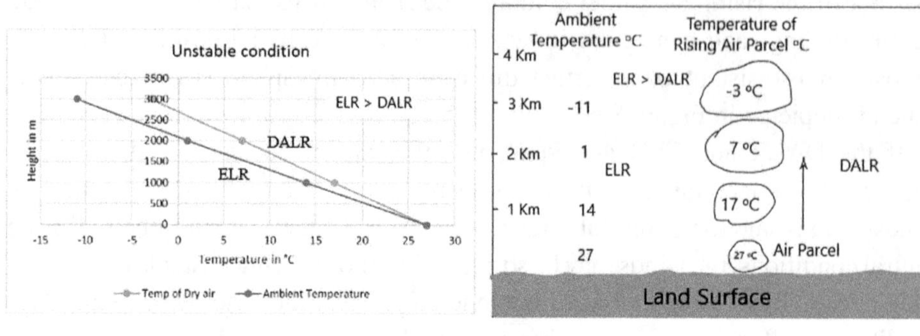

a) Unstable Atmosphere

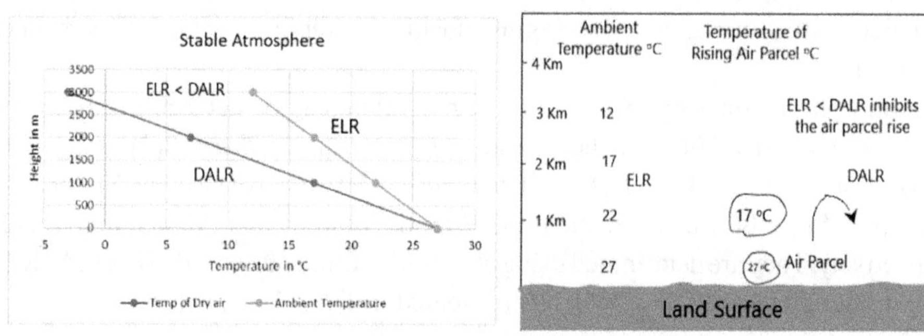

b) Stable Atmosphere

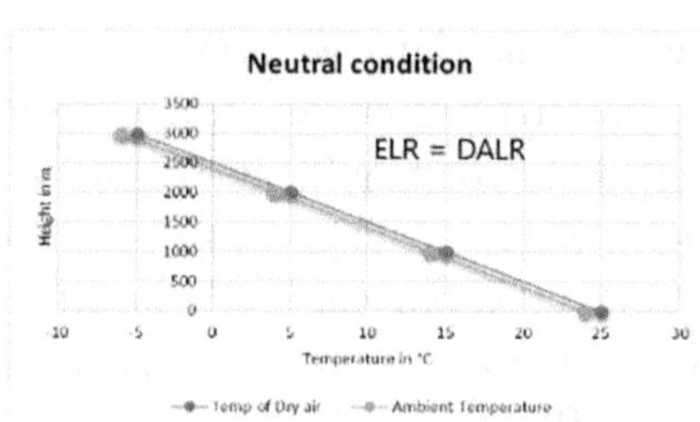

c) Neutral Atmosphere

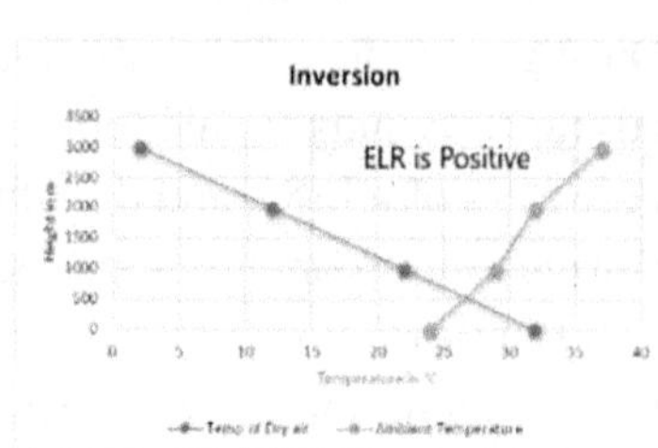

d) Inversion

During the evening, after sunset, the inversion layer will build up from the ground, keeping the plume above the inversion layer, and this is known as the Lofting Plume (Figure 9 e). Pollutants will diffuse vertically above the inversion layer. However, this condition will exist for a short duration and as the inversion completely builds up.

Table 1. Pasquill stability classes

Surface Wind at 10 m in m/s	Daytime Insolation			Night Time Cloud Cover	
	Strong	Moderate	Slight	≥ 4/8	≤ 3/8
< 2	A	A - B	B	--	--
2 – 3	A – B	B	C	E	F
3 - 5	B	B - C	C	D	E
5 – 6	C	C – D	D	D	D
> 6	C	D	D	C	D

Source: (B. Padmanabha Murty, 2004)
Note:
Strong insolation - solar elevation is more than 60°
Weak Insolation – solar elevation is less than 35°
Moderate Insolation – solar elevation is between 35° and 60° under clear sky
Solar elevation is to be obtained from Astronomical Tables for a given date, time and latitude

Sometimes, there may be an occasion when there may be inversion layers below and above and in between either neutral or unstable. Under this circumstance, the plume from a stack will get trapped, called, Trapping (Figure 9 f). However, once the ground-based inversion breaks up, the trapped plume will fumigate.

The general plume behavior discussed above is for over plains of flat terrain. The coastal environments where sea-land breeze system exists and the mountain-valley environments where the mountain-valley breeze exist, the pollutants are transported and diffused bit differently.

In a coastal environment, as the winds blow from sea to the land during daytime, pollutants from the sources near the shoreline will be carried towards the interior land. During sea-breeze, near the shoreline the condition will be stable as the air is cold near the surface and warmer above. Consequently, the pollutants from the sources near the shoreline will travel as a fanning plume, but after traveled several kilometers interior of the land, they fumigate due to existence of instability over that place. During nighttime, the pollutants may be carried towards the sea from the land.

Similarly, for a source located in a valley and surrounded by mountains, the pollutants released from the source will keep circulating within the valley due to mountain – valley breeze system. During daytime, valley floor and mountain walls

Urban Heat Island Effect

In urban areas, the presence of buildings and paved roads will absorb more energy when compared to the rural areas. Due to this, the atmosphere over urban areas will be warmer (few °C) than its rural counterparts during nighttime. The relatively

Figure 9.

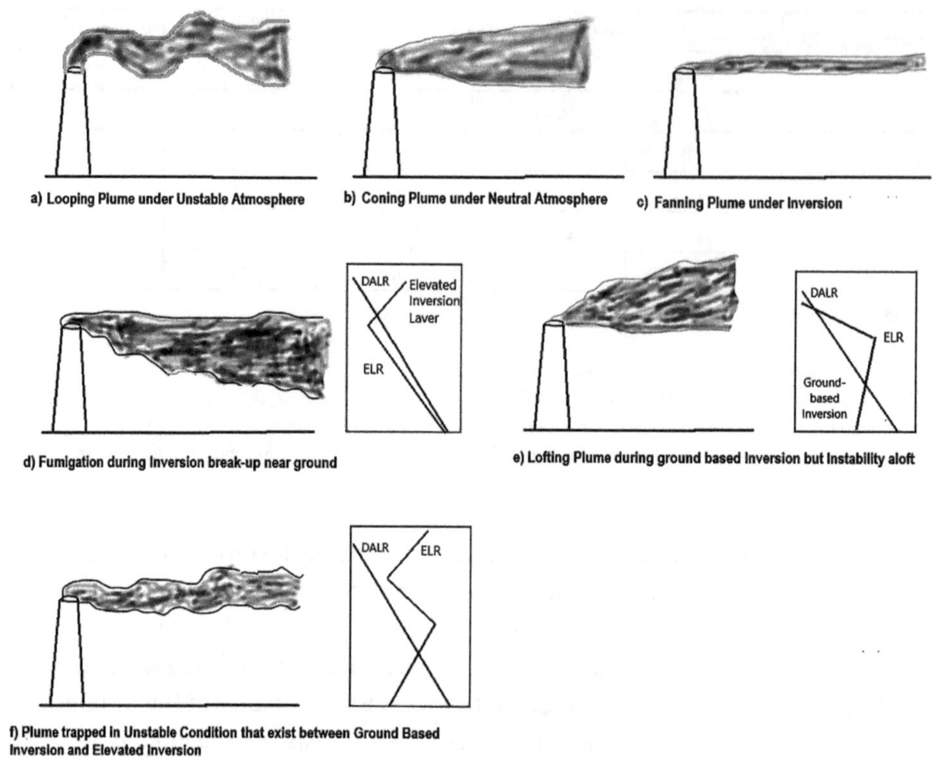

warmer urban environment is called, Urban Heat Island; it is depicted in Figure 10. Consequently, the air rises in the atmosphere and moves towards rural areas. This air becomes cooler and then started sinking in the adjoining rural areas. From rural areas, the cold air moves towards urban areas completing a cycle. This phenomenon is called, Urban Heat Island Effect. As a result, the pollutants released from urban areas, move with the air upwards, sink in rural areas and then move towards urban areas at surface level. Thus, the pollutants get circulated in the urban heat island system.

Mixing Height

In the preceding section, it is made clear that unstable and neutral conditions favor the vertical diffusion of pollutants. However, this vertical extent is limited to a certain height. This height is known as, the Mixing Height. Again, ELR and DALR determine the mixing height of a location at a given point of time. The mixing height also varies over time in a day, on a daily basis, monthly basis and seasonal basis of

Figure 10.

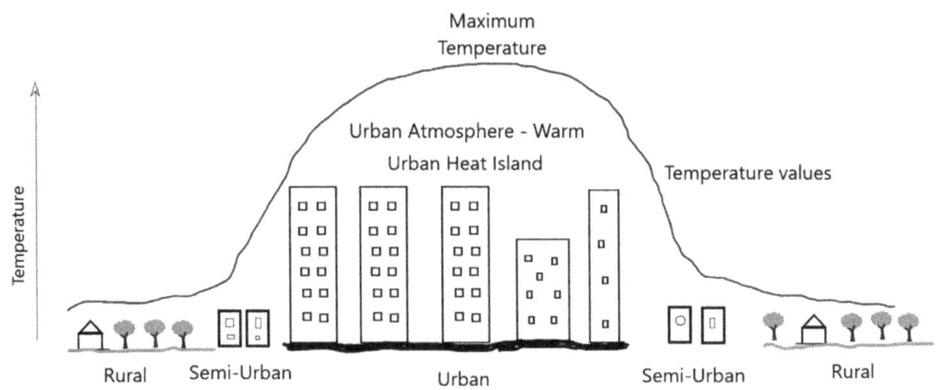

a year. While DALR is a linear, ELR may not be the same throughout the vertical height of the troposphere. The ELR may be a certain value up to a certain height and may be different from there up to a certain height and, again, may vary. That is, ELR is variable at the different altitudes of a place at the same time. The atmosphere may be unstable from the ground and, after a certain height, the atmosphere may become less unstable, neutral or stable (need not be inversion). The height at which instability (unstable condition) changes to stability (stable condition) is the mixing height. Graphically speaking, the intersecting point (height) of DALR and ELR is the mixing height of that location at the given time. It is illustrated in Figure 11.

Generally, the mixing height may be low during the night time and the lowest early in the morning and be high during daytime reaching its peak in the evening time. During nighttime, and the early morning hours, the existence of inversion limits the vertical dispersion resulting in the lowest mixing height. In the evening hours, mostly highly unstable condition exists which result in the highest mixing height. The mixing height can be determined directly by measuring the vertical temperature at various height intervals. But, direct measurements are not carried out due to financial constraints. Hence, the mixing height is determined indirectly using the surface temperature value plus 5 °C intersecting point at and the morning vertical temperature profile (Holzworth, 1971). 5 °C is added to the surface temperature to include the urban heat island effect in urban areas. In India, 00Z (0000 GMT) radiosonde data is used for morning temperature profile. Evening Mixing Height value can be obtained similarly but without adding the 5 °C value to the surface temperature. Nevertheless, the mixing height calculated in this way is an approximate value only.

Figure 11.

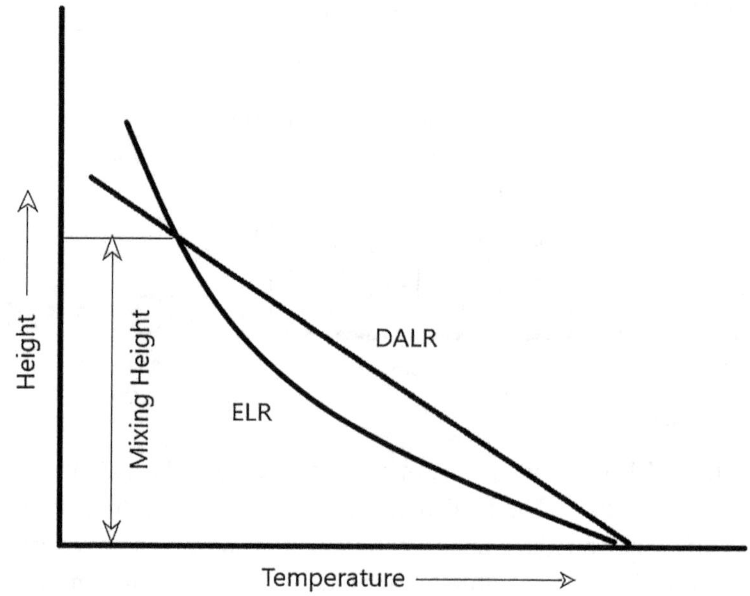

Air Pollution Potential

The mixing height and mean wind speed are multiplied to get the ventilation coefficient in m^2/s. If the ventilation coefficient is less than 6000 m^2/s. then the atmosphere is said to have a high air pollution potential. By determining time / season of occurrence of the high air pollution potential episodes, the emission sources may be controlled / regulated in such a way not to result in high air pollution. However, regulating emission is not that easy to be implemented.

Air Quality Modeling

Modeling may be defined as the development of a system, procedure or process that imitate the natural process. Air quality modeling is a field in which the processes that affect the quality of air is simulated or modeled to imitate those processes. By running a model, the resultant air quality may be predicted without the real observations of actual processes. This will help in understanding the impact of a proposed pollution source on air quality. It may be said that air quality modeling is a tool in the field of air pollution and in Environmental Impact Assessment studies.

In general, air quality models are classified into:

- Physical Models and
- Mathematical Models

Physical models are small-scale systems prepared in a laboratory to represent the original natural process / phenomena. Mathematical models represent the original physical and chemical processes / phenomena through a set of mathematical equations. Physical modeling is expensive while mathematical modeling is affordable one.

For air quality modeling, meteorological data, and source characteristics such as emission rate and stack height are required as inputs to these models. According to USEPA (2016), there are three kinds of models:

1. Dispersion modeling
2. Photochemical modeling
3. Receptor modeling

Dispersion modeling determines the concentration of a given pollutant at any downwind distance from the source at ground level, or at any height or at any lateral distance from the center of the plume. In this, the pollutant is assumed to be non-reactive.

Photochemical modeling includes the chemical reactions that take place in the atmosphere to form various secondary pollutants from primary pollutants. They simulate pollutant concentrations and deposition rates of both inert and reactive pollutants, over a large geographical area.

Receptor models are used to identify and quantify contributions by the source to the concentrations at the receptor.

Details of each kind of model can be obtained from the literature of various sources, based on one's interest. In this section, the discussion is limited to the general Guassian dispersion modeling for plains only. Gaussian dispersion modeling is done with the following assumptions (NPTEL, 2018):

1. Steady-state conditions - implying that the emission rate from the point source is constant.
2. The wind speed is constant in time and with height - Homogeneous flow.
3. The pollutant is conservative and no fallout due to gravity.
4. Perfect reflection of the plume at the underlying surface (no ground absorption).
5. The turbulent diffusion in the x-direction is neglected relative to advection in the transport direction – the model should be applied for average wind speeds of more than 1 m/s.

6. The coordinate system is directed with its x-axis into the direction of the flow, and the v and w (lateral and vertical) components of the time averaged wind vector are set to zero.
7. The terrain under the plume is flat.
8. All variables are ensemble averaged, which implies long-term averaging with stationary conditions.

The Gaussian dispersion modelling is based on the normal distribution (Gaussian Distribution) of the observations in a data set. In this approach, the concentration values are considered as the observations of a normal distribution. The centreline of the distribution is assumed to have the maximum concentration, while areas away from it will have lesser and lesser concentrations. That is, the concentration will decrease as the distance away from the centreline increases. A similar distribution of concentration is assumed both laterally and vertically (y and z coordinates), as the plume travels in the downwind direction. It is illustrated in Figure 12. The concentration of a pollutant emitted from a point source is determined using the following equation:

$$\chi_{(x,y,z,H)} = \frac{Q}{2\pi u \sigma_y \sigma_z} \times \left[\exp - \left(\frac{y^2}{2\sigma_y^2}\right)\right]\left[\exp\left\{-\frac{1}{2}\left(\frac{z-H}{\sigma_z}\right)^2\right\} + \exp\left\{-\frac{1}{2}\left(\frac{z+H}{\sigma_z}\right)^2\right\}\right]$$

Where,

χ = concentration of a pollutant in

$Q = Average emission rate in \frac{g}{s}$ = average emission rate in $\frac{g}{m^3}$

u = average wind speed in m/s

σ_y, σ_z = diffusion coefficients in y and z direction in m

H = effective stack height in m

$\exp = e$ = natural log equal to 2.71828

In this model, the diffusion coefficients, σ_y and σ_z are standard deviation values in the lateral and the vertical directions, respectively. These values cannot be determined without observations by conducting experiments. However, they can be obtained from Pasquill – Gifford curves (Turner, 1970) for a given downwind distance and atmospheric stability condition. These curves were prepared empirically and

Table 2. Diffusion coefficients

Rural Environment			Urban Environment		
Pasquill Stability Class	$\sigma_y(m)$	$\sigma_z(m)$	**Pasquill Stability Class**	$\sigma_y(m)$	$\sigma_z(m)$
A	$0.22\times(1+0.0001x)^{-1/2}$	$0.20x$	A-B	$0.32\times(1+0.004x)^{-1/2}$	$0.24\times(1+0.001x)^{-1/2}$
B	$0.16\times(1+0.0001x)^{-1/2}$	$0.12x$			
C	$0.11\times(1+0.0001x)^{-1/2}$	$0.08\times(1+0.0002x)^{-1/2}$	C	$0.22\times(1+0.0004x)^{-1/2}$	$0.20x$
D	$0.08\times(1+0.001x)^{-1/2}$	$0.06\times(1+0.0015x)^{-1/2}$	D	$0.16\times(1+0.0004x)^{-1/2}$	$0.14\times(1+0.0003x)^{-1/2}$
E	$0.06\times(1+0.001x)^{-1/2}$	$0.03\times(1+0.003x)^{-1}$	E-F	$0.11\times(1+0.0004x)^{-1/2}$	$0.08\times(1+0.00015x)^{-1/2}$
F	$0.04\times(1+0.001x)^{-1/2}$	$0.016\times(1+0.003x)^{-1}$			

(*Adapted from: CPCB, 1998*)
x *is the downwind distance in m.*

Figure 12.

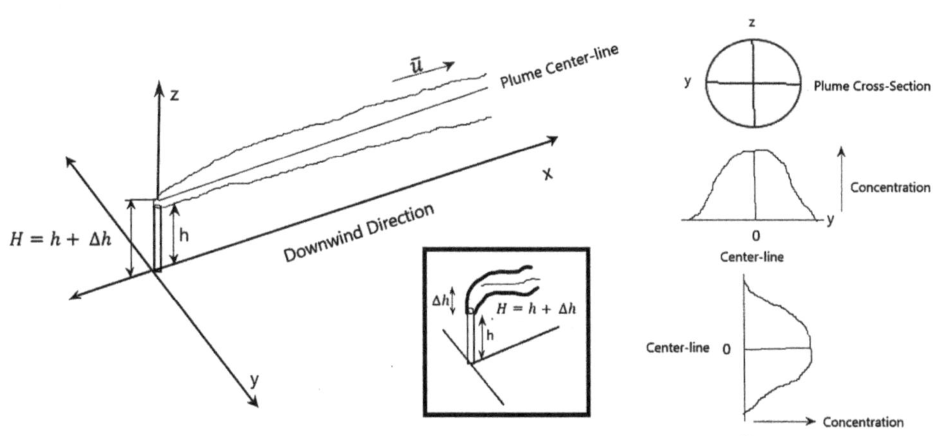

hence, yield the approximate values for other places / geographical locations. The diffusion coefficients prescribed by Briggs (1973) are recommended by CPCB (1998). Briggs' diffusion coefficients are presented in Table 2.

In the Gaussian equation, H is an important factor. H is the effective stack height which is physical stack height plus plume rise (Figure 12).

Effective stack height is calculated as follows:

$$H = h + \Delta h$$

where,

H = effective stack height in m
h = physical stack height in m
Δh = plume rise in m

Plume rise is also calculated from various plume rise models developed by many workers (Zannetti, 1990). However, no model is accurate; all of them give approximate values. The US EPA models use Briggs' (1979) formulae for plume rise calculations. Briggs' plume rise equations are used by US EPA due to their simplicity and give the values with minimum deviation from accuracy. Nevertheless, despite their wide usage, it is reported that Briggs' equations overestimate the plume rise at large distances in downwind direction. Many plume rise $\left(\Delta h\right)$ equations have the following form:

$$\Delta h\left(x\right) = const.Q_h^a x^b u^c$$

Where,

a, b, c are constants
Q_h = heat emission rate of source
x = downwind distance
u = wind speed at Z_s (at stack height)
Q_h = the heat emission rate; it is given by:

$$Q_h = Q_m C_p \left(T_s - T_a\right)$$

Where,

C_p = specific heat at constant pressure
T_s = gas exit temperature
T_a = ambient air temperature at Z_s (at stack height)

Q_m = total mass rate; it is given by:

$$Q_m = \rho_s \pi r_s^2 v_s$$

Where,

ρ_s = density of flue gases

r_s = exit radius; for a non-circular exit, $r_s = \sqrt{\frac{A}{\pi}}$

v_s = exit velocity

Of all the available methods, Briggs (1969) method is known widely. It defines the buoyancy flux parameter F_b, by

$$F_b = g v_s r_s^2 \frac{(T_s - T_a)}{T_s}$$

The critical downwind distance, x^*, (Zannetti, 1990) is given by

$$x^* = 2.16 F_b^{2/5} Z_s^{3/5}$$

for

$$Z_s < 305m$$

$$x^* = 67 F_b^{2/5} \text{ for } Z_s \geq 305m$$

For buoyant plumes, Briggs equation under neutral and unstable atmospheric conditions become (Padmanabhamurty, 2004):

$$\Delta h = \left\{ \frac{1.6 F_b^{1/3} 3.5 x^{*2/3}}{u} \right\}$$

Where,

$$F_b = \frac{gQ_h}{\pi C_p T_a \rho}$$

Where,

g = acceleration due to gravity, m/s^2
Q_h = heat emission, cal/s
C_p = specific heat at constant pressure of air, $cal/gm°K$
ρ = density of air, g/m^3
T_a = air temperature, $°K$
u = wind speed at stack level, m/s

$$x^* = 14F^{5/8}\ when\ F_b < 55\, m^4/s^3 = 34F^{2/5}\ when\ F_b > 55\, m^4/s^3$$

The equation is modified for buoyant plumes under stable conditions as

$$\Delta h = 2.6\left(\frac{F_b}{us}\right)^{1/3}$$

Where,

s = stability parameter = $\frac{g}{\theta}\frac{\partial \theta}{\partial z}$, where θ is air temperature in $°K$

For stable and calm conditions $\left(u < 1\, m/s\right)$ for buoyant plumes, plume rise becomes

$$\Delta h = 5.0F_b^{1/4}s^{-3/8}$$

By determining the stability condition of the atmosphere at a given location at a given time, the appropriate plume rise equation may be used to calculate the plume rise value. This value can be substituted in the Gaussian dispersion model to compute the concentration of a pollutant at any desired downwind distance. The model may be further simplified to calculate only ground level concentration at any downwind distance.

In the Gaussian dispersion equation, $Z - H$, accounts for the source which is above the height of H, and the $Z + H$ accounts for the "reflected source" from the ground. Reflected source is the imaginary source for the plume's re-entry after touching the ground. As after touching the ground, the plume can no longer diffuse vertically downwards, it diffuses vertically upwards from the ground. This is an additional source to the plume. Hence, $Z + H$ is included in the equation.

For ground level concentration calculations, the equation becomes:

$$\chi_{(x,y,0,H)} = \frac{Q}{\pi u \sigma_y \sigma_z} \times \left[\exp - \left(\frac{y^2}{2\sigma_y^2}\right)\right]\left[\exp\left\{-\frac{1}{2}\left(\frac{H}{\sigma_z^2}\right)^2\right\}\right]$$

Ground level concentration at the centreline of the plume is calculated as:

$$\chi_{(x,0,0,H)} = \frac{Q}{\pi u \sigma_y \sigma_z} \times \exp\left\{-\frac{1}{2}\left(\frac{H}{\sigma_z^2}\right)^2\right\}$$

CONCLUSION

In this chapter, Air Pollution Meteorology, fundamentals and basic concepts have been discussed. As the field of "Air Pollution Meteorology" has grown well and advanced and still growing, an exhaustive coverage is not possible. It requires a dedicated book / volume to cover all the aspects of air pollution meteorology and air quality modelling. However, for a reader who aims in acquiring basic knowledge, the contents in this chapter is enough. The reader interested in research may refer to the books listed in Additional Reading.

ACKNOWLEDGMENT

The Author expresses his gratitude towards Prof. Natarajan, Retired Professor of English, Bishop Heber College for his timely help in correcting the manuscript.

REFERENCES

Bagavath Singh, A., Srinivas, C.V., Maran, S., Baskaran, R., & Venkatraman, B. (2016). Wind Direction Dependent Vertical Wind Shear and Surface Roughness Parameterization in Two different Coastal Environments. *IJGAES, 4*(3), 1-9.

Blij de, H.J., & Muller, P.O. (1993). *Physical Geography of the Global Environment.* New York: John Wiley & Sons Inc.

Briggs, G. A. (1973). *Diffusion estimation for small emissions.* USAEC Report ATDL-106. Oak Ridge: NOAA.

Briggs, G. A. (1979). Some recent analysis of Plume Rise observations. In *Proc. Of 2nd Intl. Clean Air Congress*. New York: Aca. Pres.

CPCB. (1998). *Assessment of Impact to Air Environment: Guidelines for conducting Air Quality Modelling. PROBES/70/1997-98.* Delhi: Central Pollution Control Board.

Davis, M. L., & Cornwell, D. A. (1991). *Introduction to Environmental Engineering* (2nd ed.). New York: McGraw-Hill, Inc.

Encyclopaedia Britannica. (2012). *Greenhouse effect.* Retrieved from https://www.britannica.com/science/electromagnetic-radiation/-continuous-spectra-of-electromagnetic-radiation/media/183228/85040

Encyclopaedia Britannica. (2018). *Meteorology-Science.* Retrieved from https://www.britannica.com/science/meteorology

HEI & IHME. (2018). *State of Global Air/2018. A Special Report on Global Exposure to Air Pollution and its Disease Burden.* Retrieved from http://stateofglobalair.org/sites/default/files/soga-2018-report.pdf

Heidorn, K. C. (1970). A Chronology of Important Events in the History of Air Pollution Meteorology to 1970. *Bulletin of the American Meteorological Society, 59*(12), 1589–1597. doi:10.1175/1520-0477(1978)059<1589:ACOIEI>2.0.CO;2

Holzworth, C. G. (1971). Meteorological Potential for Urban Air Pollution in the Contiguous United States. *Proc. 2nd Intl. Clean Air Congress*, 1076-1080. 10.1016/B978-0-12-239450-8.50192-1

Manahan, S. E. (2011). *Environmental Science and Technology – A Sustainable Approach to Green Science and Technology*. Boca Raton, FL: CRC Press, Taylor & Francis.

NPTEL. (2018). *Air Quality Modelling*. Retrieved from https://nptel.ac.in/courses/105102089/air%20pollution%20(Civil)/Module-5/1.htm

Padmanabha Murty, B. (2004). *Environmental Meteorology*. New Delhi: I.K. International Pvt. Ltd.

Pasquill, F. (1961). The Estimation of the Dispersion of Windborne Material. *The Meteorological Magazine*, *90*(1063), 33–49.

Pidwirny, M., & Jones, S. (2009). *Atmospheric Effects on Incoming Solar Radiation*. Retrieved from http://www.physicalgeography.net/fundamentals/

Turner, B. D. (1970). *Workbook of Atmospheric Dispersion Estimates*. Environmental Protection Agency.

USEPA. (2016). *Air Quality Models*. Retrieved from https://www3.epa.gov/scram001/aqmindex.htm

Visscher, A. D. (2014). *Air Dispersion Modeling – Foundations and Applications*. John Wiley & Sons Inc.

WHO. (2018). *Air pollution and health: Summary*. Retrieved from http://www.who.int/airpollution/ambient/about/en/

Zannetti, P. (1990). *Air Pollution Modelling – theories, Computational Methods and Available Software*. New York: Van Nostrand Reinhold. doi:10.1007/978-1-4757-4465-1

ADDITIONAL READING

Arya, S. P. (1998). *Air Pollution Meteorology and Dispersion*. Oxford: Oxford University Press.

Lazaridis, M. (2011). First Principles of Meteorology and Air Pollution. In B. J. Alloway & J. T. Trevors (Eds.), *Environmental Pollution* (Vol. 19). New York: Springer.

Mahajan, S. P. (2009). *Air Pollution Control*. New Delhi: TERI Press.

Scorer, R. R. (2002). *Air Pollution Meteorology. Sawston*. Woodhead Publishing Limited.

Spellman, F. R. (2013). *The Handbook of Meteorology*. Maryland: Scarecrow Press Inc.

Vallero, D. A. (2008). *Fundamentals of Air Pollution* (4th ed.). Amsterdam: Elsevier Inc.

Chapter 8
The Climate Change and Air Pollution:
The Correlated Actions

A. S. Maheshwari
Anna University, India

ABSTRACT

Environmental pollution states the toxification by different forms of pollutants in the surroundings. This causes the natural system imbalance and affects the life on earth in different ways. The emissions of pollutants into air result in climate changes. The rising levels of CO2 and other air-polluting gases increase the greenhouse effect which results in temperature elevations and affects global weather patterns. Bill Gates also warned "Climate change is a terrible problem, and it absolutely needs to be solved; it deserves to be a huge priority". Though climate change and air pollution are closely linked, these are treated as separate problems. Hence, it is aimed to understand the aspects of association between climate change and air pollution in this chapter with the objectives: (i) To know the basics of climate change and air pollution, (ii) To review the active background studies on climate change as well as on air pollution, (iii) To identify the issues, controversies, problems which include the effects of global warming and greenhouse gases followed by the types of air pollutants and its harmful effects, (iv) To find the solutions and recommendations, the discussions are made on climate laws and policies, (v) To retrieve, analyze and store the climate change related information, the description about models, tools and databases are given, (vi) To have a flawless insights on the association of climate change and air pollution, (vii) To exemplify species indicators for the assessment of the effect of air pollution on climate change are also included. As Bernie Sanders said "Climate change is a planetary crisis. We've got to act, and we have to act boldly".

DOI: 10.4018/978-1-5225-7289-3.ch008

"We've been given a warning by science, and a wake-up call by nature; it is up to us now to heed them" - Bill McKibben, speaking of Hurricane Sandy on November 1st, 2012.

BASICS OF CLIMATE CHANGE AND AIR POLLUTION

Weather is the state of the atmosphere over a short period, whereas, climate is for extended period of time. The change in the statistical distribution of weather patterns that persists for several decades at least for 30 years is called climate change. The predictions about human-induced climate change over centuries are possible if it depends on the human activities. James Hansen stated "The evidence for human-made climate change is overwhelming". The more unstable climate has impacts on extreme weather events, stressed ecosystems, health issues in human, decreased agriculture production and unavailability of water supplies.

Due to these, some parts of the world, for example, sub-Saharan Africa becomes more vulnerable. In addition, the bleaching of coral reefs, decline of Arctic sea ice and occurrence of natural disasters and the mass migration of species are being the other proofs for change in climates. The U.S. Environmental Protection Agency (EPA) risk assessor, Neal Fann used computer models to forecast climate's impact on future air quality in 2030 for the US. He explained that there will be an increase in the number of warmer days and stagnant days for the easy formation of ozone.

Environmental Protection UK addresses two issues: (i) the local pollutants such as Ozone and black carbon are the major climate change drivers, (ii) the sources of both CO_2 and local air pollutants would cause severe adverse effects. The climate change can make the ozone worse particularly in cities. Hillary Clinton pointed "I personally believe climate change is a challenge of such magnitude and urgency that we need a president who will set ambitious goals". The IPCC (Intergovernmental Panel on Climate Change) rated that greenhouse gas is the third most significant after carbon dioxide and methane due to its increasing background concentrations of ozone.

The greenhouse gases are the hallmarks of climate changes like rising sea levels, more extreme weather conditions, heat-related health issues and increasing transmission of infectious diseases like Lyme. Peter Ewins from British Meteorological Office worried "Ignoring climate change will be the costliest of all possible choices, for us and our children". According to EPA 2014 study, carbon dioxide was responsible for 81% and methane 11% of the country's total greenhouse gas emissions. The carbon dioxide and methane raises the earth's temperature trapped in the atmosphere.

The hydrofluorocarbon (HFC) is thousands of times more powerful than carbon dioxide to trap heat that becomes another greenhouse gas emission. The United

Nations Economic Commission for Europe (UNECE) stated that black particles, ozone and its precursors as short-lived climate pollutants contribute for warming. Whereas reduction in greenhouse gas emissions may co-benefit lessen climate change. The climate change deviates the way of dispersion and transportation of pollutants and influences the effects of atmospheric pollutants.

The emissions of reactive nitrogen which end up in ecosystems is another example for how air pollution can enhance climate change. The significant impacts on climate change such as weather pattern changes, sea level rises, extreme weather measures and greenhouse gas emissions from human activities. If preventative measures would not be given importance, the average surface temperature of our world may exceed 3° C and some areas turn even more warmer. Moreover, the air pollutants settle into waterways and oceans initiate water pollution which causes adverse effects on the surrounding atmosphere.

BACKGROUND

Air pollution is caused by the harmful substances like sulphur dioxide, nitrogen oxide, ozone, particulate matter, carbon monoxide, carbon dioxide, volatile organic compounds and other greenhouse gases (Ritchie and Roser, 2017). John Walke, the Director of the Clean Air Project, part of the Climate and Clean Air program at Natural Resources Defense Council (NRDC) stated as energy use and production is being the main cause for much air pollution. The World Health Organization (WHO) reported that the air pollutants arise from different sources and its by-products be the greatest environmental threat to human health.

The emission of sulphur dioxide was peaked in North America in 1970, Europe in 1980 and in South America a decade later. Currently, the annual emission of sulphur dioxide in Europe and America fall down but in Asia and Africa it rises. Patrick Kinney, Professor of Urban Health and Sustainability, Boston University School of Public Health said "Air pollution and climate change are connected in both directions". Since moisture is drawn out of the soil into the air by higher temperature, the soil and the vegetation dries out that becomes more vulnerable to wildfires (Miller, 2017).

The air pollution-related health impacts highly depend on the climate change scenario and the projections of future emission of air pollutant are with relatively high uncertainty. The regional assessments are required in the places with higher air pollution concentration (Orru et al., 2017). David Doniger, Director of NRDC's Climate and Clean Air program agreed HFC phase-down will avoid the equivalent of more than 80 billion tons of CO over the next 35 years (Espanol, 2016). To restore and protect areas endangered by non-point source pollutants, millions of dollars per year have been spent.

In addition, the irons circulated across the ocean affect ecosystems thousands of kilometers away (Fears, 2016). National Oceanic and Atmospheric Administration (NOAA) works with several agencies to develop ways to control nonpoint source pollution which result from natural and human activities (Vikas and Dwarakish, 2015). The climate changes improve air quality in some regions like northern Europe and Poland through the changes in long-range atmospheric air pollution transportation (Orru et al., 2013; Tainio et al., 2013).

The climate scientist, Amato Evan at the University of Virginia said that human activity has a huge influence on climate processes around the Atlantic Ocean. Whereas another climate scientist, Ben Booth at Britain's Met Office Hadley Center said that aerosols create clouds. The new pollution control measures reduce the amount of aerosols, but the ocean loses its sunblock (Joyce, 2012). The frequency and intensity of forest fires will also effect air pollution and climate change. The climate change may alter energy consumption and the emissions from combustion (Pleijel, 2009).

According to the National Academic Reports in Ocean Science Series, the pollutants originate far inland and transported via rivers or air to the ocean. There is a link between climate, ecosystems and biogeochemical cycle, so, the coupling among these different systems directly affects air quality. Hence, for better understanding of these different systems behavior, the developmental science of coupling is must for the better future of earth (Brasseur, 2009).

ISSUES, CONTROVERSIES, PROBLEMS, SOLUTIONS

The encyclopedia of the atmospheric environment explored the air pollution and climate change issues. The acid rain related issues, along with air pollutants, trans-boundary pollutants and their chemicals, its effects on buildings, freshwaters, trees, soil, health and control measures need to be considered. Generally, air quality depends on the types of air pollution and its sources. The impacts of air pollution on health and the environment due to poor air quality being an another issue.

In atmosphere science, the inclusion of information on gases, atmospheric layers and the ozone hole becomes necessitate. For instance, the climate changes from savannah to tropical, the information on wind patterns and the interactions between land and weather around the world is must. Then, how measurements are made on climate changes over millions of years need to be considered. Moreover, global warming due to human influences and the impacts for the future also to be known. James Hansen said "Global warming isn't a prediction. It is happening".

Effects of Global Warming

The global warming is a serious environmental issues caused by natural (depends on the intensity of sunlight) and human influences (industrial revolution, mining, deforestation) that leads to climate change refers the observed rise in the average temperature of the earth's climate system. The effects include heat waves, droughts, heavy rainfall with floods, heavy snowfall, ocean acidification, species extinctions due to shifting temperature regimes.

As per the readings recorded by the National Science Academies of the Major Industrialized Nations, the global surface temperature rises further 0.3 °C to 1.7 °C in the lowest emissions scenario and 2.6 °C to 4.8 °C in the highest emissions scenario. The future climate change and its impacts differ from region to region. Andrew Dlugolecki, General Insurance Development believed "In the year 2065, on current trends, damage from climate change will exceed global GDP".

"Global warming is too serious for the world any longer to ignore its danger or split into opposing factions on it", this quote was collected from Tony Blair speech in Sept. 27, 2005. Whereas "The warnings about global warming have been extremely clear for a long time. We are facing a global climate crisis. It is deepening. We are entering a period of consequences" was from Al Gore, speech at National Sierra Club Convention in Sept. 9, 2005. The solutions for global warming and its effects will be achieved by using alternative fuels instead of gasoline, reducing garbage by recycling, avoiding open burning, discouraging deforestation.

The solutions for deforestation are by conserving the remainder natural resources and restoring degraded areas by replanting native tree species. The necessary efforts need to be taken to prevent the loss of biodiversity and protecting against poaching and wildlife trade. The solutions for soil degradation are adopting suitable soil conservation and existence of restoration techniques. However, only solution to control population is by women empowerment to control their own reproduction.

Effects of Greenhouse Gases

The Greenhouse gases absorb and emit infrared radiation in the wavelength range emitted by Earth. The most abundant greenhouse gases in earth's atmosphere are: water vapor, carbon dioxide, methane, nitrous oxide, ozone, chlorofluorocarbons, hydrofluorocarbons. Atmospheric concentrations are determined from sources (gas emissions by human activities and natural systems) and sinks (gas removal from the atmosphere).

The Intergovernmental Panel on Climate Change (IPCC) is the United Nations body for assessing the science related to climate change. It prepares Comprehensive Assessment Reports about knowledge on climate change, its causes, potential impacts

Figure 1. The correlated actions of climate change and air pollution

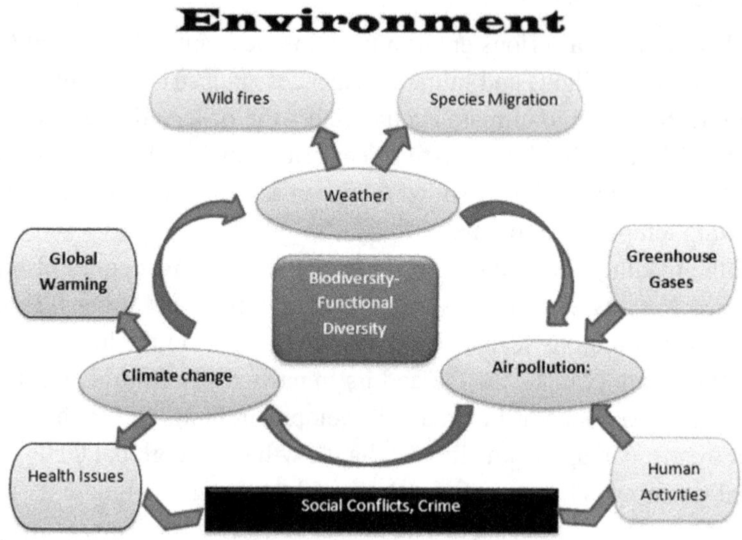

and response options. It also produces special Reports about the assessment of specific issues and methodology reports that provides practical guidelines for the preparation of greenhouse gas inventories.

The Airborne Fraction (AF) is the proportion of an emission remaining in the atmosphere after a specified time. The annual airborne fraction increased at a rate of $0.25 \pm 0.21\%$ per year in 1959–2006. An increasing AF and the related changes characterize a carbon cycle that is generating stronger-than-expected and sooner-than-expected climate forcing (Canadell et al., 2007).

The impacts of ozone depletion, ozone hole on human health are the other major issues. The sustainability on energy and renewables includes information on biodiversity, the different types of energy with fossil fuels, hydro-electric power, solar power, nuclear power, geothermal, wind power, biofuels and the importance of planning and conservation are also to be taken care of. In addition, a complete set of resources on weather, types of precipitation, clouds, weather fronts and weather measurements are similarly important.

In 2017, ScienceDaily reported a study from University of North Carolina at Chapel Hill that reveals the unaddressed future climate change by air pollution alone will cause roughly 60,000 deaths globally in 2030 and 260,000 deaths in 2100. In DW AKADEMIE 2016, Zimmermann reported the problems and solutions for five biggest environmental problems related with climate change and air pollution.

Effects of Air Pollutants

The air pollutants are mainly caused by burning coal, oil, gas and wood due to carbon overloading. The effect of carbon overloading will be reduced by replacing fossil fuels with renewable energy, reforestation, reducing emissions from agriculture and changing industrial processes. As per World Health Organization (WHO), 2012 report, one in nine deaths were caused by the presence of carcinogens and other poisons in air. About 15% of Tropical forests of the planet land are down to 6% or 7%, the remainder was degraded by logging or burning.

Species extinction on land is due to hunting wild animals for meat, ivory, medicinal purposes. At sea, fishing boats equipped with bottom-trawling may be the reason to clean out entire fish populations. Soils degradation is due to overgrazing, monoculture planting, erosion, soil compaction, overexposure to pollutants and land-use conversion etc. Human overpopulation increases from 1.6 billion to about 7.5 billion which may reach to 10 billion by 2050 worldwide. At the same time, increasing population put greater pressure on essential natural resources such as water.

SOLUTIONS AND RECOMMENDATIONS

The WHO involves tracking rapidly evolving threats, quantifying the harm to health and sounding the alarm. It also raises political awareness and advices on the best protective strategies for safeguarding public health. The protective strategies require collaboration with multiple non-health sectors.

Climate Laws and Policies

The shading of clean renewable energy and energy efficiency policies, clean air implementation and enforcement, and clean transportation innovation reflects the number of relevant policies or pieces of legislation in each country. Throughout

Table 1. The common air pollutants at elevated levels and its impacts on health

Air Pollutant	Harmful Effects
Nitrogen Dioxide, Sulphur Dioxide, Ozone	Irritate the airways of the lungs which causes lung diseases.
Fine Particles	Inflammation, heart and lung diseases.
Carbon Monoxide	Significant reduction of oxygen supply to heart that causes heart diseases.

worldwide, there are more than 1,500 climate laws and policies on environment and climate change, Climate Change Legislation, Local Climate Change Law, Climate Change Action Plan which are to be strictly given importance.

THE MODELS/ TOOLS/ DATABASES IN CLIMATE CHANGE RESEARCH

Many models, tools and databases have been developed to assist in climate change related research. Given below are some of these used for such purposes.

Models

- *BASINS-CAT:* It is a Climate Change and Watersheds Model. It provides a flexible set of capabilities for exploring the potential effects of climate change on streamflow and water quality using different watershed models in BASINS. The tools that are integrated into the BASINS system allow users to create climate change scenarios by applying change factors to modify historical daily/hourly weather data. Such data are available as meteorological input to different BASINS watershed models.
- *CMAQ:* The Community Multiscale Air Quality (CMAQ) model is a numerical air quality model which relies on scientific first principles. It predicts the concentration of airborne gases, particles and the deposition of these pollutants back to Earth's surface. Because, it contains information about the emissions and properties of compounds and classes of compounds. It also informs the users about the chemical composition of a mixture of pollutants that is particularly useful when measurements only give insights into aggregate details like total particulate mass.

It provides fast, technically sound estimates of ozone, particulates, toxics, and acid deposition. It is designed to meet the needs of the scientific community and concerned community leaders by combining current knowledge in atmospheric science and air quality modeling, multi-processor computing techniques and an open-source framework into a single modeling system.

- *GLIMPSE:* GLIMPSE is a decision support modeling tool being developed by EPA that will assist states with energy and environmental planning through the year 2050. The users can explore the impacts of energy technologies and policies on the environment. The additional analytical technologies are included for electric and hydrogen fuel cell vehicles, wind and solar power, and carbon capture and sequestration. The users can also specify energy, air quality, and water use goals within GLIMPSE, then to identify cost-effective strategies for meeting those goals.

It is built on the Global Change Assessment Model (GCAM), developed by the Pacific Northwest National Laboratory (PNNL). The model simulates the interactions between human systems (such as economy, energy, agriculture, land use, and buildings) and earth systems (such as water and carbon cycles). PNNL and EPA researchers are working together to improve GCAM's ability to support environmental analyses.

- *WEPPCAT:* Water Erosion Prediction Project Climate Assessment Tool: It is a web-based erosion simulation tool that allows for the assessment of changes in erosion rates as a consequence of user-defined climate change scenarios. It is based on the USDA-ARS Water Erosion Prediction Project (WEPP) model. WEPP has a number of customized interfaces developed for common applications such as roads, managed forests, forests following wildfire, and rangelands. This tool has the capability of taking into account all of the erosion-affecting processes. WEPP was created for agricultural modeling, but it is now widely used in forestry, fisheries, rangeland and mining studies and projects.

Tools

- *ICLUS:* EPA's Integrated Climate and Land-Use Scenarios (ICLUS) project explores future changes in human population, housing density, and impervious surface for the United States. It is consistent with peer-reviewed storylines of population growth and economic development that are useful for the climate change impacts community. The articles, reports, Web Viewer and ArcGIS and map-based tools available here help to visualize and understand these potential changes.
- *ONA:* Optimized Noise-Reduction Algorithm (ONA) Program Improves Black Carbon Particle Measurements Using Aethalometers. It reduces noise in real-time black carbon data obtained using Aethalometers. The Aethalometers optically measure the concentration of light absorbing or "black" particles that accumulate on a filter as air flows through it. Such particles are produced by incomplete fossil fuel, biofuel and biomass combustion whereas under polluted conditions, they appear as smoke or haze.

Databases

- *EPAUS9R:* It is a 9-region Energy Systems Database for use with the Market Allocation (MARKAL) Model which helps the decision makers understand the complex system interactions related to energy use. MARKAL is an energy system optimization model of the U.S. energy system used in over 35 countries and fully transparent, publicly available database. It gives thorough representation of greenhouse gas and air pollutant emissions. By using MARKAL, the EPA (United States Environmental Protection Agency) researchers analyse the environmental impacts of potential changes in the way the U.S. produces and uses energy.

EPAUS9r represents energy supply, technology, and demand throughout the major sectors of the U.S. energy system. The sectors include commercial, industrial, residential, transportation, and electricity generation. Energy resources in this database include coal, natural gas, petroleum, and renewables (e.g., wind, solar, biomass, geothermal, and hydropower).

- *Freshwater Biological Traits Database:* It currently contains nearly 4000 North American macro-invertebrate taxa which include habitat, life history, mobility, morphology and ecological trait data. It was compiled for a project on climate change effects on river and stream ecosystems. The data were collected from multiple sources such as published or well-documented by trustworthy sources, appropriate for the regions being studied.

ASSOCIATION OF CLIMATE CHANGE AND AIR POLLUTION

Air pollution and climate change are closely related whereas the CO_2 emissions, the extraction and burning of fossil fuels are the major sources of air pollutants besides key drivers of climate change. Some pollutants are warming and others cooling the earth that contribute to climate change by affecting sunlight reflection or absorption. Such Short-Lived Climate-forcing Pollutants (SLCPs) include methane, black carbon, ground-level ozone and sulfate aerosols which affect ecosystems including agriculture adversely and responsible for roughly half of current global warming.

It was documented that the complex environmental interactions contribute to the recovery of red spruce forest ecosystems from pervasive acidic air pollution beginning in 1989, about 15 years after acidic pollutants started to decline in the United States (Mathias and Thomas, 2018). The extreme climate variability in the past few years could have an equivalent impact as a consequence of large emission reduction during winter air pollution in the North China Plain (Zhao et al., 2018).

Approximately 5%–9% of exacerbated air-pollution-related mortality will be due to increase in power sector emissions from heat-driven building electricity demand. This study highlights the need for cleaner energy sources, energy efficiency and energy conservation to mitigate climate change (Abel et al., 2018). It was argued that climate change will increase strain, reduce social control, weaken social support, foster beliefs favorable to crime, contribute to traits conducive to crime, increase certain opportunities for crime and create social conflict.

Even though neglected by criminologists, there is a good reason to believe that climate change will become one of the major forces driving crime as the century progresses (Agnew, 2017). The worsening meteorological conditions are the main reasons that unusually increase air pollutant concentration. The emission control measures taken during this period contribute to mitigate the air pollution (~9%) in the regions (Liu et al., 2017).

In future, attention must be given to functional diversity measures for air pollution and climate change studies. This will establish the link between biodiversity and the provision of ecosystem services in Mediterranean ecosystems (Ochoa-Hueso et al., 2017). During the recent 15 years, the increased prevalence of Chronic actinic dermatitis is correlated with sunshine duration and ozone concentration and inversely related to air pollutant concentration (Kim et al., 2017).

It has been suggested that reducing the occurrence of "airpocalypse" events improve social welfare whereas the facemasks can only partially reduce the negative health effects of air pollution (Zhang, and Mu, 2017). The targets for interventions and recommendations for pollution reduction was highlighted through industrial upgrading, vehicle and fuel renovation, improvements in public transportation, lowering of personal exposure, mitigation of the direct effects of air pollution through healthy city development, intervention at population-based level and improvement in air quality (Guan et al., 2016).

Among the all-possible climate factors, variability of the autumn Arctic sea ice extent, local precipitation and surface wind during winter would be the most influential to the haze pollution change (Wang and Chen, 2016). The maintenance of potentially stress-tolerant genotypes by protecting existing biodiversity is crucial to promote rapid adaptation to present-day climate change ecosystems including coral reefs (Hume et al., 2016). Around 30% of mammal species have potential spread rates slower than the global mean velocity of climate change (Santini et al., 2016).

Increasing the temperature and decreasing the relative humidity promote soil drying and increases dust emissions, aerosol optical depth (AOD) for which the increase may be due to climate change (Klingmüller et al., 2016). The urbanization and high emissions from vehicle are correlated with increase in the frequency of pollen-induced respiratory allergy of people in urban and rural areas. Apart from these, measures of mitigation are needed for reducing future impacts of climate change on our planet, then, addition and adaptation to the impacts of future climate variability is also required (D'Amato et al., 2016).

Effect of Pollen Allergy Among Local Residents

Pollen allergy is used to study the relationship between air pollution and respiratory allergic diseases, such as rhinitis and bronchial asthma. Climatic factors such as temperature, wind speed, humidity, thunderstorms, etc. can affect both biological and chemical components interaction (D'Amato et al., 2016). The airports are the largest sources of air pollution in US and they experience large changes in daily air pollution and its emissions depending on the amount of time planes spend idling on the tarmac.

The idiosyncratic variation in daily airplane taxi time affects the health of local residents, largely driven by increased levels of carbon monoxide exposure. The one standard deviation increase in daily pollution levels leads to an additional $540 thousand in hospitalization costs for respiratory and heart-related admissions of 6 million individuals living within 10 km of the airports in California. It was suggested that there may be sizable morbidity benefits from lowering the existing CO standard (Schlenker and Walker, 2016).

The primary heating with a low-polluting fuel and more frequent kitchen ventilation could reduce personal PM 2.5 exposures in winter. Whereas the gaseous fuel or electricity for cooking and reduces exposure to outdoor PM 2.5 that will show its impacts in summer (Nia et al., 2016). Though the sources, abundance, composition and the effects of biological aerosols are not yet well characterized, the atmospheric microbiome constitute a large gap in the scientific understanding of the interaction and co-evolution of life and climate in the earth system (Fröhlich-Nowoiskya et al., 2016).

Maternal exposure to traffic-related air pollutant, NO_2 during pregnancy is associated with an increased risk of developing asthma, rhinitis, and eczema in children. The childhood allergic diseases originate in fetal life and are triggered by traffic-related air pollution in sensitive trimesters (Deng et al., 2016). The long-term ambient O_3 contributes to risk of respiratory and circulatory mortality. The substantial health and environmental benefits will be achieved by implementing further measures to control O_3 concentration (Turner et al., 2016).

Projections of future climate need to be considered these uncertainties for estimating the atmospheric lifetime and concentrations of key greenhouse gases.

The air pollutants on a local scale when •OH radicals respond rapidly to changes in UV radiation and on global scales, models differ in their predictions by nearly a factor of two. Though no new negative environmental effects of substitutes for ozone depleting substances and breakdown of products have been identified, some substitutes will contribute to global climate change if concentration rises above current levels (Madronich et al., 2015).

Global warming is expected to affect the start, duration and intensity of the pollen season on one hand; and the rate of asthma exacerbations due to air pollution, respiratory infections, cold air inhalation and other conditions on the other hand (D'Amato et al. 2015). As per growing evidences, the global climate change affects the indoor air quality and the pollutant exposure level of occupants. The findings indicated that exposure to air pollution induces sperm aneuploidy (Jurewicz et al., 2015). Additional studies are necessary for the understanding of how climate change will affect the health of residents in homes (Leung, 2015).

Interactive Effects

The projected changes in wildfire activity increase daily summertime surface ozone above the 95th percentile by 1 ppbv in the northwestern US, 5 ppbv in the high latitudes of Canada and 15 ppbv in Alaska suggest a greater frequency of pollution episodes in the future atmosphere (Yue et al., 2015). It was suggested that improving basic education, climate literacy and public understanding of the local dimensions of climate change are vital to public engagement and support for climate action (Lee et al., 2015).

The climatic and geographic variables do not influence the color morph proportions in *P. cinereus* populations. The bio-indicator paradigms are challenged by new ecological research and representative databases (Moore and Ouellet, 2015). It has been demonstrated that the mortality risks of air pollution exposure differ by weather type, with increased accuracy obtained when accounting for interactive effects through adjustment for dependent pollutants using a DLNM (Distributed Lag Nonlinear Modeling) (Vanos et al., 2015).

The tropospheric ozone is a globally effective agent of climate change concerns forest productivity particularly at hot spots of air pollution in the southern hemisphere, prognosticated for the upcoming decades (Matysseket al., 2014). As per recent evidences, a variety of air pollutants, such as environmental tobacco smoke, volatile organic compounds, formaldehyde, toluene, nitrogen dioxide and particulate matter act as risk factors for the development of Atopic Dermatitis. Such air pollutants may induce oxidative stress in the skin, leading to skin barrier dysfunction or immune dysregulation (Ahn, 2014).

In total, 61% of the respiratory-related mortality relative risk estimates are significantly higher than for cardiovascular-related mortality. The combined effect of weather and air pollution is greatest when tropical-type weather is present in the spring or summer (Vanos et al., 2014). Gains from addressing regional air pollution could counter expected future yield losses from direct climate change effects of LLGHGs (Long-Lived Green House Gases) (Burney and Ramanathan, 2014).

Ecological and Socioeconomic Consequences

By 2050, warming reduces the global crop production by >10%. It was found that depending on region some crops are primarily sensitive to either ozone or heat alone which provides a measure of relative benefits of climate adaptation versus ozone regulation for food security in different regions (Tai et al., 2014). The tourism industry faces bigger challenges of reduce investment, less resources, and minor importance from the government agencies because of the climate change and air pollution, putting them in a dismal state (Sajjad et al., 2014).

The projected warming for the coming decades raises even more concern about its ecological and socioeconomic consequences (Walther et al., 2012). The thermophilization of mountain plants mirror the degree of recent warming and the observed transformation shows the progressive decline of cold mountain habitats and their biota (Gottfried et al., 2012). When assessing climate change, the direct measures of extinction risk and measures of change in habitat area have impacts on biodiversity. In addition, direct estimation of extinction risk incorporates demographic and ecosystem processes that potentially influence species vulnerability to extinction due to climate change (Fordham et al., 2012).

INDICATOR SPECIES

Indicator species such as animals, plants, or microorganism are used to monitor changes in environment which show the status of an environmental condition and monitor pollution or climate change. These can be used as an "early warning system" by biologists and conservation managers (McDonough et al., 2012). The common cuckoo is an intercontinental bio-indicator of hotspots of bird richness under climate change scenarios and in areas where the species co-occur with other cuckoo species (Morelli et al., 2017).

The investigation on global patterns of biodiversity showed that climate change would lead to local extinction in the sub-polar regions, the tropics and semi-enclosed seas. In addition, the species invasion is to be the most intense in the Arctic and

Southern Ocean implying that ecological disturbances potentially disrupt ecosystem services (Cheung et al., 2009). The climate change will affect particulate matter (PM) concentrations in polluted environments by $\pm 0.1 - 1.0\ \mu g\ m^{-3}$ over the coming decades. Wildfires fueled by climate change would become an important PM source (Jacob and Winner, 2009).

Studies were also carried out European shelf seas, which are warming faster than the adjacent land masses. It explores the deepening of bottom-dwelling (demersal) fish assemblage in response to temperature could be used as a biotic indicator (Dulvy et al., 2008). The tropical insects are relatively sensitive to temperature change and species at higher latitudes have broader thermal tolerance. In the absence of migration and adaptation, the greatest extinction risks from global warming are in the tropics where biological diversity is more (Deutsch et al., 2008).

Phenology is the science of natural recurring events and the recorded dates provide a high-temporal resolution of ongoing changes that become preferred indicators. The species' phenology is responsive to temperature of the preceding months and the observed pattern in spring efficiently matches measured national warming across 19 European countries (Menzel et al., 2006). Some 'southern' species including *Osilinus lineatus* da Costa and *Gibbula umbilicalis* da Costa have undergone north and north-eastern range extensions. Increased populations and decreased adult size are synchronous throughout Britain for which climate is responsible. The intertidal species are used as the climate change indicators (Mieszkowska et al., 2006).

Species-energy theory indicates the recent climate warming increases in species richness, in cool and species-poor parts of the Northern Hemisphere. It may take decades or centuries before the species richness and composition of biological communities adjust to the current climate (Menéndez et al., 2006). As per models and analyses of paleo ecological records, tree populations are capable of rapid migration when climate warms. The past migration rates were slower than the rates to track 21st century warming (McLachlan et al., 2005).

The climate change reflects the changes in the distribution and abundance of species and communities. Migrant pests respond more quickly to climate change than plants that may colonize newly available crops/ habitats. Climatic phenomena, ecosystem processes and human activities are interactive, interdependent and making long-term predictions extremely tenuous (Cannon, 2004).

The high mountain vegetation is generally vulnerable to climate change. The Alps face drastic changes due to human-induced climate warming. The glaciers of the Alps respond to the temperature rise of about 1-2°C and it can be used as a sensitive "ecological indicator" for climate change effects. The risks of climate-induced upward migration processes of plants impact on the stability of high mountain ecosystems (Pauli et al., 2003).

The complexity of the natural system will present fundamental limits to predictive modelling, the bio-climate envelope approach provides a useful first approximation to the potentially dramatic impact of climate change on biodiversity. Whereas, the hierarchical modelling framework is proposed through limitations within a broader scale-dependent context (Pearson and Dawson, 2003).

As per available reports, the forest plant species dispersal limitation prevents net colonization at northern range boundaries required for range conservation (Honnay et al., 2002). Migrants and hibernators experience problems as a consequence of these changes in phenology. An earlier formation of permanent snowpack and longer period snow cover have implications for hibernating species (Inouye et al., 2000).

The potential effects analysis showed that roughly 30 species expand their range at least 10%, additional 30 species decrease by at least 10% following equilibrium after climate change. Half of the species assessed showed that the ecological optima shift at least 100 km to the north. The actual species redistributions will be controlled by migration rates through fragmented landscapes (Iverson and Prasad, 1998).

Meiofauna With Climate Change

Studies are needed to address both basic and applied research that bridge the spatial and temporal scales connecting local emissions vs air pollution and local weather vs global atmospheric chemistry and climate (Baklanov et al., 2016). Meiobenthos also called as meiofauna are small benthic invertebrates that live in both marine and fresh water environments.

The reviews on meiofauna with climate change and anthropogenic impacts concluded that foraminifers, nematodes, copepods and ostracods groups are valuable for predicting global changes. The multi-stressor experiments are more informative than single stressor ones, revealing complex ecological and biological interactions. The global change has negative effect on meiofauna with important consequences on benthic food webs. Some meiofaunal species are favored by the extreme conditions due to physiological adaptations (Zeppilli et al., 2015).

Lichen Communities and Air Pollution

The cleanest areas may have air pollution above a critical load to protect lichen communities and these indices help to track lichen responses using other data (Will-Wolf et al., 2015). The map of an independent lichen community-based bioclimatic index is strongly related to gradients in moisture availability and temperature variability. Lichen communities in the driest climates are more eutrophic than those in wetter climates at the same levels of nitrogen air pollution (Root et al., 2015).

CONCLUSION

The study about the exacerbated air-pollution-related mortality highlights the need for cleaner energy sources to mitigate climate change. The global climate change affects the indoor air quality and the pollutant exposure level of occupants induces sperm aneuploidy. Since, the air pollution-related health impacts highly depend on the climate change scenario, the regional assessments are required in the places that are highly air polluted. The absence of migration and adaptation of species, the climate change indicators reveal the greatest extinction risks. Moreover, as per the study reports, the climate change will become one of the major forces for driving crimes as the century progresses. As Bill McKibben said, the science and the impacts of climate change has given a warning about the aggravated correlation of air pollutants. It is our responsibility to obey the wake-up call by nature and take necessary actions for the betterment of life on earth.

REFERENCES

Abel, D. W., Holloway, T., Harkey, M., Meier, P., Ah, D., & Limaye, V. S. (2018). Air-quality-related health impacts from climate change and from adaptation of cooling demand for buildings in the eastern United States: An interdisciplinary modeling study. *PLoS Medicine*, *15*(7), e1002599. doi:10.1371/journal.pmed.1002599 PMID:29969461

Agnew, R. (2017). Dire forecast: A theoretical model of the impact of climate change on crime. *Transnational Environmental Crime*, *5*, 22.

Ahn, K. (2014). The role of air pollutants in atopic dermatitis. *The Journal of Allergy and Clinical Immunology*, *134*(5), 993–999. doi:10.1016/j.jaci.2014.09.023 PMID:25439225

Australian Academy of Science. (2018). Retrieved from https://www.science.org.au/learning/general-audience/science-booklets-0/science-climate change/1-what-climate-change

Zhang, J., & Mu, Q. (2017). (in press). Air pollution and defensive expenditures: Evidence from particulate-filtering facemasks. *Journal of Environmental Economics and Management*.

Baklanov, A., Molina, L. T., & Gauss, M. (2016). Megacities, air quality and climate. *Atmospheric Environment*, *126*, 235–249. doi:10.1016/j.atmosenv.2015.11.059

Benjamin, C. C., Hume, C. R., Voolstra, C. A., D'Angelo, C., Burt, J. A., Eyal, G., ... Wiedenmann, J. (2016). Ancestral genetic diversity associated with the rapid spread of stress-tolerant coral symbionts in response to Holocene climate change. *Proceedings of the National Academy of Sciences of the United States of America*, *113*(16), 4416–4421. doi:10.1073/pnas.1601910113 PMID:27044109

Brasseur, G. P. (2009). Implications of climate change for air quality. *World Meteorological Organization Bulletin*, *58*(1).

Burney, J., & Ramanathan, V. (2014). Air pollution impacts on Indian agriculture. *Proceedings of the National Academy of Sciences*, 172-75.

Canadell, J. G., Le Quere, C., Raupach, M. R., Field, C. B., Buitenhuis, E. T., Ciais, P., ... Marland, G. (2007). Contributions to accelerating atmospheric CO2 growth from economic activity, carbon intensity, and efficiency of natural sinks. *Proceedings of the National Academy of Sciences of the United States of America*, *104*(47), 18866–18870. doi:10.1073/pnas.0702737104 PMID:17962418

Cannon, R. J. C. (2004). The implications of predicted climate change for insect pests in the UK, with emphasis on non-indigenous species. *Global Change Biology*, *4*(7), 785–796. doi:10.1046/j.1365-2486.1998.00190.x

Carter, K. E., Schauer, J. J., Ezzati, M., Zhang, Y., Niu, H., Lai, A. M., ... Baumgartner, J. (2016). Seasonal variation in outdoor, indoor, and personal air pollution exposures of women using wood stoves in the Tibetan Plateau: Baseline assessment for an energy intervention study. *Environment International*, *94*, 449–457. doi:10.1016/j.envint.2016.05.029 PMID:27316628

Cheung, W. W. L., Lam, V. W. Y., Sarmiento, J. L., Kearney, K., Watson, R., & Pauly, D. (2009). Projecting global marine biodiversity impacts under climate change scenarios. *Fish and Fisheries*, *10*(3), 235–251. doi:10.1111/j.1467-2979.2008.00315.x

D'Amato, G., Holgate, S. T., & Pawankar, R. (2015). Meteorological conditions, climate change, new emerging factors, and asthma and related allergic disorders. A statement of the World Allergy Organization. *The World Allergy Organization Journal*, *8*(1), 25. PMID:26207160

D'Amato, G., Pawankar, R., Vitale, C., Lanza, M., Molino, A., Stanziola, A., ... D'Amato, M. (2016). Climate change and air pollution: Effects on respiratory allergy. *Allergy, Asthma & Immunology Research*, *8*(5), 391–395. doi:10.4168/aair.2016.8.5.391 PMID:27334776

D'Amato, G., Vitale, C., Lanza, M., Molino, A., & D'Amato, M. (2016). Climate change, air pollution, and allergic respiratory diseases: An update, 1414 total citations on Dimensions. *Current Opinion in Allergy and Clinical Immunology*, *16*(5), 434–440. PMID:27518837

Deng, Q., Lu, C., Li, Y., Sundell, J., & Norbäck, D. (2016). Exposure to outdoor air pollution during trimesters of pregnancy and childhood asthma, allergic rhinitis, and eczema. *Environmental Research*, *150*, 119–127. doi:10.1016/j.envres.2016.05.050 PMID:27281689

Deutsch, C. A., Tewksbury, J. J., Huey, R. B., Sheldon, K. S., Ghalambor, C. K., Haak, D. C., & Martin, P. R. (2008). Impacts of climate warming on terrestrial ectotherms across latitude. *Proceedings of the National Academy of Sciences of the United States of America*, *105*(18), 6668–6672. doi:10.1073/pnas.0709472105 PMID:18458348

Drupal - IASS Potsdam. (n.d.). *Air Pollution and Climate Change*. Retrieved from https://www.iass-potsdam.de/en/output/dossiers/air-pollution-and-climate-change

Dulvy, N. K., Rogers, S. I., Jennings, S., Stelzenmüller, V., Dye, S. R., & Skjoldal, H. R. (2008). Climate change and deepening of the North Sea fish assemblage: A biotic indicator of warming seas. *Journal of Applied Ecology*, *45*(4), 1029–1039. doi:10.1111/j.1365-2664.2008.01488.x

Espanol. (2016). *Air pollution: Everything you need to know*. New York: Natural Resources Defense Council.

Fears, D. (2016). Air pollution doesn't just make breathing harder for humans, fish are choking too. *Chicago Tribune*

Fordham, D. A., & Akçakaya, H. R. (2012). Plant extinction risk under climate change: Are forecast range shifts alone a good indicator of species vulnerability to global warming? *Global Change Biology*, *18*(4), 1357–1371. doi:10.1111/j.1365-2486.2011.02614.x

Fröhlich-Nowoiskya, J., & Kamp, C. J. (2016). Bioaerosols in the Earth system: Climate, health, and ecosystem interactions. *Atmospheric Research*, *182*, 346–376. doi:10.1016/j.atmosres.2016.07.018

Gottfried, M., Pauli, H., Futschik, A., Akhalkatsi, M., Barančok, P., Benito Alonso, J. L., ... Grabherr, G. (2012). Continent-wide response of mountain vegetation to climate change. *Nature Climate Change*, *2*(2), 111–115. doi:10.1038/nclimate1329

Guan, W., Zheng, X., Chung, K. F., & Zhong, N. (2016). Impact of air pollution on the burden of chronic respiratory diseases in China: Time for urgent action. *Lancet, 388*(10054), 1939–1951. doi:10.1016/S0140-6736(16)31597-5 PMID:27751401

Honnay, O., Verheyen, K., Butaye, J., Jacquemyn, H., Bossuyt, B., & Hermy, M. (2002). Possible effects of habitat fragmentation and climate change on the range of forest plant species. *Ecology Letters*, *5*(4), 525–530. doi:10.1046/j.1461-0248.2002.00346.x

Inouye, D. W., Barr, B., Armitage, K. B., & Inouye, B. D. (2000). Climate change is affecting altitudinal migrants and hibernating species. *Proceedings of the National Academy of Sciences of the United States of America*, *97*(4), 1630–1633. doi:10.1073/pnas.97.4.1630 PMID:10677510

Iverson, L. R., & Prasad, A. M. (1998). Predicting abundance of 80 tree species following climate change in the eastern United States. *Ecological Monographs*, *68*(4), 465–485. doi:10.1890/0012-9615(1998)068[0465:PAOTSF]2.0.CO;2

Joyce, C. (2012). *Pollution playing a major role in sea temperatures*. National Public Radio.

Jurewicz, J., Radwan, M., Sobala, W., Polańska, K., Radwan, P., Jakubowski, L., ... Hanke, W. (2015). The relationship between exposure to air pollution and sperm disomy. *Environmental and Molecular Mutagenesis*, *56*(1), 50–59. doi:10.1002/em.21883 PMID:24989325

Kim, H., Cho, Y., Seo, J., Kim, T., Song, K., & Kim, K. (2017). *The increased prevalence of chronic actinic dermatitis in accordance with climate change and air pollution during the recent 15 years*. Academic Press.

Klingmüller, K., Pozzer, A., Metzger, S., Stenchikov, G. L., & Lelieveld, J. (2016). Aerosol optical depth trend over the Middle East. *Atmospheric Chemistry and Physics*, *16*(8), 5063–5073. doi:10.5194/acp-16-5063-2016

Lee, T. M., Markowitz, E. M., Howe, P. D., Ko, C., & Leiserowitz, A. A. (2015). Predictors of public climate change awareness and risk perception around the world. *Nature Climate Change*, *5*(11), 1014–1020. doi:10.1038/nclimate2728

Leung, D. Y. C. (2015). Outdoor-indoor air pollution in urban environment: Challenges and opportunity. *Frontiers in Environmental Science*, 2.

Liu, T., Gong, S., He, J., Yu, M., Wang, Q., Li, H., ... Zhao, Q. (2017). Attributions of meteorological and emission factors to the 2015 winter severe haze pollution episodes in China's Jing-Jin-Ji area. *Atmospheric Chemistry and Physics, 17*(4), 2971–2980. doi:10.5194/acp-17-2971-2017

Madronich, S., Shao, M., Wilson, S. R., Solomon, K. R., Longstreth, J. D., & Tang, X. Y. (2015). Changes in air quality and tropospheric composition due to depletion of stratospheric ozone and interactions with changing climate: Implications for human and environmental health. *Photochemical & Photobiological Sciences, 14*(1), 149–169. doi:10.1039/C4PP90037E PMID:25380416

Mathias, J. M., & Thomas, R. B. (2018). Disentangling the effects of acidic air pollution, atmospheric CO2, and climate change on recent growth of red spruce trees in the Central Appalachian Mountains. *Global Change Biology, 24*(9), 3938–3953. doi:10.1111/gcb.14273 PMID:29781219

Matyssek, R., Kozovits, A.R., Schnitzler, J. P., Pretzsch, H., Dieler, J., & Wieser G. (2014). Trees in a Changing Environment: Forest Trees Under Air Pollution as a Factor of Climate Change. *Plant Ecophysiology*, 9.

McDonough, C., Jaffe, D., Watzin, M., & McGinley, M. (2018). Indicator species. In C. J. Cleveland (Ed.), *Encyclopedia of Earth*. Washington, DC: Environmental Information Coalition, National Council for Science and the Environment.

McLachlan, J. S., Clark, J. S., & Manos, P. S. (2005). Molecular indicators of tree migration capacity under rapid climate change. *Ecology, 86*(8), 2088–2098. doi:10.1890/04-1036

Menéndez, R., Megías, A. G., Hill, J. K., Braschler, B., Willis, S. G., Collingham, Y., ... Thomas, C. D. (2006). Species richness changes lag behind climate change. *Proceedings. Biological Sciences, 273*(1593), 1465–1470. doi:10.1098/rspb.2006.3484 PMID:16777739

Menzel, A., & Sparks, T. H. (2006). European phenological response to climate change matches the warming pattern. *Global Climate Change, 12*(10), 1969-1976.

Mieszkowska, N. (2006). Changes in the range of some common rocky shore species in Britain - A response to climate change? Marine Biodiversity. *Developments in Hydrobiology*, 241-251.

Miller, S. G. (2017). *Here's how climate change can cause more air pollution (All by itself)*. LiveScience.

Moore, J., & Ouellet, M. (2015). Questioning the use of an amphibian colour morph as an indicator of climate change. *Global Change Biology*, *21*(2), 566–571. doi:10.1111/gcb.12744 PMID:25230362

Morelli, F., Møller, A., Nelson, E., Benedetti, Y., Liang, W., Šímová, P., ... Tryjanowski, P. (2017). The common cuckoo is an effective indicator of high bird species richness in Asia and Europe. *Scientific Reports*, *7*(1), 4376. doi:10.103841598-017-04794-3 PMID:28663558

Ochoa-Hueso, R., Munzi, S., Alonso, R., Arróniz-Crespo, M., Avila, A., Bermejo, V., ... Theobald, M. R. (2017). Ecological impacts of atmospheric pollution and interactions with climate change in terrestrial ecosystems of the Mediterranean Basin: Current research and future directions. *Environmental Pollution*, *227*, 194–206. doi:10.1016/j.envpol.2017.04.062 PMID:28460237

Orru, H., Andersson, C., Ebi, K. L., Langner, J., Astrom, C., & Forsberg, B. (2013). Impact of climate change on ozone-related mortality and morbidity in Europe. *The European Respiratory Journal*, *41*(2), 285–294. doi:10.1183/09031936.00210411 PMID:22743679

Orru, H., Ebi, K. L., & Forsberg, B. (2017). The interplay of climate change and air pollution on health. *Current Environmental Health Reports*, *4*(4), 504–513. doi:10.100740572-017-0168-6 PMID:29080073

Pauli, H., Gottfried, M., & Grabherr, G. (2003). Effects of climate change on the alpine and nival vegetation of the Alps. *The Journal of Mountain Ecology*, 7.

Pearson, R. G., & Dawson, T. P. (2003). Predicting the impacts of climate change on the distribution of species: Are bio-climate envelope models useful? *Global Ecology and Biogeography*, *12*(5), 361–371. doi:10.1046/j.1466-822X.2003.00042.x

Pleijel, H. (2009). Air pollution & climate change, two sides of the same coin? In Common roots of air pollution and climate change. Swedish Environmental protection Agency.

Reyes-García, V., Fernández-Llamazares, A., Guèze, M., Garcés, A., Mallo, M., Vila-Gómez, M., & Vilaseca, M. (2016). Local indicators of climate change: The potential contribution of local knowledge to climate research. *Wiley Interdisciplinary Reviews: Climate Change*, *7*(1), 109–124. doi:10.1002/wcc.374 PMID:27642368

Ritchie, H, & Roser, M. (2018). *Air Pollution.* OurWorldInData.org.

Root, H. T., Geiser, L. H., Jovan, S., & Neitlich, P. (2015). Epiphytic macro lichen indication of air quality and climate in interior forested mountains of the Pacific Northwest, USA. *Ecological Indicators, 53*, 95–105. doi:10.1016/j.ecolind.2015.01.029

Sajjad, F., Noreen, U., & Zaman, K. (2014). Climate change and air pollution jointly creating nightmare for tourism industry. *Environmental Science and Pollution Research International, 21*(21), 12403–12418. doi:10.100711356-014-3146-7 PMID:24938808

Santini, L., Cornulier, T., Bullock, J. M., Palmer, S. C. F., White, S. M., Hodgson, J. A., ... Travis, J. M. J. (2016). A trait-based approach for predicting species responses to environmental change from sparse data: How well might terrestrial mammals track climate change? *Global Change Biology, 22*(7), 2415–2424. doi:10.1111/gcb.13271 PMID:27073017

Schlenker, W., & Walker, W. R. (2016). Airports, Air Pollution, and Contemporaneous Health. *The Review of Economic Studies, 83*(2), 768–809. doi:10.1093/restud/rdv043

Shimizu, Y. (n.d.). *A global health guardian: climate change, air pollution, and antimicrobial resistance.* World Health Organization.

Tai, A. P. K., Martin, M. V., & Heald, C. L. (2014). Threat to future global food security from climate change and ozone air pollution. *Nature Climate Change, 4*(9), 817–821. doi:10.1038/nclimate2317

Tainio, M., Juda-Rezler, K., Reizer, M., Warchałowski, A., Trapp, W., & Skotak, K. (2013). Future climate and adverse health effects caused by fine particulate matter air pollution: Case study for Poland. *Regional Environmental Change, 13*(3), 5–15.

Turner, M. C., & Jerrett, M. (2016). Long-term ozone exposure and mortality in a large prospective study. *AJRCCM Issues, 193*, 10. PMID:26680605

Vanos, J. K., Cakmak, S., Kalkstein, L. S., & Yagouti, A. (2015). Association of weather and air pollution interactions on daily mortality in 12 Canadian cities. *Air Quality, Atmosphere & Health, 8*(3), 307–320. doi:10.100711869-014-0266-7 PMID:26052369

Vanos, J. K., Hebbern, C., & Cakmak, S. (2014). Risk assessment for cardiovascular and respiratory mortality due to air pollution and synoptic meteorology in 10 Canadian cities. *Environmental Pollution, 185*, 322–332. doi:10.1016/j.envpol.2013.11.007 PMID:24355413

Vikas, M., & Dwarakish, G. S. (2015). Coastal Pollution: A Review, International conference on water resources, coastal and ocean engineering (ICWRCOE 2015). *Aquatic Procedia, 4*, 381–388.

Walther, G. R., Post, E., Convey, P., Menzel, A., Parmesank, C., Beebee, T. J. C., ... Bairlein, F. (2002). Ecological responses to recent climate change. *Nature, 416*(6879), 389–395. doi:10.1038/416389a PMID:11919621

Wang, H. J., & Chen, H. P. (2016). Understanding the recent trend of haze pollution in eastern China: Roles of climate change. *Atmospheric Chemistry and Physics, 16*(6), 4205–4211. doi:10.5194/acp-16-4205-2016

Will-Wolf, S., Jovan, S., Neitlich, P., Peck, J. E., & Rosentreter, R. (2015). Lichen-based indices to quantify responses to climate and air pollution across northeastern U.S.A. *The Bryologist, 118*(1), 59–82. doi:10.1639/0007-2745-118.1.059

Yue, X., Mickley, L. J., Logan, J. A., Hudman, R. C., Martin, M. V., & Yantosca, R. M. (2015). Impact of 2050 climate change on North American wildfire: Consequences for ozone air quality. *Atmospheric Chemistry and Physics, 15*(17), 10033–10055. doi:10.5194/acp-15-10033-2015

Zeppilli, D., Sarrazin, J., & Leduc, D. (n.d.). Is the meiofauna a good indicator for climate change and anthropogenic impacts? *Mar Biodiv., 45*, 505.

Zhang, J. (in press). Air pollution and defensive expenditures: Evidence from particulate-filtering facemasks. *Journal of Environmental Economics and Management.*

Zhao, S., Feng, T., Tie, X., Long, X., Li, G., Cao, J., ... An, Z. (2018). Impact of climate change on Siberian high and wintertime air pollution in China in past two decades. *Earth's Future, 6*(2), 118–133. doi:10.1002/2017EF000682

Chapter 9
Non-Point Source Pollutants From Motor Vehicles:
Effects and Control

Rajagopalan Varadarajan
Anna University, India

Abdul Majeeth Bathusha
Anna University, India

ABSTRACT

Motor vehicles emit gaseous pollutants from incomplete carbon reactions, unburned hydrocarbons, or other elements present in the fuel or air during combustion of fossil fuels. Atmospheric pollution is caused by multiple sources, making it a non-point source for the pollutants. The adverse effects of vehicular pollution are physical, chemical, and socio-economic in nature and are to be mitigated by the process of education, rules, and policies. A study has been done with the activated carbon made from Proposis cineria for mitigation.

INTRODUCTION

Motor Vehicle Pollution

Vehicle pollution is the process of releasing harmful material into the environment by motor vehicles. Transportation involves the combustion of fossil fuels to produce energy translated into motion. Pollution is created from incomplete carbon reactions, unburned hydrocarbons or other elements present in the fuel or air during combustion

DOI: 10.4018/978-1-5225-7289-3.ch009

(Shivaji Bhandarkar, 2013). The pollutants are emitted during the combustion and the primary pollutants such as carbon monoxide, soot, various gaseous and liquid vapour hydrocarbons, oxides of sulphur and nitrogen, sulphate and nitrate particulates, ash and lead are released in the atmosphere. These primary pollutants that react in the atmosphere to form ozone, secondary particulates, and other damaging secondary pollutants (Anshumala Sharmaa et al., 2018).

Non-Point Source Pollution

Atmospheric pollution are caused from multiple sources such as industrial facilities, vehicular transport, etc. Thus a point source, due to the distributional nature, long-range transport, and multiple sources of the pollution, it is considered a nonpoint source.

CAUSE OF POLLUTION

Various factors are identified and influenced in the vehicular pollution, the factors to be taken in to attribute of such in the particular area (Gang Wang et al, 2016). This also consent with up to the extent to which vehicles are properly maintained, the availability of appropriate fuels and the extent to which they are used properly; and Atmospheric, climatological and topological conditions (Julie A. Kenkel et al, 2016). Vehicle pollution is mainly caused by these factors:

1. The release of burnt fuel by-products into the atmosphere.
2. Use of Non-Eco Friendly Vehicles
3. Particulate matter
4. Evaporation of fuel
5. Lack of Vehicle Maintenance
6. Un-burnt Hydrocarbons
7. Noise from moving vehicles

POLLUTANTS RELEASED FROM VEHICLES

- *Crankcase Emission*: Crankcase Emission (also called running loss emissions) is unburnt or partially burned fuel components that, under pressure, escape from the combustion chamber, pass the pistons and enter the crankcase. This mixture is called blow-by. The main constituent of blow-by emission is HCs. If uncontrolled, it may constitute 13– 25% of total emissions. Since, diesel engines compress only air, levels of pollutants in the blow-by is very low. Evaporative Emissions Evaporative Emissions HC vapours, lost constantly and directly to the atmosphere due to volatile nature of petrol, mainly from the fuel lines, fuel tank and carburettor depending upon fuel composition, engine operating temperature and ambient temperature. Losses from the carburettor, called Hot Soak Emissions, occur when a hot engine is stopped. It should be noted that, out of total emissions, which is much more in case of petrol than diesel, 20-32% of the total emissions are due to evaporation losses, of which the HCs happen to be the chief constituents (Shivaji Bhandarkar, 2013). Exhaust Emission Automotive exhaust is the major source constituting about 60% of the total emission. Automobile exhaust releases wide range of pollutants such as (1) Hydrocarbons (Unburnt), (2) Carbon monoxide, (3) Oxides of nitrogen (NO_x), (4) Lead oxides, (5) Particulate matters e.g. lead, carbon, alkaline earth compounds, iron oxide, tar, oil, mist (6) Traces of aldehydes, esters, ethers, sulphur dioxide, peroxides, ketones benzene (C_6H_6), 1, 3 butadiene, Poly Aromatic Hydrocarbons (PAH), metal dust, asbestos fibre, dioxin, furon, ammonia, organic acids, chlorofluorocarbons (CFCs) etc (Ziyi Li et al, 2017 and Anshumala Sharmaa et al., 2018). Hydrocarbons and CO appears in the exhaust gas due to incomplete combustion of fuel. Due to the reaction of nitrogen and oxygen contained in the combustion at high temperature during combustion produces the oxides of nitrogen. Secondary pollutants are formed due to the reaction of primary pollutants released into the atmosphere. Chief among these includes HC, CO, NO_x when mixed with atmospheric water vapours in the presence of sunlight form ozone and variety of complex organic gases and resultant particulates known as Photochemical Smog (Linbo Qin et al, 2016). The major pollutants emitted from exhaust emissions of gasoline fuelled vehicles are CO, HC, NO_x and Pb while pollutants from diesel-fuelled vehicles are particulate matter (including smoke), NO_x, SO_x, PAH (Weiqing Huang et al, 2016). The detailed information of these pollutants is as given below. Carbon Monoxide (CO) - colourless and odourless gases slightly denser than air. Residence time and turbulence in the combustion chamber, flame temperature and excess O_2 affect CO formation. Conversion of CO to CO_2 in the atmosphere is slow and takes 2 to 5 months.

- *Hydrocarbon Compounds (HC):* Compounds consist of carbon and hydrogen and include a variety of other volatile organic compounds (VOCs). Most HCs are not directly harmful to health at concentrations found in the ambient air. Through chemical reactions in the troposphere, they play an important role in forming NO_2 and O_3 which are health and environmental hazards (Gang Wang et al, 2016). Among various HC, methane (CH_4) is absent from these reactions. Remaining HC, non-methane hydrocarbons (NMHC) are reactive in the formation of secondary air pollutants. NMHC are photo chemically reactive. Benzene and Polyaromatic Hydrocarbons (PAH) - Motor vehicles emit toxic HC including benzene, aldehydes and Polyaromatic hydrocarbons (PAH). About 85 to 90% benzene emissions come from exhaust and the remainder comes directly from gasoline evaporation and through distribution losses (Qijun Zhang et al, 2016). Toluene and xylene HC compounds are present in the gasoline whereas aldehydes, 1, 3 butadiene are absent from gasoline, diesel fuel, ethanol or methanol but in their exhaust emissions as partial combustion products. Comparatively the PAH emission is higher in the diesel powered engine than petrol powered engine (Anshumala Sharmaa et al., 2018).

- *Nitrogen Oxides (NOX):* includes nitric oxide (NO), nitrous oxide (N_2O), nitrogen dioxide (NO_2), dinitrogen trioxide (N_2O_3) and nitrogen pent oxide (N_2O_5). NO and NO_2 collectively represented as NOX, are the main nitrogen oxides emitted by vehicles. About 90% of these emissions are in the form of NO produced in the vehicle engine by combustion of nitrogen at high temperatures. NO_2 formed by oxidation of NO, has a reddish brown colour and pungent odour. In the atmosphere, NO_2 involved in a series of reactions Vehicle Engineering (VE) Volume 1 Issue 2, June 2013 www.seipub.org/ve 35 in the presence of UV radiation that produces photochemical smog, reducing visibility. It may also react with moisture to form nitric acid (HNO_3) aerosols. In the lower atmosphere (troposphere), NO_2 forms O_3 by means of reaction with HC, while in the upper atmosphere, it reacts with chlorine monoxide to form chlorine nitrates. In developing countries, the transport sector accounts for 49% of NO_x emissions and the power sector, 25%; the industrial sector, 11%; the residential and commercial sectors, 10% and other sources 5% (Anshumala Sharmaa et al., 2018).
- *Sulphur Dioxide (SO_2):* is a stable, non-flammable, non-explosive, colourless gas. In the atmosphere, SO_x may be converted to sulphur trioxide (SO_3) by means of reaction with O_2. SO_2 and SO_3 react with moisture in air to form sulphurous (H_2SO_3) and sulphuric (H_2SO_4) acids may precipitate to earth as acid rain. Sulphates may also be produced through reaction of these sulphur compounds with metals present in particulate matter (Richa Singh et al, 2017).
- *Ozone (O_3):* in the lower (troposphere) layer, ground level ozone (GLO) is formed by the reaction of VOCs and NO_x with ambient O_2 in the presence of sunlight and high temperatures (Anshumala Sharmaa et al., 2018). GLO is a major constituent of smog in urban areas and motor vehicles are the main emission source of its precursors. The small quantities of organic and inorganic compounds such as peroxyacetyl nitrate (PAN) and nitric acid is

formed due to the reaction of GLO. It mainly depends on the absolute and relative concentrations of its precursors and the intensity of solar radiation, which exhibits diurnal and seasonal variations. Also, the thermal inversion increases the GLO concentrations (Richa Singh et al, 2017).

- *Particulate Matter (PM):* consists of fine solids and liquid droplets other than pure water dispersed in air. Total suspended particulates are particles with an aerodynamic diameter greater than 70 µm. PM with an aerodynamic diameter greater than 10 µm known as suspended inhalable particulate matter / Respirable Suspended Particulate Matter (RSPM) or PM_{10}, remains in the atmosphere for longer periods because of its low settling velocity. PM_{10} can penetrate deeply into the respiratory tract and cause respiratory illness in humans. PM with an aerodynamic diameter of 2.5-10 µm or less is defined as fine particles ($PM_{2.5}$), while the larger PM is called coarse particles. Nearly all PM emitted by motor vehicles consists of fine particles and a large fraction of these particles has an aerodynamic diameter less than 1µm (Gang Wang et al, 2016). $PM_{2.5}$ can also be formed in the atmosphere as aerosols from chemical reactions that involve gases such as SO_x, NO_x and VOC. Sulphates which are commonly generated by conversion from primary sulphur emissions make up the largest fraction of $PM_{2.5}$ by mass. $PM_{2.5}$ can be formed as a result of solidification of volatile metals salts as crystals following cooling of hot exhaust gases from vehicles in ambient air as well. Gasoline fuelled vehicles have lower PM emission rates than diesel–fuelled vehicles. PM emissions from gasoline fuelled vehicles result from unburned lubricating oil and ash-forming fuel and oil additives. PM emitted by diesel-fuelled vehicles consists of soot formed during combustion, heavy HC condensed or adsorbed on the soot and sulphates. These emissions contain PAH (Anshumala Sharmaa et al, 2018; Burr et al 2004). With the advancement of emission control measures in engines, however, the contribution of soot has been reduced considerably. Black smoke, associated with the soot portion of PM emitted by diesel-fuelled vehicles, results from the deficiency of O_2 during the full combustion or expansion phase. Blue, grey and white smokes are caused by the condensed HC in the exhaust of diesel fuelled vehicles. Blue or grey smoke results from vaporized lubricating oil and white smoke occurs during engine start-up in cold weather. Diesel fuel additives such as Ba, Ca and Mg reduce smoke emissions but increase PM sulphate emissions. These additives may also increase PAH emissions. Dioxins- Cu based additives can reduce PM emissions but may catalyze the reaction between HC and trace amounts of chlorides in diesel fuel to form dioxins which are emitted in the exhaust (Qijun Zhang et al, 2016).

- *Chlorofluorocarbons (CFCs):* The Freon gases which are used in air conditioners is the main source of CFC emissions from motor vehicles. CFC emitted into the atmosphere rise to the stratosphere layer within 10 years and is estimated to remain there for 400 years. CFC molecules struck by UV radiation release chlorine atoms, which destroy O_3 by the formation of chlorine monoxide. Furthermore, when a free O_2 atom reacts with a chlorine molecule, an O2 molecule is formed and a chlorine atom is released to destroy more O_3 (Richa Singh et al, 2017).
- *Carbon Dioxide (CO_2):* It is a greenhouse gas associated with global warming, resulting mainly from increased combustion of fossil fuels including motor vehicle fuels (Anshumala Sharmaa et al., 2018).

EFFECTS OF VEHICLE POLLUTION

The consequences and the major effects of vehicular pollution are

- *Global Warming:* Emission from vehicles releases greenhouse gases into the atmosphere leads to depletion of the ozone layer and causes global warming. The result of this is adverse weather that more often than not results in loss of life and property. Global warming is a major problem for the people and deliberate efforts have been made to reduce it. Due to the global warming, the ozone layer gets depleted and the harmful ultraviolet rays of the sun can reach the lower surface of the earth and harm humans and other living organisms. The greenhouse effect has naturally heat the earth for over 4 billion years. Raising the concentrations of greenhouse gases to toxic levels, with possibly dangerous environmental and human health consequences. (Adeyanju and Manohar, 2017)
- *Poor Quality of Air*: Now a days many countries have very poor the quality of air that people wear masks to reduce the amounts of harmful substances inhaled. The major problem is that many countries have a high number of old vehicles generally and has absolute rules for the manufacture of new vehicles.
- *Health:* The main effect of vehicular pollutants that it leads to lung infection and cancer. As we know, hydrocarbons are not very good for human health which can cause heart disease, aggravate asthma, damage the central nervous system and make breathing difficult. Fuel spillage also a major pollutant that affect the health of plants and marine life.
- *Tourism:* Most people would not willingly go to visit countries that they know are bad for their health. This will, in turn, result in a reduction in the number of tourists to those countries and subsequently loss of foreign exchange income.

- *Smog and Acidic Rain:* Nitrogen oxides reacts in the atmospheric gases and causes highly corrosive smog that speeds up rusting of vehicles. Also nitrogen oxide dissolves in rain forms acid rain and make the rain water not fit for human, plant or animal consumption.

HOW TO REDUCE VEHICLE POLLUTION

Though vehicle pollution cannot be completely eliminated, it can still be reduced to manageable levels. There are some of the things that individuals, governments, and organizations can engage in to reduce the effects of pollution caused such as

- *Civic Education:* Ignorance is not bliss and education is definitely the key to success. Many people do not care about the effects of pollution because they are not aware of them. Civic education by government departments and non-governmental organizations plays a great role in awakening the society to the realities of pollution and the counter-productive methods to reduce vehicle pollution. Responsibility is sense that should be cultivated in everyone so that there is a desire and willingness to do what is right.
- *Progressive Policies:* Creating strict regulations that anticipate the challenges of the modern world that reduce the vehicular pollution can be very useful in mitigation process. Drafting of legislations that makes the people do the necessary as far as bringing down the levels of vehicle pollution. Such laws are to be included by placing a cap on the age of vehicles that are imported and the guidelines are to be made on the conditions of roadworthy vehicles, and creating agencies that will look into alternative fuels such as green energy.
- *Vehicle Maintenance:* This is more of a personal responsibility than it is universal. One should ensure that his or her car is in good condition and does not release a lot of harmful substances into the atmosphere. If the vehicle performs optimally and less amount of pollutants are released into the air, thus the regular car maintenance can help in repairing or replacing worn out parts. Also, things like replacing oil filters, changing the engine oil and greasing the moving parts should be done on a regular basis. Carelessness is the main reason that some vehicles release dark harmful smoke while they are moving on the roads. This is especially harmful to those outside because they inhale the smoke and the dangerous matter.
- *Discard Old Vehicles:* A change in thinking would lead to better practices. Thus, old vehicles should be either discarded or returned to the manufacturing company so that new ones can be acquired.

- *Carpooling:* is a method followed by the people from same neighbourhood and who work in the same area. This is an effective and efficient method to reduce traffic jams, save money used on car fuel and maintenance and contribute towards having a clean environment.
- *Alternative Means of Transport*: Alternate means of transport can be used to conserve energy and also saves lots of money. Walking and bike ridings are also an option when your workplace is not that far off from your place of residence. When compared to using the car it need fuelling, maintenance and using these alternative means of transport is a lot cheaper.

CASE STUDY

One of the mitigation measures to control the vehicular pollution are study on a laboratory scale. This is the process of using adsorptive material to reduce the vehicular pollution from the source level.

Introduction

A study is made by using activated carbon produced from the branches of *Proposis cineria* as an adsorptive material. *Proposis cineria* is a fast growing small trees in India which are mainly cultivated and consumed for timber and furniture industries of the developing nations. These trees are the most tolerant species for saline, alkaline soils and also capable of growing in water logged areas and are are widely grown in dry lands, arid regions and other parts of India.

Materials and Methods

- Raw Materials: The branches of *Proposis cineria* are cut down into small pieces and are sun dried for 5 days. This leaves the moisture content to the range of 5-7% of the pieced branches of *Proposis cineria.* The barks of the dried branches of *Proposis* are removed and the branches are made into smaller chips of length 5-7 cm.
- Preparation of Activated Carbon: Adsorption is preferred by its potential to remove organic and inorganic constituents, even at small concentrations (Basim Abussaud et al 2016; Camila Fernández et al 2016; Changjoo Yeom and Younghun Kim, 2017; Evi Mutiara Dewi et al 2016). Activated carbon are prepared from the dried branch chips of *Proposis cineria*. They were collected in and around the college campus of Anna University, Trichy. The materials was sun dried and are made into smaller chips of length 5-7 cm.

Then, the two branch chips were impregnated in 1:1 sulphuric acid for 24 hrs to make the branches charred in nature. Then, the impregnated char is washed with distilled water till the pH of the materials neutral and oven dried at 150°C to remove the moisture content in the char. The dried chars are placed in a muffle furnace at 850°C for 30 min to complete the carbonization of the materials. Then the activated carbon are ground and sieved with 1.4 mm and 0.9 mm size standard sieve.

- Testing of Vehicular Exhaust**:** The vehicle used for testing is a four wheeler light motor vehicle (LMV) with a petrol powered engine. Testing is carried-out by checking the exhaust emission at idle using gas analyser at the end of tail pipe. The emission test is conducted with and without the adsorbent respectively. The emission gases from the exhaust system are determined for CO, CO_2 and HC. The emission levels are calculated by the gas analyser in terms of percentage by volume for CO and CO_2 and in terms of ppm by volume for HC releasing form the exhaust of the vehicles (Mylaudy et al 2016).
- Working Process for Testing of Vehicular Exhaust

Result and Discussion

The vehicle emission check is the mandatory fitness which are tested for emission compliance for all vehicles needed to get emission check done once in every six months, at an Authorised Emission check Centres. The State Transport Department has authorised private entrepreneurs to operate emission test centres and issue test certificates after necessary emission check. These Emission Check Centres are primarily located at fuel dispensing stations and some at private garages.

- Testing for emission with and without catalyst adsorbent

Figure 1. Process flow diagram for the testing of vehicular exhaust

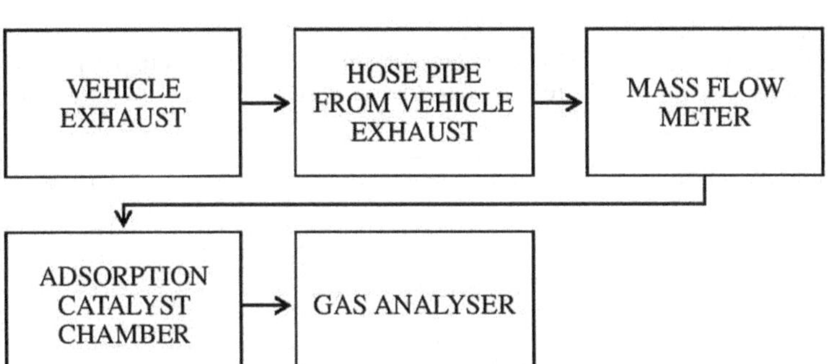

The vehicular exhaust emission test are carried out at an authorized emission testing centre to determine the exhaust efficiency to reduce the pollutants using the activated carbon prepared from *Proposis cineria*.

CONCLUSION

The study shows the significant reduction in the vehicular exhaust pollutants CO_2, CO and HC. The reduction percentage of the pollutants CO_2, CO and HC are 42.86%, 50% and 39.58% respectively. This process of adsorption won't affect the increase in emission of other gases and doesn't leave any residue to pollute the environment. Due to the adsorption process we can recycle the catalyst used. Thus, the activated carbon produced from *Proposis cineria* are comparatively cheaper from the synthetic activated carbon produced and has a significant reduction in the pollutant.

REFERENCES

Abussaud, B., Asmaly, H. A., Ihsanullah, T. A., Saleh, T. A., Gupta, V. K., laoui, T., & Atieh, M. A. (2016). Sorption of phenol from waters on activated carbon impregnated with iron oxide, aluminum oxide and titanium oxide. *Journal of Molecular Liquids*, *213*, 351–359. doi:10.1016/j.molliq.2015.08.044

Adeyanju, A. A., & Manohar, K. (2017). Effects of Vehicular Emission on Environmental Pollution in Lagos. *Sci-Afric Journal of Scientific Issues, Research and Essays, 5*(4), 34-51.

Bhandarkar, S. (2013). Vehicular Pollution, Their Effect on Human Heatlh and Mitigation Measures. *Vehicle Engineering*, *1*(2), 33–40.

Table 1. Concentration of pollutants with and without adsorption materials

Pollutants from Vehicular Exhaust	Concentration of Pollutant Without Adsorptive Catalyst	Concentration of Pollutant With Adsorptive Catalyst
CO_2	0.7%	0.4%
CO	0.1%	0.07%
HC	960 ppm	580 ppm

Burr, M. L., Karani, G., Davies, B., Holmes, B. A., & Williams, K. L. (2004). Effects on respiratory health of a reduction in air pollution from vehicle exhaust emissions. *Occupational and Environmental Medicine*, *61*(3), 212–218. doi:10.1136/oem.2002.003244 PMID:14985515

Dewi, E. M., Suwardi, Suryaningtyas, D. T., & Anwar, S. (2016). Utilization of Natural Zeolites as Cu (Ii) and Zn (Ii) Adsorbent. *J Trop Soils*, *21*(3), 153–160. doi:10.5400/jts.2016.v21i3.153-160

Fernández, C., Bion, N., Gaigneaux, E. M., Duprez, D., & Ruiz, P. (2016). 2016, Kinetics of hydrogen adsorption and mobility on Ru nanoparticles supported on alumina: Effects on the catalytic mechanism of ammonia synthesis. *Journal of Catalysis*, *344*, 16–28. doi:10.1016/j.jcat.2016.09.013

Huang, W., Fan, H., Qiu, Y., Cheng, Z., Xu, P., & Qian, Y. (2016). Causation mechanism analysis for haze pollution related to vehicle emission in Guangzhou, China by employing the fault tree approach. *Chemosphere*, *151*, 9–16. doi:10.1016/j.chemosphere.2016.02.024 PMID:26923237

Kenkel, J. A., Sisk, T. D., Hultine, K. R., Sesnie, S. E., Bowker, M. A., & Johnson, N. C. (2016). Nancy Collins Johnson; Indicators of vehicular emission inputs into semi-arid roadside Ecosystems. *Journal of Arid Environments*, *134*, 150–159. doi:10.1016/j.jaridenv.2016.06.007

Li, Z., Liu, Y., Yang, X., Xing, Y., Tsai, C.-J., Meng, M., & Yang, R. T. (2017). Performance of mesoporous silicas and carbon in adsorptive removal of phenanthrene as a typical gaseous polycyclic aromatic hydrocarbon. *Microporous and Mesoporous Materials*, *239*, 9–18. doi:10.1016/j.micromeso.2016.09.027

Qin, L., Han, J., Zhan, Y., Chen, W., & Kim, H. (2016). 2016, Reduction of Polycyclic Aromatic Hydrocarbon Emission by Porous Alumina Bed Material during Sewage Sludge Incineration. *Energy & Fuels*, *30*(1), 544–550. doi:10.1021/acs.energyfuels.5b01942

Rajdurai, Rao, & Kamalakkannan. (2016). CO_2 Capture Using Activated Alumina in Gasoline Passenger Vehicles. *Int. Journal of Engineering Research and Applications, 6*(5-4), 73-77.

Sharmaa, A., Masseya, D. D., & Tanejab, A. (2018). A study of horizontal distribution pattern of particulate and gaseous pollutants based on ambient monitoring near a busy highway. *Urban Climate*, *24*, 643–656. doi:10.1016/j.uclim.2017.08.003

Singh, R., Sharma, C., & Agrawal, M. (2017). Emission inventory of trace gases from road transport in India. *Transportation Research Part D, Transport and Environment*, *52*, 64–72. doi:10.1016/j.trd.2017.02.011

Wang, Cheng, Lang, Li, & Tian. (2016). On-board measurements of gaseous pollutant emission characteristics under real driving conditions from light-duty diesel vehicles in Chinese cities. *Journal of Environmental Sciences, 46*, 28–37.

Yeom, C., & Kim, Y. (2017). Adsorption of ammonia using mesoporous alumina prepared by a templating method. *Environmental Engineering Research*, *22*(4), 401–406. doi:10.4491/eer.2017.045

Zhang, Q., Wua, L., Yang, Z., Zou, C., Liu, X., Zhang, K., & Mao, H. (2016). Characteristics of gaseous and particulate pollutants exhaust from logistics transportation vehicle on real-world conditions. *Transportation Research Part D, Transport and Environment*, *43*, 40–48. doi:10.1016/j.trd.2015.09.005

Chapter 10
Assessment of the Biogas Technology Potential in Reducing Indoor Air Pollution:
A Review Through Cas

Thilivhali Eugene Rasimphi
University of Venda, South Africa

ABSTRACT

Many of the rural communities in developing countries are forced to rely on the traditional energy sources such as firewood, dung, and crop residues. These traditional methods are often expensive and/or time-consuming. Today, more than 90% of energy requirements globally are met by burning coal and natural gases with resultant release of harmful pollutants with a serious threat to the environment and human health. Twentieth century has witnessed great challenge to the progress of human development because of scarcity of energy and the principal infrastructure for the development of technology. The chapter reviews the potential of biogas technology to solve the problem of indoor air pollution in rural households. Useful information on the project impact was collected through villages' questionnaires, interviews, field trips, and community engagement workshops, and all the viability demonstrated from rigorous socioeconomic and environmental benefits considerations. The results show that biogas technology is sustainable and provides energy security while combating climate change.

DOI: 10.4018/978-1-5225-7289-3.ch010

INTRODUCTION

In rural areas, increased desire for improved quality of life and growing population has led to increased energy demands from all sectors, for example; transport, industrial and domestic. Indoor air pollution (IAP), a significant proportion generated from traditional cooking stoves, is thought to be responsible for 2.7% of the total global burden of disease (WHO, 2011). According to Smith *et al.,* (2013) indoor air pollution (IAP) is considered not only to be a threat in the cases of related disease, but also in subsequent death. The reduction in lives lost due to reduced incidence of IAP related disease has financial and economic implications. Indoor air pollution is considered to be a result of solid fuels use in household.

South Africa currently relies almost entirely on fossil fuels (approx. 90%) to satisfy its energy demand, with coal providing 75% of this energy supply (Amigun and Von Blottnitz, 2009). This dependence is due to its vast coal resources, making it an attractive and low-cost source of energy for South Africa (Austin & Blignaunt, 2008). The majority of total greenhouse gas (GHG) emissions in the country (85% in 2010) are produced by the Energy sector (World Energy Assessment, 2000).

In many rural areas there is still inadequate access to clean energy, thus negatively affecting household health, for example, indoor air pollution caused by burning firewood or charcoal on open fires and simple stoves (WHO, 2006; Singh and Sooch, 2004). Indoor air pollution levels in rural households are often much higher than outdoor air pollution because of burning especially firewood. For instance, typical levels of particulate matter i.e. PM^{10} in rural households range from 300 to 3000 micrograms per cubic metre (WHO, 2002) and these causes serious threat to the users thereof. World Health Organization has reported that almost 40% of acute respiratory infection and about 20% of chronic obstructive pulmonary illnesses are caused by indoor air pollution from the burning of fuelwood (Arcenas *et al.,* 2010). This is a problem in rural areas because of indoor air pollution caused by using fuel wood. Also the use of fuel wood for cooking also causes deforestation.

Poor access to recent energy supplies and unsustainable practices of energy generation in rural areas of South Africa has been causing environmental destruction and health risk to users. According to Onguntoke *et al., (*2010) some of the prevalent health problems caused by the smoke of traditional fuels, particularly fuel wood include; coughing, headache, and lung illness. For example, WHO reported about deaths in South Africa due to indoor air pollution.

The Sub-Saharan countries have low electrification rates and have high poverty levels compared to countries from other continents highlighting the link between energy access and poverty (WEO, 2009). Thus, the development of new efficient methods of producing energy and the choice of the renewable energy technologies that suit the economic and geographical conditions for particular communities particularly in developing countries is required in order to solve the escalating energy problems. The problem requires urgent strategies to protect the environment, achieve economic development and to increase social well-being.

Biogas technology is associated with social and environmental benefits such as improving human well-being through better sanitation, reduced indoor smoke, and employment generation as well as conservation of resources particularly trees and reduced greenhouse gas emission (Amigun, 2008). The use of biogas in households has the ability to uplift the quality of life in rural areas (Amigun and Von Blottnitz, 2011). Cooking with biogas is less harmful than cooking with fuel wood, and dramatically reduce risk of health related problems such as eye problems, coughs and pneumonia, in addition, less fuel needs to be collected and this mean that less trees will be cut down thus promoting a natural and sustainable environment.

OBJECTIVES

To assess the potential of domestic biodigester to reduce indoor air pollution in Vhembe district through a case study of Marubini Cooperative to reduce indoor air pollution and promote a natural and sustainable environment.

CASE STUDY AREA PRESENTATION

The study focused on assessing the potential of biogas to reduce indoor air pollution in Vhembe district around Makhado municipality in the villages of Maila (Marubini Cooperative).

Materials and Methods

Use of Existing Case Studies

Although the use of extrapolation and benefit transfer models, especially with regard to stated preference results is generally not favoured (Pearce and Ozdemirogh, 2002), however desk research is easy and quick. In the case of this research, where time and resources are not without limits, it appears to be acceptable to make provision for extrapolation, use of aggregate regional data when site specific details are not available (European Commission, 2006). Information was gathered from journals, research studies, library database and the study sites. Census 2011 rated Limpopo province poorest in the country with a headcount poverty level of about 63.8% (Lehohla, 2014) and yet with one of the best agricultural climate and vast natural resources.

In villages (Vleifontein and Fefe), site visits sought information on energy demand and use pattern, electrification of homes, waste management, energy sources used, how they obtain the energy sources, agricultural activities involved in households, animals owned in the household and the availability of water. In agricultural lands data on the management of agricultural waste, whether they burn or leave the crops to act as fertilisers, any composting of the waste or any other way of dealing with waste. The villages visited include Vleifontein i.e. Maila and Nthabalala where organic wastes are generated around Vhembe.

Questionnaire

In order to conduct a comprehensive feasibility study of the potential benefits of rural household biodigester use (Smith *et al.,* 2013), it was necessary to first bring information about the current characteristic and energy demands of rural household in the study area. For this purpose a questionnaire was designed to gather detailed information about the household, their energy requirement, their livestock keeping practice, water usage and the production of crops and vegetables at their homesteads. The questionnaire design, survey and analysis process is discussed in the preceding section.

Questionnaire Design

The questionnaire used in this research was based on and adapted from previously used questionnaire: a questionnaire used in a biomass energy audit conducted by Rhodes University and a biogas perception and behaviour questionnaire conducted by Agama Energy and Jabezi in development of the South African National Biogas feasibility study (Austin and Blignuant, 2008; Smith, 2013). The questionnaire was used as a basis for this research project as they had been tried and tested in the field and had contributed to save meaningful studies (Smith, 2013), and such was relevant to the study.

RESULTS AND DISCUSSION

Household Energy

Energy is crucial for meeting our most basic necessities, i.e. cooking, boiling water, lighting and heating. World's poorest people dwell in rural areas where there is no access of energy services (IEA, 2011). The reliance on fuel wood to overcome cooking and heating needs is a driver of cutting down of trees in rural areas. Women and their children use substantial amount of time collecting fuel wood for cooking and heating. Their efforts spent in collecting firewood have negative effects on their lives, especially the educational prospects of children. Cooking with fuel wood, cow dung, and coal is a big risk for lung diseases amongst children and chronic respiratory disease among adults.

Rural communities need to improve their access to current energy patterns (World Bank, 2000). Fuel wood use has an influence on the social and economic status of a rural community as they depend mainly on it for cooking and other needs. Most people still depend on fuel wood and paraffin as their main sources of energy (Shackleton *et al.,* 2007). However fire wood is collected from far away forest hence consuming a lot of time.

The collection procedure of fuel wood is costly activity in a poor household, both in time and energy. Daily activities of women revolve around subsistence farming and much of their time is devoted to collecting fuel wood, water, food preparation, child care, cleaning and household maintenance. A study by Masekoang *et al.,* (2005) in the villages of Giyani, Limpopo, showed that women and children spend between 5 to 6 hours per day collecting fuel wood. Thus about 45% of a day is spent on fuel wood gathering. Similarly, another study in Kwa-Zulu Natal by Eberhand (1996) reported that in a village of Mahlabathini, the total time spent collecting fuel wood is about 9 hours per household per week. Masekoang *et al.,* (2005), stated that in rural areas of Giyani more than 80% of energy related work is done by women. Women play a significant role in the domestic energy system by obtaining and handling fuel wood for their families. The implication of this is that women's participation in other development activities is impaired.

The study further revealed that about 72% of the household indicated that fuel wood is very limited and when wood becomes scarce, people resort to collecting very inconvenient kinds of energy sources which create yet even more environmental and social problems (Masekoang *et al.,* 2005). The 2008/2009 poverty line Living Conditions Survey (LCS) results, state that Limpopo is the poorest province (48.5%) in South Africa. With regard to energy poverty the province is seen to be on the dominating in the use of fossil fuels like wood.

Fuelwood use has an influence on the social and economic characteristics of a rural community as the main energy source. Many of rural households in South Africa are still dependent on firewood and paraffin for cooking (Shackleton *et al.*, 2007). For example, in Limpopo more than 50% of the households depends on firewood and paraffin for cooking (SSA, 2007). According to Statistic South Africa (2011) the prevalent energy sources for household energy use in Vhembe district are presented in Figure 1 with fuel wood at 63% for cooking which is a large percentage thus it reveals that firewood is the dominant energy source.

Cooking represents one of the most energy-intensive applications. For example, figure1 shows that in rural areas of Vhembe district fuel wood is the significant energy source for cooking at 63%, followed by electricity at 34% and other between 1 and 2%. Concerns caused by reliance on traditional biomass for cooking are now well known. The smoke emitted by burning of fire wood in traditional cook-stoves contains harmful pollutants including particulate matter (WHO, 2002).

Traditional biomass use for cooking and heating is principal in most rural communities (AGECC, 2010). Rural communities are dominated by the poor households living there with limited or no access to energy services. This deficiency of access to energy services not only affects economic production but is also a

Figure 1. Household energy usage percentages for cooking in Vhembe district (SSA, 2011)

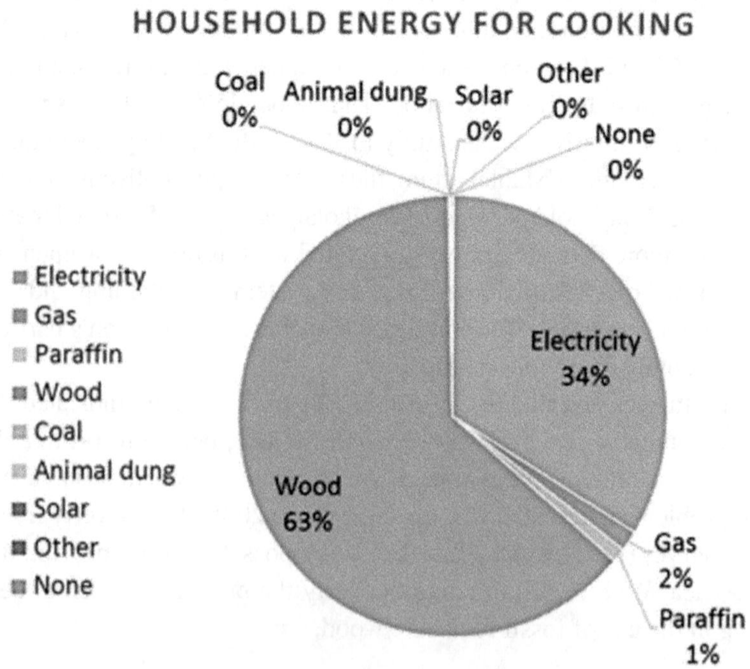

stumbling block to the proper provision of other crucial basic services such as health care and education (IEA, 2011). The over dependence on fuel wood to meet cooking and heating needs is a primary driver for deforestation in disadvantaged communities (WHO, 2002). Biogas technology can provide much cleaner access to energy for rural communities, thus helping to create new economic opportunities, generate more income and bring about rural development (FAO, 2009). In figure 1, the people in the community showed that they neither use any of the energy sources and thus it was accounted as none, as they were not using any of the sources for cooking.

Energy Forms Used for Cooking and Heating

The graphical presentation of energy use profile for cooking and heating is presented on Figure 2. As can be seen from the histogram (Figure 2) the household energy sources for cooking and heating, 63,4% of the respondents use firewood for cooking, 55,1% for heating. This is because firewood is considered the primary source of energy in this rural community. Electricity is the second most important energy source for cooking and heating as it is seen at 32,2% for cooking and 10,2% for heating. People in rural areas mostly depend on electricity when they can't access fuel wood as such electricity is used least from fuel wood.

The area of concern is biogas use, from Figure 2; the respondents clearly do not use biogas technology as it is evident from the histogram because it has 0% both for cooking and heating. The knowledge about the technology is poorly disseminated and this is not surprising when biogas is not known as an energy source for cooking and heating. Biogas technology is not known in the area of study, almost 99% of the people never heard of the technology. However, in some parts of the province the technology is slowly gaining popularity, where the shortage of clean and sustainable energy source represent destruction to the environment and its people (Amigun and Von Blottnitz, 2009). For example, in Giyani the technology is slowly growing. Biogas as a renewable energy has a massive potential to be one of the energy sources for cooking. Another widely used source of energy is the LPG though it has negative impacts. From Figure 2, about 4,6% and 0,9% of the respondents use liquid petroleum gas both for cooking and heating. This shows clearly that fuelwood is considered the primary energy source for cooking and heating in the study area though it has adverse effects on human well-being and the environment as a whole.

Social Benefits of Biogas Technology

The use of biogas has the ability to develop the welfare of rural people. This includes health, sanitation, and better living conditions. Fuel wood is a dirty source of energy, that when fuel wood is burned it leaves behind ash which can later contribute to

Figure 2. Households cooking and heat energy use profile in the study area
(Tinarwoet al, 2014)

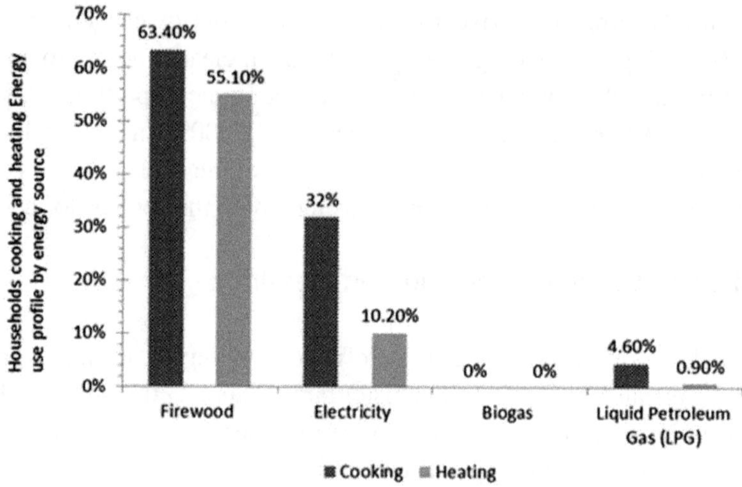

pollution, whereas organic waste either animal dung and/ food wastes are breeding grounds for disease carrying insects. However cooking with biogas promotes clean cooking area, improve sanitation, and reduce health associated problems with fuel wood like eye irritation, and coughing (Amigun & Von Blottnitz, 2011). Fuel wood collection is a tedious process and consumes time especially for women and children. It is estimated that they can spend about 2-6 hours daily in collecting fuel wood (DFID, 2002) and that depends on the country and region. Fuel wood collection takes away time that could have been better utilized in productive activities such as income generation or education.

A baseline study in Uganda (2010) showed that women spend about six hours daily on fuel wood collection. Because of their multiple roles in the house, women have only 2 hours to spend for other social commitments. Similarly, a user's survey was conducted one year later in Uganda (Africabiogas, 2014) amongst users' of biogas. The results indicated that the respondents (61%) agree strongly that biogas plant greatly reduced their work load. Time saved as a result is creatively put to things like attending community development activities, supporting children in their homework, and making handcrafts. Women are able to participate in home based initiatives (such as selling vegetables, poultry keeping, and other activities) to generate additional income (Africabiogas, 2014). In addition, incidence of disease (respiratory and eye irritation) decreased by 56%.

Biogas plants implementation in rural areas could see a huge change from reliance on traditional biomass to a renewable energy technique that improves social status of rural community taking the example of Uganda. Such benefits can help in decreasing the workload of women and girls in collecting firewood. Social benefits of biogas technology include better sanitation, decreased workload for women in firewood collection and cooking, reduced volume of disposed waste products, improved hygiene conditions through reduction of pathogens and flies, and environmental benefits such as protection of soil, water and woody vegetation. Other benefits are that biogas is a renewable source of energy and provides reliable power supply and is an environmental-friendly technology.

The cooperative uses a lot of energy in the form of heat to cook food for children. They get firewood supplies from local entrepreneurs who charges exorbitant prizes and the health of users is always at risk. On the other hand the cooperative produce a lot of waste from the egg layers (egg laying chickens) which can be used to feed the biodigester and the waste produced poses disposed challenge in addition to the health issues.

A study in Chinas rural areas, most villages do not have electricity, and commercial fuels like kerosene and coal are expensive. Thus the rural population still relies heavily on biological sources of energy. Fuelwood and crop residues (mostly straw) account for 80% of rural household energy use and 54% of total rural energy use. The amount of heat actually used is only about 10% for traditional stoves, so a lot of fuel is required for such simple tasks as preparing food and tea. This has serious environmental consequences, such as deforestation, reduced soil fertility, erosion, and even localized desertification. It also causes severe indoor air pollution, with resultant health effects such as upper respiratory infections and eye problems.

Besides better hygiene, the most important human health benefit from biogas is a reduction in indoor air pollution. Households that burn coal for cooking have 74% more carbon monoxide, 84% more sulfur dioxide, 27% more carbon dioxide, and 77% more suspended particulates than households that use biogas. Exposure to these high concentrations causes a variety of ailments—pneumonia, chronic obstructive lung disease, cancer and tuberculosis, as well as heart disease and low birth weight. According to Smith (2007), 380,000 premature deaths per year are blamed on indoor smoke from solid fuels in China. It is also bad for the eyes. In an interview with the UN International Fund for Agricultural Development, which helped build 30,000 biogas digesters in China, farmer Liu Chun Xian is quoted as saying, "We used to cook with wood. The smoke made my eyes tear and burn and I always coughed. The children too were often sick and had to go to the clinic, which was expensive. Now that we're cooking with biogas, things are much better" (IFAD n.d.).

As a result of the fuel substitution in the household, and fewer crop residues being burned in open fields, ambient (outdoor) air quality is improved as well, especially particulate matter and SO_2, which causes acid rain.

Biogas technology played an important role in improving the quality of life for the cooperative since it was introduced. In addition to being a renewable source of energy that is locally available, it had a lot on benefits and some include:

1. More hygienic conditions, thanks to the elimination of indoor air pollution from fuel wood, which is particularly harmful to women who are responsible for cooking.
2. Preventing deforestation by reducing the detrimental harvesting of trees as fuelwood. Maila cooperative used fuel wood for cooking food for the children and the introduction of biogas has lessen the reliance on wood as energy source.
3. Burning biogas is cleaner than burning fuel wood. Apart from being smokeless, the only emissions are carbon dioxide and water, thus carbon emission to the atmosphere is reduced.

Evidence from the monitoring process and expression by the biogas users at the cooperative for cooking suggests that biogas if used to its full capacity can provide enormous benefits. The cooperative members revealed that their lives have dramatically changed ever since the introduction of the biogas digesters. In addition to saving on fuelwood purchases, they identified that they no longer needed to constantly manage cooking fires and that they were saving time as they were simply able to turn the gas on, light it and begin cooking without the need to watch over the process. Time saving, as a result of more efficient cooking practices and reduced fuelwood purchases and collection, reveals that a biodigester could contribute to promoting gender equality and empowerment of women as women are relieved of time-consuming domestic duties, while efficient cooking fuel and reduced wood harvesting also have the potential to ensure environmental sustainability.

By replacing firewood, kerosene, and chemical fertilisers emissions of greenhouse gases (GHGs) and so mitigating global warming (Amigun *et al, 2009*). For example, based on the research in India, a family size biogas plant substitutes 316 L of kerosene, 5,535 kg firewood and 4,400 kg cattle dung cake as fuels every year. It means a family size biogas plant reduces NO_x of 16.4 kg, SO2 of 11.3 kg, CO of 987.0 kg and volatile organic compounds of 69.7 kg per year (Amigun *et al.* 2009).

REFERENCES

Africa Biogas. (2012). *Domestic Biogas in Africa: A first Assessment of the Potential and Need*. Author.

AGECC. (2010). *Advisory group on Energy and Climate Change*. New York: Energy for a Sustainable Future.

Amigun, B., & Von Blottnitz, H. (2009). Capital Cost Prediction for Biogas Installation in Africa. *Environmental Progress & Sustainable Energy*, *28*, 134–142.

Austin, G., & Blignaunt, J. (2008). *South African National Rural Domestic Biogas Feasibility Assessment: Prepared for Ministry for Development Co-Operation, The State of Netherlands*. Cape Town: Agama Energy.

Bhat, P., Chanakya, H., & Ravindranth, N. (2001). Biogas Plant Dissemination; Success Story of Sirsi, India. *Energy for Sustainable Development*, *1*(5), 39–46.

Bond, T., & Templeton, M. (2011). History and Future of Domestic Biogas Plants in the Developing World. *Energy for Sustainable Development*, *15*, 347–354.

Department of Energy (DoE). (2015). State of Renewable Energy in South Africa. Department of Energy, South Africa.

FAO. (2009). *How to Feed the World in 2050: Background paper for the High-Level Forum on How to feed the World in 2050*. Rome: FAO.

Greben, H., & Oelofse, S. (2009). *Unlocking the Resource Potentail of Organic Waste: A South African Persepctive*. Pretoria: CSIR- Natural Resource and the Environment, Waste Management Research.

IEA. (2011). *Energy For All: Financing Access for the Poor, Special Early Except of the World Energy Outlook*. Oslo, Norway: IEA.

Jiang, X., Sommer, S., & Chritensen, K. (2011). A Reveiw of the Biogas Industry in China. *Energy Policy*, *39*, 6073–6081.

Khoiyangbam, R. (2011). Environmental Implication of Biomethanation in Conentional Plants. *Energy & Environment*, *2*, 181–187.

Lehohla, P. (2014). *Poverty Trends in South Africa: An examination of absolute poverty between 2006 and 2011*. Pretoria: Statistics South Africa.

Nijaguma, B. (2002). Biogas technology. New Technology International, 46-48.

NRDC. (2007). *Medium and Long-term Develoment Plan for Renewable Energy in China*. Beijing: National Development and Reform Commission.

Pandey, B., Subedi, P., Sengendo, M., & Monroe, I. (2007). *Report on the Feasibility for a national Household Commercialization program in Uganda*. Brussels: Winrock International.

Pearce, D., & Ozdemiroglu, E. (2002). *Economic Valuation with Stated preference Techniques*. London: Renwick, Department of Tranport, Local Government and the Regions.

Renwick, M., Suebi, S. & Hutton, G. (2007). *A Cost-Benefit Analysis of National and Regional Integrated Biogas and Sanitation Programmes in sub-Saharan Africa*. Winrock International.

Signh, J., & Sooch, S. (2004). Comparative Studies of Economies of Different Models of Family Size Biogas Plants for State of Punjab. *Energy Conversion and Management*, *45*, 1329–1341.

Smith, M., Goebel, J., & Blignaut, J. (2013). *The Economic and Fiancial Feasibility of Rural Household Biodigesters for Poor Communities in South Africa*. Waste Management.

Tinarwo, D., Mugera, W. M., Rasimphi, T. E., Nekhubvi, V., & Machaba, H. (2014). Assessment of biogas Technology, A Case Study of a rural village in Vhembe District, Limpopo, South Africa. Paper Presented at *a Conference, Nairobi*, Kenya.

World Energy Assessment. (2000). *Energy and the Challenge of Sustainability*. United Nations Development Programme.

World Health Organization. (2006). Fuel for Life Household Energy and health. Geneva: WHO.

Chapter 11
System Dynamics Modelling for Analyzing CO_2 Emissions:
A Case Study in Turkey

Arzu Eren Şenaras
https://orcid.org/0000-0002-3862-4551
Uludag University, Turkey

Şahin İnanç
Uludag University, Turkey

Hayrettin Kemal Sezen
Uludag University, Turkey

ABSTRACT

System dynamics is an interdisciplinary problem-solving methodology that utilizes several significant thinking skills such as dynamic thinking and cause-and-effect thinking. System dynamics is a disciplined collaborative approach that could accelerate learning by combining a multifaceted perspective that provides insight into complex and interactive issues. This chapter aims to develop a system dynamic model. The amount of CO_2 oscillations were investigated from the production of electrical energy depending on the rate ofpopulation growth up 2030. Model results, which are developed using Vensim package program, have been observed related to the variables of the adaptations.

DOI: 10.4018/978-1-5225-7289-3.ch011

INTRODUCTION

System dynamics approach was developed by Jay Forrester from MIT during 1950's to analyze especially the complex behavior in administration with computer simulation in social sciences. System dynamics is a form of systems approach as a methodology to understand the dynamic behavior of complex systems. The basis of system dynamics is to understand how system structures cause system behavior and system events (Sezen, 2009: 298).

Jay Forrester initially constructed his first dynamic model upon his meeting with the management of General Electric Corporation. Big fluctuations in production, inventory, labor force and profitability were compelling GE management. Despite hard efforts of the management, these fluctuations were mostly associated with external factors. Especially, the fluctuations in the business were related to received orders. Forrester interacted with the management to observe the system operations in other departments. In the first model he developed, he observed that simulations were necessary since the system could not be monitored analytically. He demonstrated that the corporation could experience serious fluctuations due to management policies even when the demand is considered constant with the weekly simulation he ran. Later on, he designed the computer simulation for the same problem. In his later studies, Forrester demonstrated how the feedback control theory could be adapted for complex administration and human systems. He published his initial findings in an article in Harvard Business Review. Later on, he developed this study to write his famous work "Industrial Dynamics" (Lane and Sterman, 2011; Ramage and Shipp, 2009: 100-101).

LITERATURE REVIEW

Some of the studies in the literature made by system dynamics approach are as follows.

A systematic dynamical study in China focuses on whether the future electricity demand is not high enough and whether it is right to conduct long-term power plants (Qian, 2008).

In a system dynamics study for South Africa, it is worth considering whether the increase in energy supply is due to enlargement of nuclear power generation, or whether it should be closed down with stringent measures to be taken in the use of energy. As a result of the work done, it turns out that the strict measures to be taken in energy use are more appropriate (Musango, Brent and Bassi, 2009).

In a system dynamics study in Germany, the factors affecting electricity prices and measures to reduce CO2 emissions are covered (Jäger, Schmidt and Karl, 2008).

A system dynamics study was conducted on how to reduce the impact of intercity passenger transport in China on CO2 emissions. This study will be higher than expected CO2 emissions in 2020 it will be emphasized and the development of rail transport is the best solution accordingly, (Han and Yoshitsugu, 2008).

United Arab another system dynamics work in the Emirates will also reduce CO2 emissions outside to meet the required energy for future electricity and clean water requirement seeks to identify the most appropriate technology (Choucr of Goldsmith and Mezher 2008).

A systematic dynamics study in Italy is about renewable energy sources. After the banning of nuclear energy in Italy in 1987 and the restriction of CO2 emissions by the Kyoto Protocol, there has been no shift towards renewable energy sources. In this study, the extent to which a renewable energy resources 2030 and fossil fuels to the costs arising from its use (together with additional costs for the excess CO2) extent advantageous or be disadvantaged by modeled (Giorgio, et.al, 2008).

Agent based simulation (ATS) is a new model technique used for modeling complex systems besides Forresterr's system dynamics (SD) approach model. In many studies models can be modeled by both approaches. In this study (SD), the specifications for deriving from the model (ATS) are presented (Macal, 2004).

The system dynamic model in Iran has been developed to show how it affects electricity price changes. Since electricity storage is unlikely to be practical in practice, it is necessary to protect supply demand balance. For this purpose, three groups, low, medium and high income according to income level, have been produced and distributed according to these conditions (Esmaeeli et al., 2006).

Alternative Energy Technologies (AET) in a study on (EEC) 'The development of a number of macroeconomic and geological forces, is uncertain because of a negative situation such as national oil companies. In this study, it is aimed to invest by using the system dynamics approach (EEC) for less risk (EEC) (Tan et al., 2007).

They have developed the SD model related to the global warming effect of oil use and renewable energy investments (Kars and Altunbay, 2010).

Liu et al.(2015) developed a system dynamics model to analyze the traditional energy consumption, alternative energy consumption and the CO2 emission performance until 2020 in China. A number of scenarios are simulated to examine the impact of different economic growth rates and policies on different combinations of energy uses and CO2 emissions in the future. They also examine whether the target of CO2 emissions per GDP in 2020 could be achieved.

Rafieisakhaei et.al. (2016). studied the main variables and causes behind the emissions, and investigate the mechanisms on the carbon market, particularly the European Union ETS. They also provide an updated oil market model to include the recent data and developments in the global oil market. They connect the two models to each other so that the effects of the oil price on the climate change can be investigated.

Capelo et al.(2017). studied to develop a sustainable EPC industry from an energy service venture perspective. A system dynamics model representing an ESCO venture in the Portuguese market was created. The model simulation provides a helpful basis for accelerating learning regarding the policies and managerial processes that are critical for the success of the venture. The simulation of the base case produces an overall insignificant market value added and a low probability of success, mainly due to long sales cycles. The model is sensitive to the wordof-mouth (WOM) parameter, however, which suggests that effective policy interventions should consider initiatives that accelerate WOM among EPC adopters and prospects.

SYSTEM DYNAMICS

As the systems grow more complicated, inter-system interactions have increased. Thus, the problems that emerge within the system and the system status continuously interact. The models that would be used to examine and assess these systems, which include novel concepts and change continuously, should be designed to

- scrutinize the system as a whole of the related processes,
- enable the system to monitor the system behavior and changes in that behavior,
- reflect the phenomena that change within the system and conditions that trigger these phenomena,
- to provide flexibility in redesign of the system (Erkut, 1995).

The conventional analytical approach implemented to estimate the behavior of a system or to explain the underlying problem concentrates all our attention on the immediate state of the system. In reality, however, events do not work in this way. Events do not determine the system behavior, but they occur as a consequence of the system behavior. If we are to examine how behavior occurs, then it is necessary to determine the system structure that the interactions between the elements of the system create. As it could be observed, the main factor in this approach is to be able to see and understand the whole created by the interactions between the elements of the system. Briefly, we need to make sure that interdependence is fundamental in our approaches rather than linear causality, so that we could better understand the problems we face today. According to Peter Senge, the properties of a system could not be defined by examining the isolated functions of the parts forming the system. Primarily, the behavior of a system is not about the function of each part, but it is about how each piece interacts with the others (Ayanoğlu and Gökçe, 2007).

System dynamics as a field of system theory is a method developed to understand the dynamic behavior of complex systems. The basic principle of system dynamics is that the behavior of the system is rooted in its structure and that the structure of the system (the relations between components) is as significant as the components themselves. Because, the behavior of the whole is often not explained by the singular scrutiny of the components; the behavior of the whole may be quite different from the behavior of the individual components (Öğüt and Şahit, 2012: 5).

The systems approach could be expressed as a collection of methods, tools, and principles that examine the relationships between the forces that form organizational decisions and behavior and perceive the forces forming the systems as a part of a common function. The idea underlying systems approach is that all system behavior follow certain common principles. As a new paradigm and method of analysis, systems approach aims to enable us to use our thinking methodology on how to improve organizational operations further more efficiently. Thus, the two keywords System and Thinking together are used to refer to Systems Thinking as a discipline. If a definition is needed for System Thinking / System Dynamics, it could be defined as scientific and artistic activity that attempts to draw reliable conclusions about the behavior of the system by continuously developing an in depth understanding of the system's infrastructure (Ayanoğlu and Gökçe, 2007).

System Thinking and System Dynamics provide computer technology and conceptual and numerical modeling techniques that could help explain the dynamics and forces underlying the complexity and change in business, politics, socio-economic and environmental systems. System dynamics, policy analysis and design are used for learning and decision making (Abdelbari et al., 2015; Maani and Cavana, 2007).

System dynamics deals with how things change over time. Almost all are interested in how the past formed the present moment and how today's actions determine the future (Forrester, 1995: 16).

The concept of dynamics indicates change over time. If something is dynamic, it changes constantly. Therefore, a dynamic system is a system in which there are interactions that promote change over time. System dynamics approach is a method used to understand how the system changes over time. The elements and variables that constitute a system that changes in time are expressed as the system behavior. The aim is to understand the basic behavior system of the variables, to discover the factors that cause this mode of behavior and to improve the system behavior. Thus, it could be argued that system dynamics is a method to explain how the systems change with time. In dynamic systems, variables influence each other simultaneously (Barlas, 2005a; Ayanoğlu and Gökçe, 2007).

Dynamic complexity arises from connections and disconnections that link social and business systems. When a change occurs in one part of the system, it causes change in another part later (sooner or later). These effects could not always be observed clearly, usually they are beyond expectations (Morecroft, 2015: 21).

Dynamic complexity arises due to the following systems characteristics (Sterman, 2000):

- **Dynamics:** Systems constantly change. Even an apparent constant would change over a long time period.
- **Close Relationship:** The components of the system have a strong interaction.
- **Management by feedback**: Actions return back to themselves. Decisions cause a change and trigger other system components to take action.
- **Nonlinearity**: Systems are mostly nonlinear, i.e. the effects are not proportional to their outcomes in general.
- **Historical dependence**: Many actions have no return. An action taken will cause irreversible effects as a result of several feedbacks.
- **Self-organization**: The dynamics of a system arises from its own internal structure.
- **Adaptability**: Rules and abilities change over time. They allow selection and multiplication as they develop over time.
- **Contradiction to intuition**: Causes and effects are different from each other in the temporal and spatial contexts. Individuals search for the reasons of events in the periphery of the events they are trying to explain.
- **Resistance to Politics**: The complexity of the system that we work on could reduce our ability to understand the system. After all, many obvious solutions fail and even worsen the situation.
- **Determination of Reconciliation**: Time delay feedbacks provide different responses in the short and long term. Often a good response occurs before the bad behavior (response).

The main principle of system dynamics is that the ongoing accumulation of the complex behavior of organizational and social systems (human, material, financial assets, information, biological and psychological states) is also the result of balancing and empowering feedback mechanisms. (Richardson, 1999).

System dynamics is a method that allows analysts to separate complex social, and behavioral systems into components, to visualize them by reconstructing them as a whole again, and to develop a simulation model (Tang and Vijay, 2001: 3).

There is never one single correct answer in system dynamics. Since system dynamics represents the relationships within the current system, it presents multiple possible approaches that could be applied, instead of providing a single correct answer. Each approach could provide some of the desired result, as well as some unexpected ones. In fact, another significant characteristic of system thinking is to become aware why we preferred a solution over others (Sterman, 2000: 127-133). System dynamics was designed to help managers make decisions similar to other information technologies. The aim of system dynamics studies is to assist decisions by analyzing old policies and designing new ones (Wheeler, 1994: 80).

System dynamics models are not derived statistically from time series data. They are statements related to policies and system structure that guide decisions. Models include admissions made about the system. A model is only as good as the expertise that underlies its formulation. A good computer model is distinguished from a bad one by the accuracy of its reflection of the system it represents. Several other mathematical models are limited due to the nonlinear nature of the real system and the lack of their inclusion of several feedback cycles. On the other hand, system dynamics computer models could reflect real system behavior (Forrester, 1995: 6).

The system dynamics approach is a simulation method used to analyze complex, nonlinear, dynamic feedback systems and to design policies to improve the system performance (Radzicki, 2007: 727).

System dynamics deals with the behavior of time-dependent management systems in order to identify and understand system behavior with qualitative and quantitative models. It investigates how system behavior is managed by information feedback and how simulation and optimization could be utilized to design healthy information feedback and control policies (Coyle, 1996: 10).

SD Model

This study aims to develop a system dynamic model. Amount of C02 oscillations were investigated from the production of electrical energy depending on the rate of population growth up 2030. Model results which is developed using Vensim package program have been observed related to the variables of the adaptations.

Figure 1 shows the pattern of energy input and output, population growth and CO2 emissions. In this model, the energy requirement is included in the model because it increases with population growth. Energy imports or exports also affect the amount of energy.

As shown in the Figure 2, the amount of CO2 increases compared to the years. Since this increase is a cumulative increase, it is seen that the annual increase rate is more rapidly. You need to look at the chart below for a consistent increase amount.

Figure 3 shows CO2 emissions over the years. According to energy demand, CO2 emissions also increase over time. In the context of the Kyoto protocol, CO2 emissions will have to be limited, especially in the coming years. Values of CO2E can be seen as Table 1.

Figure 1. SD Model

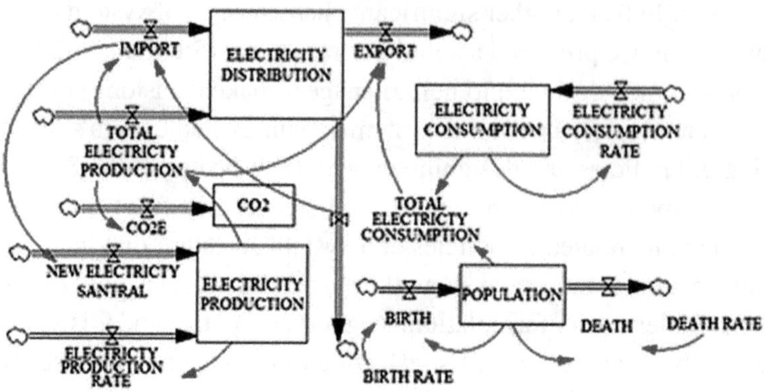

Figure 2. Graph of CO2

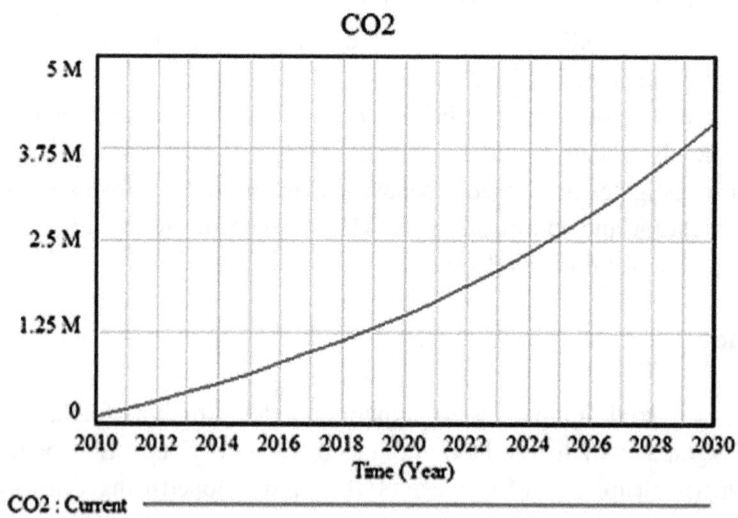

CONCLUSION

We can find this model in the following conclusions;

- Consistent estimates of CO2 emissions for the coming years can be made.
- There may be an answer to questions such as whether to restrict the power plants to reduce CO2 emissions.
- Decisions can be given more consistently, such as which bases should be initiated first and which ones should be supported to meet the energy need.

Figure 3. Graph of CO2E

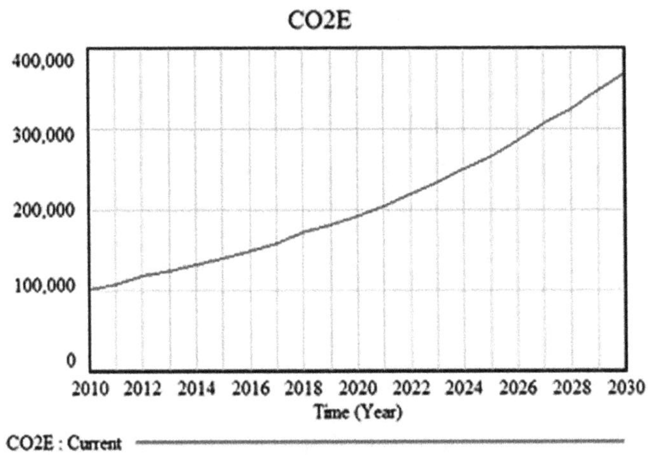

Table 1. Values of CO2E

Time (Year)	CO2E
2010	101470
2011	107484
2012	117752
2013	124859
2014	132408
2015	140376
2016	148763
2017	157765
2018	170991
2019	181155
2020	192041
2021	203452
2022	219348
2023	232434
2024	250141
2025	265130
2026	284777
2027	305678
2028	324024
2029	347387
2030	368186

- The model can be further developed to find out what measures can be taken to limit imports of energy from abroad. As a result of these measures, the current account deficit due to energy imports can be prevented and new domestic power plants and domestic employment can be provided.
- The answer to questions such as how much restriction should be made in the Kyoto protocol framework to be the most economical for clean power plants.
- Energies can be diversified and improved in different technological areas.
- Different power plants with low CO2 emissions, such as nuclear energy, can be considered. Thus, other technologies besides CO2 emissions are transferred.

REFERENCES

Abdelbari, Elsawah, & Shafi. (2015). Model Learning Using Genetic Programming Under Full And Partial System Information Conditions. *33rd International Conference of the System Dynamics Society*, Cambridge, MA.

Ayanoğlu, M., & Gökçe, M. (2007). Sistem Düşüncesinden Sistem Dinamiklerine. *Makine Teknolojileri Elektronik Dergisi*, 29-41.

Barlas. (2005a). Dinamik Sistem Yaklaşımı: Modeller, Kurumsal Öğrenme Ve Sorun Çözme – I. *Endüstri Ve Otomasyon Dergisi,* (94).

Capelo, Dias, & Pereira. (2017). *Exploring the Dynamics of an Energy Service Venture for Energy Efficiency Policy*. International Conference of the System Dynamics Society, Delft, The Netherlands.

Choucri, Goldsmith, & Mezher, . (2008). Framework for Modeling Technology Policy: Renewable Energy in Abu Dhabi. *International Conference of the System Dynamics Society*.

Coyle. (1996). *Systems Dynamics Modeling*. London: Chapman & Hall.

Esmaeeli, Shakouri, & Sedighi. (2006). Investigation Of Pricing İmpact On The Electrical Energy Consumption Behavior Of The Household Sector By A System Dynamics Approach. *International Conference of the System Dynamics Society*.

Forrester, . (1995). *Road Map 1: Counterintuitive Behaviour Of Social Systems*. MIT System Dynamics in Education Project.

Giorgio, Enzo, & Stefania, . (2008). Renewable Energy İn Italy: Scenarios To 2030. *International Conference Of The System Dynamics Society*.

Han, & Yoshitsugu, . (2008). *A System Dynamics Model Of CO2 Mitigation İn China's İnter-City Passenger Transport*. Elsevier Science Direct.

Jäger, T., Schmidt, S., & Karl, U. (2009). A System Dynamics Model For The German Electricity Market – Model Development And Application. *International Conference of the System Dynamics Society*.

Kars, C., & Altunbay, S. (2010). *Petrol Kullanımının Küresel Isınmaya Etkisi Ve Yenilebilir Enerji Yatırımları- Bir Sistem Dinamiği Modeli*. YAEM.

Lane & Sterman. (2011). Profiles in Operations Research: Jay Wright Forrester. In *Profiles in Operations Research: Pioneers and Innovators*. New York: Springer. Retrieved from http://jsterman.scripts.mit.edu/docs/Lane2011%20Profiles%20in%20Operations%20Research.pdf

Liu, Mao, Ren, Li, Guo, & Zhang, . (2015). How Might China Achieve Its 2020 Emissions Target? A Scenario Analysis Of Energy Consumption And CO2 Emissions Using The System Dynamics Model. *Journal of Cleaner Production*, *103*, 401–410. doi:10.1016/j.jclepro.2014.12.080

Maani & Cavana. (2007). Systems Thinking And Modelling Understanding Change and Complexity. Pearson Education.

Macal, C. M. (2004). To Agent-Based Simulation From System Dynamics. *International Conference of the System Dynamics Society*.

Morecroft. (2015). *Strategic Modelling And Business Dynamics: A Feedback Systems Approach*. Chichester, UK: John Wiley & Sons.

Musango, J. K., Brent, A. C., & Bassi, A. (2009). South African Energy Model: A System Dynamics Approach. *International Conference of the System Dynamics Society*.

Öğüt & Şahin. (2012). *Türkiye'de Finansal İstikrar Sorununa Sistem Dinamiği Yaklaşımı*. Türkiye Ekonomi Kurumu.

Qian, . (2008). An Experimental Study On The Construction Of Electricity Power Stations İn China. *International Conference of the System Dynamics Society*.

Radzicki, M. J. (1990). Methodologia Oeconomiae Et Systematis Dynamis. *System Dynamics Review*, *6*(2), 123–147. doi:10.1002dr.4260060202

Rafieisakhaei, Barazandeh, Moosavi, Fekri, & Bastani. (2016). *Modeling Dynamics of the Carbon Market: A System Dynamics Approach on the CO2 Emissions and its Connections to the Oil Market.* International Conference of the System Dynamics Society, Delft, The Netherlands.

Ramage. (2009). *Sytem Thinkers, The Open University Walton Hall.* Milton Keynes, UK: Springer London.

Richardson. (1999). *Feedback Throught in Social Science And Systems Theory.* Waltham: Pegasus Communications, Inc.

Sezen, H. (2009). *Günal Murat.* Bursa: Yöneylem Araştırmasında Benzetim, Ekin Yayınevi.

Sterman. (2000). Business Dynamics Systems Thinking And Modelling In A Complex World. McGraw-Hill.

Tan, Anderson, & Parker. (2007). Managing Risk in Alternative Energy Product Development. *International Conference of the System Dynamics Society.*

TangVijay, . (2001). *System Dynamics Origins, Development And Future Prospects of a Method.* Research Seminar in Engineering Systems.

Wheeler. (1994). How to Get Managers To Use System Dynamics. *International System Dynamics Conference*, Stirling, UK.

KEY TERMS AND DEFINITIONS

Decision Functions: States of the policy that determines how to convert available information in stock into a decision.

Delays: A characteristic of dynamic systems and affect both material and information flow.

Flow: It is defined as the increase or decrease in the unit time interval in stocks and denoted by f (*t*).

Information Delay: They are usually caused by the transfer of information and the delay in the process of taking an action after the information is received. Information delays could occur in vertical hierarchical administrations.

Information Flow: Creation, control, and distribution of information is the most significant task of business management. It might be difficult to model the information properly in the business process.

Material Delays: In this structure, the material moves forward from the beginning to the end of the pipeline without any change during a period of time, similar to a water running through a pipeline.

Material Flow: It includes stocks and flow rates of physical goods such as raw materials, inventories in the process, or finished goods.

Stocks: The present values of the variables that are formed by the accumulated difference between the inflow and outflow.

System Dynamic Language: This language consists of four components: stocks, flows, decision functions, and information flow.

System Dynamics: The basis of system dynamics is to understand how system structures cause system behavior and system events.

Chapter 12
Prevention of Air Pollution for Sustainable Environment

Md. Mahfuzar Rahman Chowdhury
The Legal Care, India

ABSTRACT

Air pollution is the natural processes caused by human activities through which certain substances enter the atmosphere at a sufficient concentration to cause environmental pollution. The World Health Organization reported that 80% of the world's cases of heart disease and stroke deaths were due to air pollution, and a total of 7 million people in the world died of air pollution in 2012. Unplanned urbanization, industrialization, and agricultural activities contribute to air pollution. Climate change affects air pollution in a number of factors including changes in temperature, solar radiation, humidity, precipitation, atmospheric transport, and biogenic emissions. Increasing scientific evidence shows that air pollution and climate change policies must be integrated to achieve sustainable development and a low carbon (LC) society. Combined efforts to deal with air pollution and climate issues at the urban level will be particularly important as most people are exposed to air pollution, and 75% of global GHG emissions are generated in urban areas.

INTRODUCTION

Air pollution has become the world's largest environmental health risk (Huanqiunet 2015). It is the natural processes caused by the human activities that certain substances enter the atmosphere at a sufficient concentration and endanger environmental pollution. In 2015, the World Health Organization released a report stating that

DOI: 10.4018/978-1-5225-7289-3.ch012

at least one in eight people died of air pollution globally. Rapid industrialization and urbanization are promoting development and at the same time increasing air pollution. The World Health Organization reported that 80% of the world's cases of heart disease and stroke deaths were due to air pollution, and a total of 7 million people in the world died of air pollution in 2012 (Huanqiunet, 2014).

Again, less than 1% of China's 500 major cities met the standards of the World Health Organization (Zhang, 2012). Since January 1, 2013, the Ministry of Environmental Protection has monitored the air quality index of different major cities, which are collectively called the 74 cities, in brief (China's National Environmental Monitoring Centre 2013). The concentrations of pollutants such as PM10 and SO2 have been monitored since November 2014 (China's National Environmental Monitoring Centre 2013a).

AIR POLLUTION

The vehicular pollution, industrial effluents, power plants, use of stems causes air pollution. Unplanned urbanization, industrialization, and agricultural activities contribute to the air pollution. It can have haphazard effects on the expansion, growth and durability of different creatures. Air pollution is harmful for health because of the presence of suspended particulate matter (SPM) in the air. Dust, effluents discharged from industries, and smoke from households and vehicles etc. are examples of SPM. They deteriorate air quality and causes environmental degradation. Pollutants released from vehicular pollution have adverse effects on flora, fauna, agricultural yield and human health. Moreover, with the growth of urbanization, industrial and commercial practices in the city, the requirement of transportation has been continuously increasing (Empirical Study of Air Pollution, 2014).

Vehicular emission causes adverse effects on respiratory and immune system of the human body as including causing pulmonary diseases and physiological effects. Carbon monoxide generated from vehicular emission causes sickness due to acute lack of oxygen in blood and tissues. Smokes emitted from using wood, crop residues, animal dung, and low quality coal produce dangerous particles and gases and their in-doors usage can cause serious respiratory disease and tuberculosis.

Air pollution is a slow poison and one of the greatest causes of major global diseases. Due to the effects of air pollution, the diseases like bronchitis, asthma, lung cancer and heart ailments affect human health. The state of air pollution is often expressed as air quality. The concentrations of gaseous pollutants and particulate matter are then used as a measure of air quality (Dalsøren, et al. 2016). Again, ozone is strongly coupled to meteorological variability both in terms of regional photochemical production and loss, and the contribution from intercontinental transport (UNEP, 2014).

Since the 1980s, air-quality regulations have led to a decline in air pollution (Mishchenko et al. 2007). Heavy air pollution can even lead to die-off of Sphagnum mosses, triggering for example peat erosion in blanket mires (Tallis, 1985). Contribution to air pollution and deposition from domestic and transboundary sources varies significantly subject to the pollutant, local source strength, proximity to major transboundary sources, and the geographical size of the individual countries. The transportation sector consumes about 28% (91 Exa joules) of the global energy demand and contributes to 23% of fossil fuel CO2 emissions (The World Bank, 2018).

Air pollution (smoke and noise), municipal waste problems, loss of forest areas, habitat destruction, threats to biodiversity, resource depletion, and the global green house problem are related to economic growth (Alege & Ogundipe, 2013). A critical collapse of the world economy due to environmental problems might impede growth and generates adverse effects on an economy.

AIR POLLUTION AND CLIMATE CHANGE

Climate change affects air pollution in a number of factors as including changes in temperature, solar radiation, humidity, precipitation, atmospheric transport and biogenic emissions. Since the famous London killer fog of 1952, it has been clear that air pollution may result devastating impacts, different perceptions centering air pollution are moving around the developed and developing countries. One of the effects of climate change is heat wave. The 2003 heat wave caused around 70,000 excess deaths in Europe, with about 20–38% attributed to air pollution (Jalkanen 2011). There was an overall 17% increase in death rates for England and Wales with the excess mortality most pronounced in London, with a 33% increase in the over 75-year old age group (Kovats et al. 2006).

Air pollutants can influence climate through their radioactive balance of the atmosphere. Emissions from aviation, such as carbon dioxide (CO2), nitrogen oxides (NOX), and particulate matter, contribute to climate change via radioactive forcing (European Aviation Environmental Report, 2016). Impact of aviation on GHG emissions has grown rapidly in recent years owing to the significant increase in the annual average number of passengers using air transport (Kousoulidou & Lonza, 2016).

SUSTAINABILITY AND DEVELOPMENT

National security is a very popular notion and concerned with the international peace and security. The security in the 1960s was no longer simply an issue of external aggressors but also one of internal threats to stability. While national security concerns were being dominated by the threat to a state's borders and sovereignty, the environment began to take its place within the security agenda in the 1970s at the time the Club of Rome published *The Limits to Growth (*Meadows, Randers, Meadows, & Behrens, 1972*)*. Alexandre Timoshenko rightly articulates security cannot only be political or military; it must be as well ecological, economical, and social. It must ensure the fulfillment of the aspirations of humanity as a whole. Thus, the necessity of a secure environment for the survival of mankind was duly stressed in the report.

There is an essential connection between social justice and the environment (Kassiola, 2002) that casts the essence of environmental policy. Proper application of policy and regulation in respect of environmental sustainability can help earning benefits of sustainable development. The Court of Justice of the European Union (EU) has, to a certain extent, demonstrated willingness to engage in balancing exercises where commercial economic interests are involved (Hamlyn, 2015). In fact, a balancing decision making among social, economic, and environmental aspects renders development.

In 1950s and 1960s, unsustainable resource exploitation and population increases severely affected the environmental sustainability. The utilization of natural resources pursuant to the national policies became a key concern for the nation states for their economic development. In 1977, Lester Brown published a World Watch Paper under the title of *Redefining National Security* in which he criticized the militarization of the world economy and opined that the emerging threat for the mankind would be their relationship with nature.

In the 1980s, the theme of unsustainable environmental exploitation gathered pace leading to the World conservation Union's World Conservation Strategy in 1980 (World Conservation Union, 1980) and the formation and reports of the World Commission on Environment and Development. Brundtland Report of the commission put an emphasis on an integrated approach and subsequently it became known as sustainable development.

INTERNATIONAL ENVIRONMENTAL LAW

Climate change continues to dominate international environmental efforts, and thereby, the issues of sustainability and development. Environmental law continues to change and evolve, as new regulations are adopted and existing rules are amended or challenged in courts or interpreted by agencies. Therefore, the international community is experiencing many areas of relevant international legal activity as well as the breadth and complexity of international development and economic activity.

The key international environment instruments in respect of climate change issues are the UNFCCC and the Kyoto Protocol that invoke the principles of inter- and intra-generational equity, Common But Differentiated Responsibilities (CBDR), the precautionary principle, the right of all parties to sustainable development, and the need to promote a supportive and open international economic system.

Obligation of not to causing environmental harm is widely viewed as an extension of the general international norm that prohibits a state from using its territory, or allowing others to use its territory, in a way that can harm the interests of another State. Despite the significant progress that international treaties have made on certain issues, such as regulation of the ozone layer, treaties do have particular weaknesses in effectively regulating environmental issues. States can too easily make reservations or derogations in respect of obligations as inserted in the treaty provisions that might go against the interest of individual member state(s). Therefore, the binding force or effectiveness of the treaties raises serious concern for the member states as indiscriminate activities of a single or a group of states can bring catastrophic effects for the people across the countries.

Treaty negotiations are the primary international approach to addressing climate change while the slow pace of treaty negotiations has been frustrating. Growing number of countries have integrated emission trading schemes into their national climate policies, and regional arrangements have been established to combat climate change. The essence of a functional environmental law regime for a sound global economic order has been a demand of time. Compliance with the principles as including the principle of Common But Differentiated Responsibility (CBDR) has been an urgent concern for the existence of humans and their environment in the planet.

HISTORICAL BACKGROUND

International environmental law has developed through bilateral and multilateral negotiation of treaties and "soft law" instruments such as declarations and plans of action. It can be traced back to 1351 as to the conclusion of a fishery treaty between England and Castile. Subsequently, the significant development was initiated with the emergence of 1893 Pacific Fur Seal Arbitration[1] and Trail Smelter Arbitration[2].The principle developed in Trail Smelter Arbitration is pivotal to prevent transboundary environmental harm and subsequently the same became as settled principle of customary international law. Initially, international environmental law treaties concerned protection of interests of particular states, rather than the environmental protection in common. In the late 1990s, serious concerns were expressed about growing evidence of environmental degradation that prompted to realize that environmental degradation was closely linked to unsustainable levels of economic development and population growth.

The United Nations General Assembly articulated, ...effective measures to preserve natural resources, flora and fauna should be taken at the earliest possible moment simultaneously with economic development, including industrialization, and urbanization (United Nations General Assembly, 1962). United Nations Conference on the Human Environment (UNCHE) in Stockholm in 1972 aroused an understanding that development and environmental issues are clearly interdependent.

In addition to UNCHE in Stockholm in 1972, the 1992 United Nations Conference on Environment and Development (UNCED) in Rio de Janeiro, the World Summit on Sustainable Development (WSSD) in Johannesburg in 2002 were significant events for the development of international environmental law and adoption of the notion of rights to development and protection of the environment. The Stockholm meeting produced a number of significant outcomes including the establishment of UNEP and the adoption of the Stockholm Declaration (UNCHE, 1972) setting out 26 Principles and an Action Plan containing 109 recommendations.

The Bruntland Report in 1987 emphasized on the need to ensure "sustainable development" that meets the needs of the present without compromising the ability of future generations to meet their own needs. The Brundtland Report led to the UNCED (the Earth Summit) in Rio de Janeiro in Brazil in 1992. The Earth Summit adopted three non-binding instruments: the Rio Declaration, the UNCED Forest Principles, and Agenda 21. In addition, the Convention on Biological Diversity and the UN Framework Convention on Climate Change were opened for signature at the Earth Summit.

Following the Earth Summit, the existence of several key principles of international environmental law including the precautionary principle, the "polluter pays" principle, common but differentiated responsibility, sustainable development reflect environmental and development concerns.

UNFCCC began to hold Conference of the Parties to address the adverse consequences of climate change. States gathered in Berlin, Germany for holding the 1st Conference of Parties from 28 March to 7 April 1995.

IMPACTS OF AIR POLLUTION

Air quality perturbs the physical and chemical properties of the environment, and thus, impacts on the environment and contributes to the climate change. Climate policies imply energy efficiency and technical measures for emissions of air pollutants. Air quality mitigation measures affect greenhouse gas emissions. Mitigation optimized for a climate or air quality target in isolation could have synergistic or antagonistic effects (UNEP, 2014).

The concentrations of pollutants such as PM10 and SO2 have been monitored since November 2014 (China's National Environmental Monitoring Centre 2013a). Short-term exposure to particulate air pollution was associated with increased ischemic heart disease (IHD) mortality (Xu et al. 2014). Total economic cost of the health impact was approximately 341.4 billion Yuan, 87.79% of which was attributable to premature deaths (RJ et al. 2010).

SDG 3.9 seeks to substantially reduce the number of deaths and illnesses from hazardous chemicals and air, water and soil pollution and contamination by 2030. Based on a recent report from the authoritative Lancet Commission on Pollution and Health (The Lancet, 2017), at least 9 million deaths per year are caused globally due to pollution.

Air pollution has been a threat to the human existence as this can be depicted from the following observation:

In Bangladesh, Somalia and Chad, over a quarter of all deaths are attributable to pollution because of household air pollution, unsafe water sources and inadequate sanitation. In countries such as the United States, Russia and China, most pollution-related deaths are caused by modern pollutants linked to industrial development. (The Lancet, 2017)

Environmental hazards, including exposure to air pollution, are known to have a disproportionately severe effect on people whose place in society has already compromised their health and nutritional status (Air Quality and Sustainable Development, 2018).

Air pollution has become the fourth highest risk factor for premature deaths – one in 10 deaths worldwide is attributable to air pollution exposure (The World Bank, 2018). According to the World Health Organization (WHO), an estimated 12.6 million people die from environmental health risks annually.

IMPLICATION FOR THE PREVENTION OF AIR POLLUTION

Between 2003 and 2012, there was no clear trend in the annual mean of the daily maximum eight hour average ozone concentrations, recorded at different types of station (traffic, urban, rural and other, mainly industrial), for the EU-28 as a whole (European Environmental Agency, 2018).

Due to industrial revolution in the 19th century, air quality in many parts of the world has been threatened. Greenhouse gas (GHG) emissions have rapidly increased with the rise of energy consumption worldwide. Combustion, mining, agriculture, transportation and many other societal practices have been associated with the emissions of gases and aerosols that threaten the health and quality of life of billions of people, as well as eroding the quality of ecosystems and endangering other species (Vallero, 2014). According to the Fifth Assessment Report of the Intergovernmental Panel on Climate Change (IPCC), without additional GHG reduction efforts, the average global temperature will increase by 3–5 oC by 2100 (Hong et al. 2016).

Increases in the frequency or intensity of some extreme weather events, rising sea levels, and melting snow and ice are already disrupting people's lives and damaging some sectors of the U.S. economy (Melillo, 2014). Among the various forms of pollution, air pollution is the most detrimental to health. Though the international environmental agreements and policy pronouncements are framed they have minimal level of implementation at state and local level.

Violation of air pollution legislation both in national and international level is a commonplace. The inter-relation between air pollution and sustainable development is inevitable. Uninterrupted air pollution impedes sustainable development as pronounced in international commitments and policy pronouncements.

Some urban areas and some regions in Europe, prominent for most energy consumption and most greenhouse gases (GHGs) emission, will continue to struggle with air pollution and related health effects. These areas will require ad hoc additional local actions to complement medium and long-term national and EU-wide strategies to reach EU air quality objectives (Guariso & Volta, 2017).

Air quality and related impacts are therefore influenced by local meteorological features, regional (transboundary) processes, and intercontinental transport (UNEP, 2014). Poor air quality in Europe is a serious human health issue as annual premature mortalities were over 400,000 in the EU area in 2010 and the total external costs of the health impacts were estimated at EUR 330–940 billion (EU 2013). In the European Union, ship traffic accounted for 7% of the total health effects in Europe due to air pollution in 2000 (Brandt et al. 2013). In Denmark, the relative contribution from international ship traffic is comparable to the contributions from road traffic or the domestic use of wood stoves (Brandt et al. 2013a).

The sources of air pollution that emit the Short Lived Climate Pollutants (SLCPs), causes deaths in millions, destruction of crops in millions of tons, melts snow packs and glaciers worldwide (Ramanathan, 2014). The reductions in SLCPs would cut the projected warming trends from now to 2050 by nearly half. Mitigation of SLCPs also helps save over 3.5 million lives lost to air pollution every year and save as much as hundred million tons of cop damages and significantly slowing down retreat of Himalayan glaciers (The World Bank, 2018).

The first major attempt to calculate both the human and financial cost of the continent's pollution suggests dirty air could be killing 712,000 people a year prematurely, compared with approximately 542,000 from unsafe water, 275,000 from malnutrition and 391,000 from unsafe sanitation (The Guardian, 2016). For Africa as a whole, the estimated economic cost of premature air pollution deaths in 2013 was roughly $215bn (£175bn) a year for outdoor air pollution, and $232bn for household, or indoor, air pollution (Austin et al. 1997). Africa's local air pollution is likely to contribute considerably more to climate change when its population increases to around 2.5 billion, or 25% of the world's population in 2050, and thence to around 4.4 billion, or 40% of the world's population in 2100(Austin et al. 1997).

Poor air quality is a growing challenge, especially in cities and urban centres, compromising the lives of millions worldwide. Action to reduce air pollution will save lives and provide co-benefits for the climate, ecosystem services, biodiversity, and food security (Steiner, 2014). United Nations Environment Assembly, held in Nairobi on June 23-27, 2014, unanimously agreed to encourage governments to set standards and policies across multiple sectors to reduce air pollution emissions and manage the negative impacts of air pollution on health, the economy, and overall sustainable development (UNEP, 2014).

AIR POLLUTION AND SUSTAINABLE DEVELOPMENT

Sustainable development is meant to balance and integrate the social, economic, and environmental components in a given society. Managing air quality within a sustainable development context put emphasis on balancing among environment, economy, and society.

Greenhouse gas (GHG) emissions have rapidly increased with the rise of energy consumption worldwide. According to the Fifth Assessment Report of the Intergovernmental Panel on Climate Change (IPCC), without additional GHG reduction efforts, the average global temperature will increase by 3–5 C by 2100 (Hong et al. 2016). The emissions of GHGs and other air pollutants generated by transportation (vehicles, trains, aircraft, and ships) have significantly impacted the atmosphere and contribute to climate change (Shon et al. 2011).

Science in general and environmental science specifically is beginning to make use of widely available communications and sensor applications and technologies to improve the coverage and granularity of environmental measurements (Hasenfratz, 2012). Research on the relationships between ecosystem conditions and air quality, climate change impact assessments, and pollution detection and compliance monitoring (Haklay, 2013) can be useful for the greater purposes.

The Air Pollution Model (TAPM) (Hurley et al. 2005) was used for the simulation of different mitigation scenarios. It predicts meteorology and air pollution concentrations, simulates the transport, dispersion and chemistry of atmospheric pollutants, at both local and regional scale, and thereby, suitable for long term simulations.

INTERNATIONAL COOPERATION FOR CONTROLLING AIR POLLUTION

The increase in the breadth and depth of international environmental cooperation is the reflection of the necessity of environmental protection for sustainability. The environmental agreements vary in scope, ranging from global to regional, sub-regional, and bilateral arrangements. Global agreements tend to be a framework character whereas more specific requirements are often found in other agreements at regional, sub-regional or national level. International agreements and other cooperative agreements have no independent influence on the actions of the states as they reflect the national interests of the participating states or the specific preferences of the most powerful states.

1979 UN Economic Commission for Europe Convention on Long-range Transboundary Air Pollution (LRTAP) addresses problems in Europe and North America concerning airborne pollutants and develops a framework for co-coordinating pollution control measures and common emission standards.The concerned states are required to adopt precautionary principle approach as set forth in the 1992 Rio Declaration. A monitoring system has been set up, and UN Economic Commission for Europe 1988a, volatile organic compounds 1991, UN Economic Commission for Europe 1991, UN Economic Commission for Europe 1994, UN Economic

Commission for Europe 1998a, UN Economic Commission for Europe 1998b have been negotiated.

Joint Russian–Norwegian Environmental Commission, established in 1988 comprising of the leading environmental agencies in the two states has contributed to the establishment of the Working Group on Airborne Pollution to control air pollution. Norway has developed an environmental monitoring and modeling program and established the Environmental Commission.

RECENT DEVELOPMENT

Responding to carbon emission scheme can bring sustainable changes for the lives and environment. Stern Review rightly puts, cost of unabated climate change on the global economy will equate to a loss of at least 5% of global gross domestic product (GDP) each year, potentially rising to 20%. In contrast, a shift to a low-carbon pathway would benefit the economy by US$2.5 trillion a year, with markets for low-carbon technologies worth at least US$500 billion per year by 2050. Avoiding the worst impacts of climate change would require an investment of only 1-2% of global GDP per annum (Stern, 2007).

Warming of the climate system is unequivocal, and since the 1950s, many of the observed changes are unprecedented over decades to millennia (IPCC, 2014). Again, atmospheric concentrations of carbon dioxide, methane, and nitrous oxide have increased to levels unprecedented in at least the last 800,000 years. Human influence on the climate system is clear from the increasing greenhouse gas concentrations in the atmosphere, positive radiative forcing, observed warming, and our current understanding of the climate system. Carbon dioxide concentrations have increased by over 40% since pre-industrial times, primarily from fossil fuel emissions, and from net land-use change emissions (IPCC, 2014). Singapore may experience 'climate departure' as soon as 2028 (i.e. the year when the coolest annual temperatures are predicted to exceed the warmest annual temperature recorded in a 1960-2005 baseline period) (Mora et al., 2013) and may cause sea level rise.

U.S. Climate Science Special Report: Fourth National Climate Assessment termed the current period as the warmest in the history of modern civilization clearly mentioning that human activities, especially emissions of greenhouse gases, are the dominant cause of observed warming since the mid-20th century and the current trend might cause a global sea level rise of up to 2.4 metres by the end of the century, an increased incidence of drought and flooding, more frequent wildfires, and devastating storms (Wuebbles et al., 2017). The IPCC Fourth Assessment Report in 2007 warned that due to the disintegration of global ice-sheets, notably

Greenland and Western Antarctica might cause near-term sea level rise to coastlines and low-lying islands.

The UN World Meteorological Organization (WMO) reported that 2016 was the warmest year on the instrumental record, with the yearly rise in atmospheric carbon dioxide in 2016, at 3.1 ppm, was 50% higher than the previous 10-year average (World Meteorological Organization Greenhouse Gas Bulletin, 2017). As per the latest data from the WMO, 2017 was shaped up to be the third warmest year on the instrumental record, meaning that 17 out of the 18 warmest years in the instrumental record will have all occurred in the 21st Century (Obbard, 2017).

The United Nations Emission Gap Report, in its 2017 review of global carbon emissions, reported that the emissions gap between the national pledges under the PCA to cut carbon emissions and the actual reductions needed to keep temperature rise below 2oC of warming is now "alarmingly high". Coal combustion in China contributed to 2oC rise of carbon emission in 2017(Global Carbon Project, 2017) while reluctance of USA adds negative impacts on the carbon emission trajectory worldwide. Inaction of the global leaders and states concerned simply reveal the observation that time is simply running out to keep warming well below 2oC under the PCA let alone 1.5oC.

The United Nations-led 21st Conference of the Parties (COP-21) to the United Nations Framework Convention on Climate Change held at Paris in December, 2015 agreed on some important issues that form the basis of inspiration to address the effects of climate change. The PCA established specific actions and targets for the world to reduce greenhouse gases emissions, to mitigate and adapt to climate change, and to finance efforts to alleviate impacts in developing countries. As of November, 2017, a total of 195 countries have signed the PCA, and 169 have become a party to it. Signatory countries pledge to limit global mean temperature rise to below 2oC and make strong efforts to keep the rise to no more than 1.5oC. However, states should concentrate more on achieving carbon mission targets for a sustainable world.

States concerned unanimously agreed to enhance capacities of global response to climate change, and therefore, limit global temperature increase to well below 2 degrees Celsius, possibly to 1.5 degrees. They also agreed to communicate their nationally determined contribution (NDCs) every five years and provide information necessary for clarity and transparency reflecting gradual progress with the NDCs (i.e., successive NDC will represent a progression than the previous NDC). COP-21 also asked the developed countries to take the lead by undertaking economy-wide reduction targets, while developing countries were urged on enhancing their mitigation efforts, and subsequently move toward economy-wide targets considering the typical circumstances of a given country with the aspiration of a low carbon future for the planet.

Recently, 23rd Conference of the Parties (COP23) concluded on 20 November 2017 in Bonn, Germany. Global leaders gathered in the meeting with a view to determining how best to implement the goals of the Paris Climate Agreement (PCA) of COP21 held in December, 2015. The standing of the U.S. in COP-23 was not cooperative rather its plan of withdrawal from Paris Climate Agreement cast the light of despair and uncertainties while the global temperatures are still rising, and carbon emissions have resumed upward trend. A global transformation to low-carbon technologies has now become embedded as an economic and environmental imperative.

Thus, despite the U.S. withdrawal from the PCA, the intended changes will pave the way forward to a safer world with economic prosperity and environmental sustainability. In response to the U.S. stance on coal, several countries including Canada, the United Kingdom and Mexico responded at COP23 by forming a new global alliance to move away from unabated coal combustion.

RECOMMENDATION AND CONCLUDING REMARKS

The current global response to pervasive poor air quality is inadequate. Despite this lack of a holistic response, numerous countries and regions are coming up with effective - and cost-effective - measures to improve air quality. The Sustainable Development Goals provide an opportunity to replicate those best practices globally, and bring about cleaner air, and social and economic benefits worldwide (Steiner, 2016). Global urban air pollution levels increased by 8 per cent between 2008 and 2013 and more than 80 per cent of people living in urban areas are exposed to air quality levels that exceed WHO limits, threatening lives, productivity and economies (CCAC Secretariat, 2016).

Climate impacts on air quality are substantial as being evidenced by the heat wave of 2003 in Europe. Heat waves might be more frequent in the coming decades, and therefore, reductions in the emission of air pollutants need to be achieved. Otherwise, the planet might experience catastrophic effects and humans and their environment would be put into risks. It is predicted that changes in air pollutant emissions as well as in climate will affect the future air quality

Cost-effective allocation of controls along geographic lines or according to preferences for air versus water quality improvement is necessary (Austin et al. 1997). Again, costs of reducing emissions could be lowered by reallocating emission reductions based on type of source and also on geographic location (Krupnick et al. 1998). European Commission assessed the marginal environmental costs of energy production and identified key externality issues to be addressed in future policy (European Commission, 1996).

Effective mitigation of SLCPs will require detailed assessments of the multiple impacts of emitted air pollutants together with CO_2, their sources, their atmospheric interactions and their potential for mitigation (Schmale et al. 2014). Concerted efforts at city and state level is important as more people are exposed to air pollution, and 75% of global CO_2 emissions is generated in cities. Controlling pollution can improve overall scenario and generate welfare by economizing on energy resources and minimizing unwanted residuals.

Air pollution and climate change threaten longevity and quality of life of the citizens. Thus, the integration programs for reducing and mitigating the corresponding impacts would involve substantial cost savings and important benefits for human health and the environment. Increasing scientific evidence shows that air pollution and climate change policies must be integrated to achieve sustainable development and a low carbon (LC) society (Ramanathan, 2014). Combined efforts to deal with air pollution and climate issues at the urban level will be particularly important as most people are exposed to air pollution, and 75% of global GHG emissions are generated in urban area (Schmale et al. 2014a).

Setting the target for limiting the maximum per capita emission to about 10 tons/PP/year until 2030 and reduce it to near zero by 2070 (The World Bank, 2018) can help to meet the sustainable development goals. In addition, reducing both CO_2 and the four short-lived climate pollutants-methane; black carbon; ozone and HFCs-for the reduction of warming in the coming decades can save 4 million lives, avoid hundred million tons of crop damages, and slow down the melting of Himalaya glaciers. Enabling clean energy for cooking and lighting and reducing carbon dioxide emissions will save 4 million live every year, slow the rate of warming, melting of Himalayan-Tibetan glaciers, and increase monsoon rainfall. Initiating sub-national and nation-to-nation collaborative projects can be helpful for sharing the knowledge and actions to reduce air pollution and GHGs emissions from outdated technologies in the transportation and industrial sectors in developing nations. One-third of total food is wasted each year globally that requires emission of a great deal of CO_2. Efforts to minimize the waste of food can ensure reduction of CO_2 emission for the production of food.

REFERENCES

Air quality and sustainable development. (n.d.). Retrieved February 15, 2018 from https://www.environment.gov.za/sites/default/files/docs/stateofair_airqualityand_sustainable_development.pdf

Alege, O. P., & Ogundipe, A. A. (2013). Environmental quality and economic growth in Nigeria: A fractional cointegration analysis. *International Journal of Development and Sustainability*, *2*(2), 580–596. Retrieved from https://isdsnet.com/ijds-v2n2-10.pdf

Austin, H. D., Krupnick, J, A., & McConnell, D. V. (1997). Efficiency and Political Economy of Pollution Control with Ancillary Benefits: An Application to NOX Control in the Chesapeake Bay Airshed. In *Resources for the Future, 1997.* Discussion Paper 97-34. Retrieved February 28, 2018 from http://ageconsearch.umn.edu/bitstream/10553/1/dp970034.pdf

Brandt, J., Silver, D. J., Christensen, H. J., Andersen, S. M., Bønløkke, H. J., Sigsgaard, T., ... Frohn, M. L. (2013). Assessment of past, present and future health-cost externalities of air pollution in Europe and the contribution from international ship traffic using the EVA model system. *Atmospheric Chemistry and Physics*, *13*(15), 7747–7764. doi:10.5194/acp-13-7747-2013

Brandt, J., Silver, D. J., Christensen, H. J., Andersen, S. M., Bønløkke, H. J., Sigsgaard, T., ... Frohn, M. L. (2013a). Contribution from the ten major emission sectors in Europe and Denmark to the health-cost externalities of air pollution using the EVA model system – an integrated modelling approach. *Atmospheric Chemistry and Physics*, *13*(15), 7725–7746. doi:10.5194/acp-13-7725-2013

Chen, R. J., Chen, B. H., & Kan, H. D. (2010). A health-based economic assessment of particulate air pollution in 113 Chinese cities. *Zhongguo Huanjing Kexue*, *30*, 3.

China's National Environmental Monitoring Centre. (2013). *Monthly air quality report in January 2013 in 74 cities.* Retrieved from https://www.ncbi.nlm.nih.gov/pmc/articles/PMC4816944/

Dalsøren, B. S., Jonson, E. J., Quante, M., & Colijn, F. (Eds.). (2016). *North Sea Region Climate Change Assessment*. Regional Climate Studies. doi:10.1007/978-3-319-39745-0_16

Empirical Study of Air Pollution and Sustainable Development of. (2014, September 19). Retrieved April 14, 2018 from shodhganga.inflibnet.ac.in/bitstream/10603/109618/15/15%20chapter%206.pdf

EU. (2013). *Commission staff working document impact assessment*. Retrieved April 13, 2018 from https://eurohealthnet.eu/media/news-releases/eurohealthnet-expresses-concern-european-commission-withdrawal-anticipated-eu-%E2%80%98

European Aviation Environmental Report. (2016). Retrieved October 16, 2016 from https://www.easa.europa.eu/eaer/

European Commission. (1996). ExternE — Externalities of Energy. A Research Project of the European Commission: Vol. 1. *Summary Report EUR 16520 EN*. Brussels: Author.

European Environmental Agency. (2018). *Air pollution due to ozone: health impacts and effects of climate change*. Retrieved March 15, 2018 from https://www.eea.europa.eu/data-and-maps/indicators/air-pollution-by-ozone-2/assessment

Global Carbon Project. (2017). *Carbon Budget 2017*. Retrieved on November 20, 2017, from: http://www.globalcarbonproject.org/carbonbudget/index.htm

Guariso, G., & Volta, M. (Eds.). (2017). *Air Quality Integrated Assessment*. Springer Briefs in Applied Sciences and Technology. Springer. Retrieved April 10, 2018 from https://link.springer.com/chapter/10.1007/978-3-319-33349-6_6#

Haklay, M. (2013). Citizen science and volunteered geographic information: Overview and typology of participation. In *Crowdsourcing Geographic Knowledge*. Heidelberg, Germany: Springer. doi:10.1007/978-94-007-4587-2_7

Hamlyn, O. (2015). Sustainability and the Failure of Ambition in European Pesticides Regulation. *Journal of Environmental Law*, *0*, 1–25. Retrieved from http://discovery.ucl.ac.uk/1469006/7/Sustainability%20and%20the%20Failure%20of%20Ambition%20in%20European%20Pesticides%20Regulation.pdf

Hasenfratz, D., Saukh, O., Sturzenegger, S., & Thiele, L. (2012). Participatory air pollution monitoring using smartphones. *Proceedings of the 2nd International Workshop on Mobile Sensing*.

Hong, S., Chung, Y., Kim, J., & Chun, D. (2016). Analysis on the level of contribution to the national greenhouse gas reduction target in Korean transportation sector using LEAP model. *Renewable & Sustainable Energy Reviews*, *60*, 549–559. doi:10.1016/j.rser.2015.12.164

Huanqiunet. (2014). *WHO reported that 7 million people died of air pollution in 2012*. Retrieved April 10, 2018 from http://world.huanqiu.com/exclusive/2014-03/4929416.html

Huanqiunet. (2015). *WHO: Air Pollution has become the World's largest Environmental Health risk*. Retrieved January 22, 2018 from https://www.ncbi.nlm.nih.gov/pmc/articles/PMC4816944/

Hurley, J. P., Physick, L. W., & Luhar, K. A. (2005). TAPM: A practical approach to prognostic meteorological and air pollution modelling. *Environmental Modelling & Software*, *20*(6), 737–752. doi:10.1016/j.envsoft.2004.04.006

IPCC. (2014). *Climate Change 2014: Synthesis Report. Contribution of Working Groups I, II and III to the Fifth Assessment Report of the Intergovernmental Panel on Climate Change* (R. K. Pachauri & L. A. Meyer, Eds.). Geneva, Switzerland: IPCC.

Jalkanen, L. (2011). WMO addressing climate and air quality. *Technical Workshop on Science and Policy of Short-lived Climate Forcers*. Retrieved March 31, 2018 from http://mce2.org/ SLCFWorkshop/docs/%28Jalkanen%29%20SLCF%20Mexico%20WMO.pdf

Kassiola, J. (2002). Why Environmental Thought and Action Must Include Considerations of Social Justice. In M. Gillroy & J. Bowersox (Eds.), *The Moral Austerity of Environmental Decision Making Sustainability, Democracy, and Normative Argument in Policy and Law*. Duke University Press.

Kousoulidou, M., & Lonza, L. (2016). Biofuels in aviation: Fuel demand and CO2 emissions evolution in Europe toward 2030. *Transp. Res. D.*, *46*, 166–181. doi:10.1016/j.trd.2016.03.018

Kovats, R. S., Johnson, H., & Griffiths, C. (2006). Mortality in southern England during the 2003 heat wave by place of death. *Health Statistics Quarterly*, *29*, 6–8. PMID:16523675

Krupnick, A., McConnell, V., Austin, D., Cannon, M., Stoessell, T., & Morton, B. (1998). The Chesapeake Bay and the Control of NOX Emissions: Policy Analysis. *Resources for the Future, 1998.* Discussion Paper 98-46. Retrieved February 28, 2018 from http://ageconsearch.umn.edu/bitstream/10576/1/dp980046.pdf

Meadows, D. H., Randers, J., Meadows, D. L., & Behrens, W. W. (1972). *The Limits to Growth: Report for the Club of Rome's Project on the Predicament of Mankind*. New York: Universe Books.

Melillo, J. M., Richmond, T. C., & Yohe, G. W. (Eds.). (2014). *Climate Change Impacts in the United States: The Third National Climate Assessment*. Washington, DC: U.S. Global Change Research Program; doi:10.7930/J0Z31WJ2

Mishchenko, M. I., Geogdzhayev, I. V., Rossow, W. B., Cairns, B., Carlson, B. E., Lacis, A. A., ... Travis, L. D. (2007). Long-term satellite record reveals likely recent aerosol trend. *Science*, *315*(5818), 1543. doi:10.1126cience.1136709 PMID:17363666

Mora, C., Frazier, A. G., Longman, R. J., Dacks, R. S., Walton, M. M., Tong, E. J., ... Giambelluca, T. W. (2013). The projected timing of climate departure from recent variability. *Nature*, *502*(7470), 183–187. doi:10.1038/nature12540 PMID:24108050

Obbard. (2017). *APCEL Climate Change Adaptation Platform COP23 Bonn, Germany: The U.S. Stands Alone*. Asia Pacific Centre for Environmental Law. National University of Singapore.

Ramanathan, V. (2014). The Two Worlds Approach for Mitigating Air Pollution and Climate Change. *Sustainable Humanity, Sustainable Nature: Our Responsibility*. Extra Series: 41, Pontifical Academy of Sciences. Retrieved April 13, 2018 from http://www.pas.va/content/dam/accademia/pdf/es41/es41-ramanathan.pdf

Schmale, J., Shindell, D., von Schneidemesser, E., Chabay, I., & Lawrence, M. (2014a). Air pollution: clean up our skies. *Nature, 515*, 335–337. Retrieved April 14, 2018 from https://www.nature.com/news/air-pollution-clean-up-our-skies-1.16352

Schmale, J., van Aardenne, J., & von Schneidemesser, E. (2014). New Directions: Support for integrated decision-making in air and climate policies – Development of a metrics-based information portal. E. *Atmospheric Environment*, *90*, 146–148. doi:10.1016/j.atmosenv.2014.03.016

Secretariat, C. C. A. C. (2016, May 30). *Two New UNEP Reports On Air Quality Show More Can Be Done To Tackle Air Pollution*. Retrieved February 25, 2018 from http://www.ccacoalition.org/en/news/two-new-unep-reports-air-quality-show-more-can-be-done-tackle-air-pollution

Shon, Z. H., Kim, K. H., & Song, S. K. (2011). Long-term trend in NO2 and NOx levels and their emission ratio in relation to road traffic activities in East Asia. *Atmospheric Environment*, *45*(18), 3120–3131. doi:10.1016/j.atmosenv.2011.03.009

Steiner, A. (2014, June 23). *UNEP*. Retrieved March 30, 2018 from http://drustage.unep.org/transport/astf/events/air-pollution-priority-united-nations-environment-assembly

Steiner, A. (2016, May 30). Two New UNEP Reports On Air Quality Show More Can Be Done To Tackle Air Pollution. *CCAC Secretariat*. Retrieved February 25, 2018 from http://www.ccacoalition.org/en/news/two-new-unep-reports-air-quality-show-more-can-be-done-tackle-air-pollution

Stern, N. H. (2007). *The Economics of Climate Change: The Stern Review*. Cambridge, UK: Cambridge University Press. doi:10.1017/CBO9780511817434

Tallis, J. H. (1985). Mass movement and erosion of a Southern Pennine blanket peat. *Journal of Ecology*, *73*(1), 283–315. doi:10.2307/2259784

The Guardian. (2016, October 20). *Air pollution more deadly in Africa than malnutrition or dirty water, study warns.* Retrieved January 20, 2018 from https://www.theguardian.com/global-development/2016/oct/20/air-pollution-deadlier-africa-than-dirty-water-or-malnutrition-oecd

The Lancet. (2017). *The Lancet Commission on Pollution and Health.* Retrieved March 15, 2018 from http://www.thelancet.com/commissions/pollution-and-health

The World Bank. (2018, April 5). *Reducing Pollution.* Retrieved April 14, 2018 from http://www.worldbank.org/en/topic/environment/brief/pollution

UNCHE. (1972). *Declaration of the UN Conference on the Human Environment, Stockholm, 5–16 June 1972.* Report of the UN Conference on the Human Environment, UN Doc.A/CONF.48/14.

UNEP. (2014, June 23). Retrieved March 30, 2018 from http://drustage.unep.org/transport/astf/events/air-pollution-priority-united-nations-environment-assembly

United Nations General Assembly. (1962). *Economic Development and the Conservation of Nature.* G.A. Resolution 1831 (XVII), UN GAOR, 17th Session, Supp. no. 17, UN Doc. A/RES/1831 (XVII).

Vallero, D. A. (2014). *Fundamentals of Air Pollution* (5th ed.). Waltham, MA: Elsevier Academic Press.

Vallero. (2016). Air Pollution Monitoring Changes to Accompany the Transition from a Control to a Systems Focus. Sustainability, 8, 1216. doi:10.3390u8121216

Voorhees, S. A., Sakai, R., Araki, S., Sato, H., & Otsu, A. (2001, July). Cost-Benefit Analysis Methods For Assessing Air Pollution Control Programs in Urban Environments — A Review. *Environmental Health and Preventive Medicine*, *6*, 63–73. Retrieved from https://slideheaven.com/cost-benefit-analysis-methods-for-assessing-air-pollution-control-programs-in-ur.html

World Conservation Union. (1980). *World Conservation Strategy: Living Resource Conservation for Sustainable Development*. Gland, Switzerland: International Union for Conservation of Nature.

World Meteorological Organization Greenhouse Gas Bulletin. (2017). *The state of Greenhouse Gases in the Atmosphere Based on Global Observations through 2016.* World Meteorological Organization. Retrieved on January 10, 2018, from: https://library.wmo.int/opac/doc_num.php?explnum_id=4022

Wuebbles, D. J., Fahey, D. W., Hibbard, K. A., Dokken, D. J., Stewart, B. C., & Maycock, T. K. (Eds.). USGCRP, 2017: Climate Science Special Report: Fourth National Climate Assessment (vol. 1). U.S. Global Change Research Program. doi:10.7930/J0J964J6

Xu, M. M., Guo, Y. M., Zhang, Y. J., Westerdahl, D., Mo, Y., Liang, F., & Pan, X. (2014). Spatiotemporal analysis of particulate air pollution and ischemic heart disease mortality in Beijing, China. *Environmental Health, 13*(1), 109. doi:10.1186/1476-069X-13-109 PMID:25495440

Zhang, Q. F. (2012). *Toward an Environmentally Sustainable Future: Country Environmental Analysis of the People's Republic of China*. Beijing, China: China Financial and Economic Publishing House.

ENDNOTES

1. Pacific Fur Seal Arbitration settled a dispute between the United States and the United Kingdom for the protection of fur seals in the Bering Sea from overexploitation. Award of the arbitration developed the notion that consent of the concerned states is imperative for the conservation of living resources as they have the rights to assert even if it is outside their jurisdiction.
2. Trail Smelter Arbitration developed the principle that no state has the right to use or permit the use of its territory in such a manner as to cause injury by fumes in or to the territory of another or the properties or persons therein.

Compilation of References

A blanket around the Earth, Global Climate Change by NASA. (n.d.). Retrieved from https://climate.nasa.gov/causes/

Abdelbari, Elsawah, & Shafi. (2015). Model Learning Using Genetic Programming Under Full And Partial System Information Conditions. *33rd International Conference of the System Dynamics Society*, Cambridge, MA.

Abel, D. W., Holloway, T., Harkey, M., Meier, P., Ah, D., & Limaye, V. S. (2018). Air-quality-related health impacts from climate change and from adaptation of cooling demand for buildings in the eastern United States: An interdisciplinary modeling study. *PLoS Medicine*, *15*(7), e1002599. doi:10.1371/journal.pmed.1002599 PMID:29969461

Abussaud, B., Asmaly, H. A., Ihsanullah, T. A., Saleh, T. A., Gupta, V. K., laoui, T., & Atieh, M. A. (2016). Sorption of phenol from waters on activated carbon impregnated with iron oxide, aluminum oxide and titanium oxide. *Journal of Molecular Liquids*, *213*, 351–359. doi:10.1016/j.molliq.2015.08.044

Adam, K. N. (2016). Incertezas e impactos de mudanças climáticas sobre o regime de vazões na bacia hidrográfica do rio Uruguai (Doutorado em Recursos Hídricos e Saneamento Ambiental). Universidade Federal do Rio Grande do Sul - Instituto de Pesquisas Hidráulicas. Porto Alegre.

Adaptation Sub-Committee. (2016). *UK Climate Change Risk Assessment 2017: synthesis report: priorities for the next five years*. Committee on Climate Change.

Adeyanju, A. A., & Manohar, K. (2017). Effects of Vehicular Emission on Environmental Pollution in Lagos. *Sci-Afric Journal of Scientific Issues, Research and Essays, 5*(4), 34-51.

Advantages of LED Lighting. (2016). Retrieved from http://energy.ltgovernors.com/in-depth-advantages-of-led-lighting.html

Africa Biogas. (2012). *Domestic Biogas in Africa: A first Assessment of the Potential and Need.* Author.

AGECC. (2010). *Advisory group on Energy and Climate Change.* New York: Energy for a Sustainable Future.

Agência Nacional de Águas (ANA). (2013). Conjuntura dos recursos hídricos no Brasil. Agência Nacional de Águas.

Agnew, R. (2017). Dire forecast: A theoretical model of the impact of climate change on crime. *Transnational Environmental Crime*, *5*, 22.

Ahmaruzzaman, M. (2009). Role of fly ash in the removal of organic pollutants from wastewater. *Energy & Fuels*, *23*(3), 1494–1511. doi:10.1021/ef8002697

Ahmaruzzaman, M., & Gupta, V. K. (2012). Application of Coal Fly Ash in Air Quality Management. *Industrial & Engineering Chemistry Research*, *51*(47), 15299–15314. doi:10.1021/ie301336m

Ahn, K. (2014). The role of air pollutants in atopic dermatitis. *The Journal of Allergy and Clinical Immunology*, *134*(5), 993–999. doi:10.1016/j.jaci.2014.09.023 PMID:25439225

Ainsworth, E. E., Yendrek, C. R., Sitch, S., Collins, W. J., & Emberson, L. D. (2012). The effects of tropospheric ozone on net primary productivity and implications for climate change. *Annual Review of Plant Biology*, *63*(1), 637–661. doi:10.1146/annurev-arplant-042110-103829 PMID:22404461

Air quality and sustainable development. (n.d.). Retrieved February 15, 2018 from https://www.environment.gov.za/sites/default/files/docs/stateofair_airqualityand_sustainable_development.pdf

Alege, O. P., & Ogundipe, A. A. (2013). Environmental quality and economic growth in Nigeria: A fractional cointegration analysis. *International Journal of Development and Sustainability*, *2*(2), 580–596. Retrieved from https://isdsnet.com/ijds-v2n2-10.pdf

Alexander, B., Park, R. J., Jacob, D. J., Li, Q. B., Yantosca, R. M., Savarino, J., ... Thiemens, M. H. (2005). Sulfate formation in sea-salt aerosols: Constraints from oxygen isotopes. *Journal of Geophysical Research, D, Atmospheres*, *110*(D10).

Alves, N. O., Loureiro, A. L. M., Santos, F. C., Nascimento, K. H., Dallacort, R., Vasconcellos, P. C., ... Medeiros, S. R. B. (2011). Genotoxicity and composition of particulate matter from biomass burning in the eastern Brazilian Amazon region. *Ecotoxicology and Environmental Safety*, *74*(5), 1427–1433. doi:10.1016/j.ecoenv.2011.04.007 PMID:21496924

American Medical Association. (2012). *Council on Science and Public Health Report 4, light pollution. Adverse effects of night time lighting*. American Medical Association.

Amigun, B., & Von Blottnitz, H. (2009). Capital Cost Prediction for Biogas Installation in Africa. *Environmental Progress & Sustainable Energy*, *28*, 134–142.

Anderson, H. R., Favarato, G., & Atkinson, R. W. (2011). Long-term exposure to air pollution and the incidence of asthma: Meta-analysis of cohort studies. *Air Quality, Atmosphere & Health*, *6*(1), 47–56. doi:10.100711869-011-0144-5

Anderson, T. A., Kruger, E. L., & Coats, J. R. (1994). Enhanced degradation of a mixture of three herbicides in the rhizosphere of a herbicide-tolerant plant. *Chemosphere*, *28*(8), 1551–1557. doi:10.1016/0045-6535(94)90248-8

Angelo, J., Andrade, L., Madeira, L. M., & Mendes, A. (2013). An overview of photocatalysis phenomena applied to NOx abatement. *Journal of Environmental Management, 129*, 522–539. doi:10.1016/j.jenvman.2013.08.006 PMID:24018117

Annesi-Maesano, I., Baiz, N., Banerjee, S., Rudnai, P., & Rive, S. (2013). Sinphonie Group. Indoor air quality and sources in schools and related health effects. *Journal of Toxicology and Environmental Health. Part B, Critical Reviews, 16*(8), 491–550. doi:10.1080/10937404.2013.853609 PMID:24298914

Apling, A. J., Blanchet, J. P., Charlson, R. J., Crommelynck, D., Grassl, H., Husson, N., & Rodhe, H. (1990). *Radiative Forcing of Climate. IPCC, Climate Change. The IPCC Scientific Assessment.* Cambridge, UK: Cambridge University Press.

ApSimon, H., Amann, M., Åström, S., & Oxley, T. (2009). Synergies in addressing air quality and climate change. *Climate Policy, 9*(6), 669.

Ares, A., Fernandes, J. A., Aboal, J. R., & Carballheira, A. (2011). Study of the air quality in industrial areas of Santa Cruz de Tenerife (Spain) by active biomonitoring with *Pseudoscleropodium purum. Ecotoxicology and Environmental Safety, 74*(3), 533–541. doi:10.1016/j.ecoenv.2010.08.019 PMID:20828820

ASTM C 618 94. (1994). Specification for coal fly ash and raw calcined natural pozzolan for use as a mineral admixture in Portland cement concrete. *Annual Book of ASTM Standards, 4*(2), 296–298.

Atun, G., Hisarlı, G., Kurtoglu, A. E., & Ayar, N. (2011). A comparison of basic dye adsorption onto zeolitic materials synthesized from fly ash. *Journal of Hazardous Materials, 187*(1-3), 562–573. doi:10.1016/j.jhazmat.2011.01.075 PMID:21300434

Austin, H. D., Krupnick, J, A., & McConnell, D. V. (1997). Efficiency and Political Economy of Pollution Control with Ancillary Benefits: An Application to NOX Control in the Chesapeake Bay Airshed. In *Resources for the Future, 1997.* Discussion Paper 97-34. Retrieved February 28, 2018 from http://ageconsearch.umn.edu/bitstream/10553/1/dp970034.pdf

Austin, G., & Blignaunt, J. (2008). *South African National Rural Domestic Biogas Feasibility Assessment: Prepared for Ministry for Development Co-Operation, The State of Netherlands.* Cape Town: Agama Energy.

Australian Academy of Science. (2018). Retrieved from https://www.science.org.au/learning/general-audience/science-booklets-0/science-climate change/1-what-climate-change

Avery, M. A. (2010). Increasings pringtime ozone mixing ratios in the free troposphere over western North. *American Naturalist, 463*, 344–348.

Aw, J., & Kleeman, M. J. (2003). Evaluating the first order effect of intraannual temperature variability on urban air pollution. *Journal of Geophysical Research, D, Atmospheres, 108*(D12).

Ayanoğlu, M., & Gökçe, M. (2007). Sistem Düşüncesinden Sistem Dinamiklerine. *Makine Teknolojileri Elektronik Dergisi*, 29-41.

Badman, D. G., & Jaffe, E. R. (1996). Blood and Atmospheric pollutants: State of knowledge and research needs. *Otolaryngology - Head and Neck Surgery, 114*(2), 205–208. doi:10.1016/S0194-5998(96)70166-3 PMID:8637733

Bagavath Singh, A., Srinivas, C.V., Maran, S., Baskaran, R., & Venkatraman, B. (2016). Wind Direction Dependent Vertical Wind Shear and Surface Roughness Parameterization in Two different Coastal Environments. *IJGAES, 4*(3), 1-9.

Baklanov, A., Molina, L. T., & Gauss, M. (2016). Megacities, air quality and climate. *Atmospheric Environment, 126*, 235–249. doi:10.1016/j.atmosenv.2015.11.059

Balmes, J. R., Fine, J. M., & Sheppard, D. (1987). Symptomatic bronchoconstriction after short-term inhalation of sulfur dioxide. *The American Review of Respiratory Disease, 136*(5), 1117–1121. doi:10.1164/ajrccm/136.5.1117 PMID:3674573

Barbour, D. A. (1986). *Why are there so few butterflies in Liverpool? An answer.* London: Royal Entomological Society.

Bárdossy, A., & Pegram, G. (2012). Multiscale spatial recorrelation of RCM precipitation to produce unbiased climate change scenarios over large areas and small. *Water Resources Research, 48*(9), W09502. doi:10.1029/2011WR011524

Barlas. (2005a). Dinamik Sistem Yaklaşımı: Modeller, Kurumsal Öğrenme Ve Sorun Çözme – I. *Endüstri Ve Otomasyon Dergisi,* (94).

Barrie, L. A., & Hoff, R. M. (1984). The oxidation rate and residence time of sulphur dioxide in the Arctic atmosphere. *Atmospheric Environment, 18*(12), 2711-2722.

Barry, R. G., & Chorley, R. J. (2013). *Atmosfera, Tempo e Clima (9th ed.).* Porto Alegre: Bookman.

Bartók, B., Wild, M., Folini, D., Lüthi, D., Kotlarski, S., Schär, C., ... Imecs, Z. (2017). Projected changes in surface solar radiation in CMIP5 global climate models and in EURO-CORDEX regional climate models for Europe. *Climate Dynamics, 49*(7-8), 2665–2683.

Basha, S., Jhala, J., Thorat, R., Goel, S., Trivedi, R., Shah, K., ... Jha, B. (2010). Assessment of heavy metal content in suspended particulate matter of coastal industrial town, Mithapur, Gujarat, India. *Atmospheric Research, 97*(1-2), 257–265. doi:10.1016/j.atmosres.2010.04.012

Batterman, S., Ganguly, R., Isakov, V., Burke, J., Arunachalam, S., Snyder, M., ... Lewis, T. (2014). Dispersion Modeling of Traffic-Related Air Pollutant Exposures and Health Effects among Children with Asthma in Detroit, Michigan. *Transportation Research Record: Journal of the Transportation Research Board, 2452*(1), 105–113. doi:10.3141/2452-13

Beelen, R., Hoek, G., Van den Brandt, P. A., Goldbohm, R. A., Fischer, P., Schouten, L. J., ... Brunekreef, B. (2008). Long-term effects of traffic-related air pollution on mortality in a Dutch cohort (NLCS-AIR study). *Environmental Health Perspectives, 116*(2), 196–202. doi:10.1289/ehp.10767 PMID:18288318

Bellinger, D. C. (2005). Teratogen update: Lead and pregnancy. *Birth Defects Research. Part A, Clinical and Molecular Teratology, 73*(6), 409–420. doi:10.1002/bdra.20127 PMID:15880700

Benjamin, C. C., Hume, C. R., Voolstra, C. A., D'Angelo, C., Burt, J. A., Eyal, G., ... Wiedenmann, J. (2016). Ancestral genetic diversity associated with the rapid spread of stress-tolerant coral symbionts in response to Holocene climate change. *Proceedings of the National Academy of Sciences of the United States of America, 113*(16), 4416–4421. doi:10.1073/pnas.1601910113 PMID:27044109

Bentayeb, M., Simoni, M., Norback, D., Baldacci, S., Maio, S., & Viegi, G. (2013). Indoor Atmospheric pollutants and respiratory health in the elderly. *J Environ Sci Health A Tox Hazard Subst Environ Eng, 48*, 1783-9.

Bhandarkar, S. (2013). Vehicular Pollution, Their Effect on Human Heatlh and Mitigation Measures. *Vehicle Engineering, 1*(2), 33–40.

Bhat, P., Chanakya, H., & Ravindranth, N. (2001). Biogas Plant Dissemination; Success Story of Sirsi, India. *Energy for Sustainable Development, 1*(5), 39–46.

Birch, M. B. L., Gramig, B. M., Moomaw, W. R., Doering, O. C. III, & Reeling, C. J. (2011). Why metrics matter: Evaluating policy choices for reactive nitrogen in the chesapeake bay watershed. *Environmental Science & Technology, 45*(1), 168–174. doi:10.1021/es101472z PMID:20853823

Bishoi, B., Prakash, A., & Jain, V. K. (2009). A Comparative Study of Air Quality Index Based on Factor Analysis and USEPA Methods for an Urban Environment. *Aerosol and Air Quality Research, 9*(1), 1–17. doi:10.4209/aaqr.2008.02.0007

Blij de, H.J., & Muller, P.O. (1993). *Physical Geography of the Global Environment.* New York: John Wiley & Sons Inc.

Blok, K., Afanador, A., van der Hoorn, I., Berg, T., van Vuuren, D., & PBL, O. E. (2017). *Sectoral Greenhouse Gas Emission Reduction Potentials in 2030.* Academic Press.

Bloomberg New Energy Finance. (2016). *Global trends in renewable energyfinance 2016.* Author.

Bloomer, B. J., Stehr, J. W., Piety, C. A., Salawitch, R. J., & Dickerson, R. R. (2009). Observed relationships of ozone air pollution with temperature and emissions. *Geophysical Research Letters, 36*(9).

Blunden, J., & Arndt, D. S. (2016). State of the Climate in 2015. *Bulletin of the American Meteorological Society, 97*(8), Si-S275.

Bocchi, C., Bazzini, C., Fontana, F., Pinto, G., Martino, A., & Cassoni, F. (2016). Characterization of urban aerosol: Seasonal variation of mutagenicityand genotoxicity of PM2.5, PM1and semi-volatile organic compounds. *Mutation Research, 809*, 16–23. doi:10.1016/j.mrgentox.2016.07.007 PMID:27692295

Boffetta, P., Merler, E., & Vainio, H. (1993). Carcinogenicity of mercury and mercury compounds. *Scandinavian Journal of Work, Environment & Health, 19*(1), 1–7. doi:10.5271jweh.1510 PMID:8465166

Bond, T., & Templeton, M. (2011). History and Future of Domestic Biogas Plants in the Developing World. *Energy for Sustainable Development, 15*, 347–354.

Boonen, E., Akylas, V., Barmpas, F., Boréave, A., Bottalico, L., Cazaunau, M., ... Beeldens, A. (2015). Construction of a photocatalytic De-polluting field site in the Leopold II tunnel in Brussels. *Journal of Environmental Management, 155*, 136–144. doi:10.1016/j.jenvman.2015.03.001 PMID:25863437

Bortnick, S. M., Coutant, B. W., & Eberly, S. I. (2002). Using Continuous PM2.5 Monitoring Data to Report an Air Quality Index. *Journal of the Air & Waste Management Association, 52*(1), 104–112. doi:10.1080/10473289.2002.10470763 PMID:15152670

Boucher, O. (2012). Comparison of physically-and economically-based CO2-equivalences for methane. *Earth System Dynamics, 3*(1), 49–61.

Boucher, O., Halloran, P. R., Burke, E. J., Doutriaux-Boucher, M., Jones, C. D., Lowe, J., ... Wu, P. (2012). Reversibility in an Earth System model in response to CO2 concentration changes. *Environmental Research Letters, 7*(2), 024013.

Boucher, O., Randall, D., Artaxo, P., Bretherton, C., Feingold, G., Forster, P., & Rasch, P. (2013). Clouds and aerosols. In *Climate change 2013: the physical science basis. Contribution of Working Group I to the Fifth Assessment Report of the Intergovernmental Panel on Climate Change* (pp. 571–657). Cambridge University Press.

Bouwer, L. M., Aerts, J. C. J. H., Van De Coterlet, G. M., Giesen, N., Gieske, A., & Mannaerts, C. (2004). Evaluating downscaling methods for preparing global circulation model (GCM) data for hydrological impact modelling. In Climate change in contrasting river basins: adaptation strategies for water, food and environment. Wallingford: CABI.

Brandt, J., Silver, D. J., Christensen, H. J., Andersen, S. M., Bønløkke, H. J., Sigsgaard, T., ... Frohn, M. L. (2013). Assessment of past, present and future health-cost externalities of air pollution in Europe and the contribution from international ship traffic using the EVA model system. *Atmospheric Chemistry and Physics, 13*(15), 7747–7764. doi:10.5194/acp-13-7747-2013

Brandt, J., Silver, D. J., Christensen, H. J., Andersen, S. M., Bønløkke, H. J., Sigsgaard, T., ... Frohn, M. L. (2013a). Contribution from the ten major emission sectors in Europe and Denmark to the health-cost externalities of air pollution using the EVA model system – an integrated modelling approach. *Atmospheric Chemistry and Physics, 13*(15), 7725–7746. doi:10.5194/acp-13-7725-2013

Brasseur, G. P. (2009). Implications of climate change for air quality. *World Meteorological Organization Bulletin, 58*(1).

Brasseur, G. P., Muller, J.-F., Tie, X., & Horowitz, L. (2001). Tropospheric ozone and climate: past, present and future. In T. Matsuno & H. Kida (Eds.), *Present and Future of Modeling Global Environmental Change: Toward Integrated Modeling* (pp. 63–75). Tokyo: TERRAPUB.

Brasseur, G. P., Schultz, M., Granier, C., Saunois, M., Diehl, T., Botzet, M., ... Walters, S. (2006). Impact of climate change on the future chemical composition of the global troposphere. *Journal of Climate*, *19*(16), 3932–3951.

Brazil, G. M., Kenefick, L., Callanan, M., Haro, A., De Lorenzo, V., Dowling, D. N., & O'gara, F. (1995). Construction of a rhizosphere pseudomonad with potential to degrade polychlorinated biphenyls and detection of bph gene expression in the rhizosphere. *Applied and Environmental Microbiology*, *61*(5), 1946–1952. PMID:7646029

Briggs, G. A. (1973). *Diffusion estimation for small emissions.* USAEC Report ATDL-106. Oak Ridge: NOAA.

Briggs, G. A. (1979). Some recent analysis of Plume Rise observations. In *Proc. Of 2nd Intl. Clean Air Congress*. New York: Aca. Pres.

Brown, S. G., Frankel, A., Raffuse, S. M., Roberts, P. T., Hafner, H. R., & Anderson, D. J. (2007). Source apportionment of fine particulate matter in Phoenix, AZ, using positive matrix factorization. *Journal of the Air & Waste Management Association*, *57*(6), 741–752. doi:10.3155/1047-3289.57.6.741 PMID:17608008

Buffler, P. A., Kwan, M. L., Reynolds, P., & Urayama, K. Y. (2005). Environmental and Genetic Risk Factors for Childhood Leukemia: Appraising the Evidence. *Cancer Investigation*, *23*(1), 60–75. doi:10.1081/CNV-46402 PMID:15779869

Burney, J., & Ramanathan, V. (2014). Air pollution impacts on Indian agriculture. *Proceedings of the National Academy of Sciences*, 172-75.

Burr, M. L., Karani, G., Davies, B., Holmes, B. A., & Williams, K. L. (2004). Effects on respiratory health of a reduction in air pollution from vehicle exhaust emissions. *Occupational and Environmental Medicine*, *61*(3), 212–218. doi:10.1136/oem.2002.003244 PMID:14985515

Butler, J. & Montzka, S.A. (2016). *The NOAA annual greenhouse gas index (AGGI).* Earth System Research Laboratory, Global Monitoring Division Website.

Bytnerowicz, A., Arbaugh, M., Schilling, S., Fraczek, W., & Alexander, D. (2008). Ozone distribution and phytotoxic potential in mixed conifer forests of the San Bernardino Mountains, southern California. *Environ. Pollut*, *155*(3), 398–408. doi:10.1016/j.envpol.2008.01.046 PMID:18359541

Canadell, J. G., Le Quere, C., Raupach, M. R., Field, C. B., Buitenhuis, E. T., Ciais, P., ... Marland, G. (2007). Contributions to accelerating atmospheric CO2 growth from economic activity, carbon intensity, and efficiency of natural sinks. *Proceedings of the National Academy of Sciences of the United States of America*, *104*(47), 18866–18870. doi:10.1073/pnas.0702737104 PMID:17962418

Cannon, R. J. C. (2004). The implications of predicted climate change for insect pests in the UK, with emphasis on non-indigenous species. *Global Change Biology*, *4*(7), 785–796. doi:10.1046/j.1365-2486.1998.00190.x

Capelo, Dias, & Pereira. (2017). *Exploring the Dynamics of an Energy Service Venture for Energy Efficiency Policy*. International Conference of the System Dynamics Society, Delft, The Netherlands.

Cardelino, C. A., & Chameides, W. L. (1990). Natural hydrocarbons, urbanization, and urban ozone. *Journal of Geophysical Research*, *95*(D9), 13,971–13,979. doi:10.1029/JD095iD09p13971

Carreras, H. A., Calderón-Segura, M. A., Gómez-Arroyo, S., Murillo-Tovar, M. A., & Amador-Muñoz, O. (2013). Composition and mutagenicity of PAHs associated with urban airborne particles in Córdoba, Argentina. *Environmental Pollution*, *178*, 403–410. doi:10.1016/j.envpol.2013.03.016 PMID:23624338

Carreras-Sospedra, M., Dabdub, D., Rodriguez, M., & Brouwer, J. (2006). Air quality modeling in the South Coast Air Basin of California: What do the numbers really mean? *Journal of the Air & Waste Management Association*, *56*(8), 1184–1195. PMID:16933651

Carslaw, K. S., Boucher, O., Spracklen, D. V., Mann, G. W., Rae, J. G. L., Woodward, S., & Kulmala, M. (2010). A review of natural aerosol interactions and feedbacks within the Earth system. *Atmospheric Chemistry and Physics*, *10*(4), 1701–1737. doi:10.5194/acp-10-1701-2010

Carter, K. E., Schauer, J. J., Ezzati, M., Zhang, Y., Niu, H., Lai, A. M., ... Baumgartner, J. (2016). Seasonal variation in outdoor, indoor, and personal air pollution exposures of women using wood stoves in the Tibetan Plateau: Baseline assessment for an energy intervention study. *Environment International*, *94*, 449–457. doi:10.1016/j.envint.2016.05.029 PMID:27316628

Carvalho, S. A. D., & Furtado, A. T. (2014). Os desafios da adaptação às mudanças climáticas globais. Climacom –. *Cultura Cientifica*, *2*, 1–11.

Catcott, E. J. (1961). Effects of Atmospheric pollutants on animals. In *World Health Organization Monograph series, no.41* (pp. 221–231). Geneva: WHO.

Cavanagh, J. E. (2006). *Potential of Vegetation to Mitigate Road-generated Atmospheric pollutants*. Retrieved April 21st, 2016 from https://www.niwa.co.nz/sites/niwa. co.nz/files/import/attachments/co1x0405 p05.pdf

Cecchi, L. D., Amato, G., & Annesi-Maesano, I. (2013). Climate, urban Atmospheric pollutants, and respiratory allergy. In R. A. Pielke (Ed.), *Climate vulnerability: understanding and addressing threats to essential resources* (pp. 105–113). Waltham, MA: Academic Press. doi:10.1016/B978-0-12-384703-4.00116-7

Central Pollution Control Board (CPCB). (2009). *National Ambient Air Quality Standards for India*. Available online at http://cpcb.nic.in/National_Ambient_Air_Quality_Standards.php

Chandra, S. V., & Garg, N. (1991). A cross-sectional study of pulmonary function among workers exposed to multimetals in the glass bangle industry. *American Journal of Industrial Medicine, 20*(3), 391–399. doi:10.1002/ajim.4700200311 PMID:1928114

Change, I. C. (1996). *The science of climate change*. Working Group I to the Second Assessment Report of the IPCC.

Chang, H. H., Hao, H., & Sarnat, S. E. (2014). A statistical modeling framework for projecting future ambient ozone and its health impact due to climate change. *Atmospheric Environment, 89*, 290–297. PMID:24764746

Chatterjee, A. (2014). *Technological initiatives and status of fly ash in India. In Fly ash: Sources, applications and potential environmental impacts* (pp. 315–359). New York: Nova Science Publishers.

Chauhan, A. J., Krishna, M. T., Frew, A. J., & Holgate, S. T. (1998). Exposure to nitrogen dioxide (NO2) and respiratory disease risk. *Reviews on Environmental Health, 13*, 73. PMID:9718623

Chen, E., Ng, Y. K., Tan, Y. F., & Toh, J. S. Y. (2017). Environmentally Responsible Happy Nation Index: Refinements and 2015 Rankings. *Social Indicators Research, 134*(1), 39–56.

Chen, R. J., Chen, B. H., & Kan, H. D. (2010). A health-based economic assessment of particulate air pollution in 113 Chinese cities. *Zhongguo Huanjing Kexue, 30*, 3.

Chestnut, L., & Mills, D. M. (2005). A fresh look at the benefits and costs of the US acid rain program. *Journal of Environmental Management, 77*(3), 252–266. doi:10.1016/j.jenvman.2005.05.014 PMID:16171931

Cheung, W. W. L., Lam, V. W. Y., Sarmiento, J. L., Kearney, K., Watson, R., & Pauly, D. (2009). Projecting global marine biodiversity impacts under climate change scenarios. *Fish and Fisheries, 10*(3), 235–251. doi:10.1111/j.1467-2979.2008.00315.x

China's National Environmental Monitoring Centre. (2013). *Monthly air quality report in January 2013 in 74 cities.* Retrieved from https://www.ncbi.nlm.nih.gov/pmc/articles/PMC4816944/

Choucri, Goldsmith, & Mezher, . (2008). Framework for Modeling Technology Policy: Renewable Energy in Abu Dhabi. *International Conference of the System Dynamics Society.*

Chou, S. C., Lyra, A., Mourão, C., Dereczynski, C., Pilotto, I., Gomes, J., ... And Marengo, J. (2014). Assessment of Climate Change over South America under RCP 4.5 and 8.5 Downscaling Scenarios. *American Journal of Climate Change, 3*(05), 512–527. doi:10.4236/ajcc.2014.35043

Christensen, J. H., Hewitson, B., Busuioc, A., Chen, A., Gao, X., Held, R., & Magaña Rueda, V. (2007). Regional climate projections. In *Climate Change, 2007: The Physical Science Basis. Contribution of Working group I to the Fourth Assessment Report of the Intergovernmental Panel on Climate Change* (pp. 847–940). Cambridge University Press.

Clark, N. A., Demers, P. A., Karr, C. J., Koehoorn, M., Lencar, C., Tamburic, L., & Brauer, M. (2010). Effect of early life exposure to air pollution on development of childhood asthma. *Environmental Health Perspectives*, *118*(2), 284–290. doi:10.1289/ehp.0900916 PMID:20123607

Clean Air Institute. (2013). *La Calidad del Aire en América Latina: Una Visión Panorámica.* Retrieved from http://www.cleanairinstitute.org/calidaddelaireamericalatina/cai-report-spanish.pdf

Clifford, A., Lang, L., Chen, R., Anstey, K. J., & Seaton, A. (2016). Exposure to Atmospheric pollutants and cognitive functioning across the life course–A systematic literature review. *Environmental Research*, *147*, 383–398. doi:10.1016/j.envres.2016.01.018 PMID:26945620

Climate and Environment. (2016). Retrieved from https://www.triplepundit.com/2016/04/world-air-quality-index-brings-data-life/

Climate Change Indicators. (n.d.). *Greenhouse Gases.* US Environmental Protection Agency. Retrieved from https://www.epa.gov/climate-indicators/greenhouse-gases

Climate Change Indicators: Global Greenhouse Gas Emissions, US Environmental Protection Agency. (n.d.). Retrieved from https://www.epa.gov/climate-indicators/climate-change-indicators-global-greenhouse-gas-emissions

Climate Change Measurements - Climate Measurement. (n.d.). *A review of rainfall and temperature measurement standards in Guyana.* National Institute of Standards and Technology. Retrieved from https://www.nist.gov/sites/default/files/documents/iaao/OdessaShako.pdf

Cohen, A. J., Ross Anderson, H., Ostro, B., Pandey, K. D., Krzyzanowski, M., Kunzli, N., ... Smith, K. (2005). The global burden of disease due to outdoor Atmospheric pollutants. *Journal of Toxicology and Environmental Health. Part A.*, *68*(13-14), 1301–1307. doi:10.1080/15287390590936166 PMID:16024504

Cohen, D. D., Crawford, J., Stelcer, E., & Atanacio, A. J. (2012). Application of positive matrix factorization, multi-linear engine and back trajectory techniques to the quantification of coal-fired power station pollution in metropolitan Sydney. *Atmospheric Environment*, *61*, 204–211. doi:10.1016/j.atmosenv.2012.07.037

Cohen, M. F., Williams, J., & Yamasaki, H. (2002). Biodegradation of diesel fuel by an *Azolla*-derived bacterial consortium. *Journal of Environmental Science and Health. Part A*, *37*(9), 1593–1606.

Collischonn, W., Allasia, D. G., Silva, B. C., & And Tucci, C. E. M. (2007). The MGB-IPH model for large scale rainfall-runoff modeling. *Hydrological Sciences Journal*, *52*(5), 878–895. doi:10.1623/hysj.52.5.878

Consoli, N. C., Rocha, C. G., & Saldanha, R. B. (2014). Coal fly ash–carbide lime bricks: An environment friendly building Product. *Construction & Building Materials*, *69*, 301–309. doi:10.1016/j.conbuildmat.2014.07.067

Cooking and Climate Change, Global Alliance for Clean Cookstoves. (n.d.). Accessed on March 12, 2018 from http://carbonfinanceforcookstoves.org/about-cookstoves/cooking-and-climate-change/

Cooper, O. R., Parrish, D. D., Stohl, A., Trainer, M., Nédélec, P., Thouret, V., ... Leblanc, T. (2010). Increasing springtime ozone mixing ratios in the free troposphere over western North America. *Nature*, *463*(7279), 344–348. doi:10.1038/nature08708 PMID:20090751

Cox, W. M., & Chu, S. H. (1996). Assessment of interannual ozone variation in urban areas from a climatological perspective. *Atmospheric Environment*, *30*(14), 2615–2625.

Coyle. (1996). *Systems Dynamics Modeling*. London: Chapman & Hall.

CPCB. (1998). *Assessment of Impact to Air Environment: Guidelines for conducting Air Quality Modelling. PROBES/70/1997-98*. Delhi: Central Pollution Control Board.

Crispim, B. A., Vaini, J. O., Grisolia, A. B., Teixeira, T. Z., Mussury, R. M., & Seno, L. O. (2012). Biomonitoring the genotoxic effects of pollutants on *Tradescantia pallida* (Rose) D.R. Hunt in Dourados, Brazil. *Environmental Science and Pollution Research International*, *19*(3), 718–723. doi:10.100711356-011-0612-3 PMID:21912955

Crouse, D. L., Peters, P. A., Van Donkelaar, A., Goldberg, M. S., Villeneuve, P. J., Brion, O., ... Burnett, R. T. (2012). Risk of nonaccidental and cardiovascular mortality in relation to long-term exposure to low concentrations of fine particulate matter: A Canadian national-level cohort study. *Environmental Health Perspectives*, *120*(5), 708–714. doi:10.1289/ehp.1104049 PMID:22313724

D'Amato, G., Holgate, S. T., & Pawankar, R. (2015). Meteorological conditions, climate change, new emerging factors, and asthma and related allergic disorders. A statement of the World Allergy Organization. *The World Allergy Organization Journal*, *8*(1), 25. PMID:26207160

D'Amato, G., Pawankar, R., Vitale, C., Lanza, M., Molino, A., Stanziola, A., ... D'Amato, M. (2016). Climate change and air pollution: Effects on respiratory allergy. *Allergy, Asthma & Immunology Research*, *8*(5), 391–395. doi:10.4168/aair.2016.8.5.391 PMID:27334776

D'Amato, G., Vitale, C., Lanza, M., Molino, A., & D'Amato, M. (2016). Climate change, air pollution, and allergic respiratory diseases: An update, 1414 total citations on Dimensions. *Current Opinion in Allergy and Clinical Immunology*, *16*(5), 434–440. PMID:27518837

Dahl, A., Gharibi, A., Swietlicki, E., Gudmundsson, A., Bohgard, M., Ljungman, A., ... Gustafsson, M. (2006). Traffic – generated emission of ultrafine particles from pavement – tire interface. *Atmospheric Environment*, *40*(7), 1314–1323. doi:10.1016/j.atmosenv.2005.10.029

Dallarosa, J., Teixeira, E. C., Meira, L., & Wiegand, F. (2008). Study of the chemical elements and polycyclic aromatic hydrocarbons in atmospheric particles of PM10 and PM2.5 in the urban and rural areas of South Brazil. *Atmospheric Research*, *89*(1-2), 76–92. doi:10.1016/j.atmosres.2007.12.004

Dalsøren, B. S., Jonson, E. J., Quante, M., & Colijn, F. (Eds.). (2016). *North Sea Region Climate Change Assessment*. Regional Climate Studies. doi:10.1007/978-3-319-39745-0_16

Damek-Poprawa, M., & Sawicka-Kapusta, K. (2003). Damage to the liver, kidney, and testis with reference to burden of heavy metals in yellow-necked mice from areas around steelworks and zinc smelters in Poland. *Toxicology*, *186*(1-2), 1–10. doi:10.1016/S0300-483X(02)00595-4 PMID:12604166

Dash, S. K., & Dash, A. K. (2015). Determination of Air Quality IndexStatus near Bileipada, Joda Area of Keonjhar, Odisha, India. *Indian Journal of Science and Technology*, *8*(35). doi:10.17485/ijst/2015/v8i35/81468

Davidson, E. A. (2009). The contribution of manure and fertilizer nitrogen to atmospheric nitrous oxide since 1860. *Nature Geoscience*, *2*(9), 659–662. doi:10.1038/ngeo608

Davidson, E. A. (2012). Excess nitrogen in the US environment: Trends, risks, and solutions. *ESA Issues Ecology*, *15*, 1–16.

Davis, M. L., & Cornwell, D. A. (1991). *Introduction to Environmental Engineering* (2nd ed.). New York: McGraw-Hill, Inc.

Dawson, J. P., Adams, P. J., & Pandis, S. N. (2007). Sensitivity of PM 2.5 to climate in the Eastern US: A modeling case study. *Atmospheric Chemistry and Physics*, *7*(16), 4295–4309.

Dawson, J. P., Bloomer, B. J., Winner, D. A., & Weaver, C. P. (2014). Understanding the meteorological drivers of US particulate matter concentrations in a changing climate. *Bulletin of the American Meteorological Society*, *95*(4), 521–532.

De Jong, M. F., Drijfhout, S. S., Hazeleger, W., Van Aken, H. M., & Severijns, C. A. (2009). Simulations of hydrographic properties in the northwestern North Atlantic Ocean in coupled climate models. *Journal of Climate*, *22*(7), 1767–1786.

DeAngelo, B. J. (2013). Bounding the role of black carbon in the climate system: A scientific assessment. *Journal of Geophysical Research, D, Atmospheres*, *118*(11), 5380–5552. doi:10.1002/jgrd.50171

Deng, Q., Lu, C., Li, Y., Sundell, J., & Norbäck, D. (2016). Exposure to outdoor air pollution during trimesters of pregnancy and childhood asthma, allergic rhinitis, and eczema. *Environmental Research*, *150*, 119–127. doi:10.1016/j.envres.2016.05.050 PMID:27281689

Department of Energy (DoE). (2015). State of Renewable Energy in South Africa. Department of Energy, South Africa.

Details of the Math behind the Model. (n.d.). University Corporation for Atmospheric Research (UCAR). Retrieved from https://scied.ucar.edu/activity/very-very-simple-climate-model-activity)

Deutsch, C. A., Tewksbury, J. J., Huey, R. B., Sheldon, K. S., Ghalambor, C. K., Haak, D. C., & Martin, P. R. (2008). Impacts of climate warming on terrestrial ectotherms across latitude. *Proceedings of the National Academy of Sciences of the United States of America*, *105*(18), 6668–6672. doi:10.1073/pnas.0709472105 PMID:18458348

Dewi, E. M., Suwardi, Suryaningtyas, D. T., & Anwar, S. (2016). Utilization of Natural Zeolites as Cu (Ii) and Zn (Ii) Adsorbent. *J Trop Soils*, *21*(3), 153–160. doi:10.5400/jts.2016.v21i3.153-160

Dibike, Y. B., & Coulibaly, P. (2005). Hydrologic impact of climate change in the Saguenay watershed: Comparison of downscaling methods and hydrologic models. *Journal of Hydrology (Amsterdam)*, *307*(1-4), 145–163. doi:10.1016/j.jhydrol.2004.10.012

Direct Emissions by GHG. (2015). US Environmental Protection Agency. Retrieved from https://www.epa.gov/ghgreporting/ghgrp-reported-data

Divan, A. M. Junior, Oliva, M. A., & Ferreira, F. A. (2008). Dispersal pattern of airbone emissions from aluminium smelter in Ouro Preto, Brazil, as expressed by foliar fluoride accumulation in eight plant species. *Ecological Indicators*, *8*(5), 454–461. doi:10.1016/j.ecolind.2007.04.008

Dockery, D. W., & Stone, P. H. (2007). Cardiovascular risks from fine particulate Atmospheric pollutants. *The New England Journal of Medicine*, *356*(5), 511–513. doi:10.1056/NEJMe068274 PMID:17267912

Dodd, K. T., & Gross, D. R. (1980). Ammonia inhalation toxicity in cats: A study of acute and chronic respiratory dysfunction. *Archives of Environmental Health*, *35*(1), 6–14. doi:10.1080/00039896.1980.10667455 PMID:7362273

Domene, L. A. F., & Ayres, R. U. (2001). Nitrogen's role in industrial *systems. Journal of Industrial Ecology*, *5*(1), 77–103. doi:10.1162/108819801753358517

Donham, K. J. (1991). Association of environmental air contaminants with disease and productivity in swine. *American Journal of Veterinary Research*, *52*, 1723–1730. PMID:1767997

Drupal - IASS Potsdam. (n.d.). *Air Pollution and Climate Change*. Retrieved from https://www.iass-potsdam.de/en/output/dossiers/air-pollution-and-climate-change

Dubey, K. K., & Fulekar, M. H. (2011). Effect of pesticides on the seed germination of *Cenchrus setigerus* and *Pennisetum pedicellatum* as monocropping and co-cropping system: Implications for rhizospheric bioremediation. *Romanian Biotechnological Letters*, *16*(1), 5908–5918.

Dulvy, N. K., Rogers, S. I., Jennings, S., Stelzenmüller, V., Dye, S. R., & Skjoldal, H. R. (2008). Climate change and deepening of the North Sea fish assemblage: A biotic indicator of warming seas. *Journal of Applied Ecology*, *45*(4), 1029–1039. doi:10.1111/j.1365-2664.2008.01488.x

Ebi, K., & McGregor, G. (2009). Climate change, tropospheric ozone and particulate matter, and health impacts. *Ciencia & Saude Coletiva*, *14*(6), 2281–2293. PMID:20069198

Economopoulos, A. P. (1993). *Assessment of Sources of Air Water and Land Pollution. A Guide to Rapid Source Inventory Techniques and Their Use in Formulating Environmental Control Strategies. Part One: Rapid Inventory Techniques in Environmental Pollution. Part Two: Approaches for Consideration in Formulating Environmental Control Strategies (Unpublished document WHO/YEP/93.1.)*. Geneva: WHO.

Elshout, S. (2008). Comparing Urban Air Quality in Europe in Real Time A Review of the Existing Air Quality Indices and the Proposal of a Common Alternative. *Environment International*, *34*(5), 720–726. doi:10.1016/j.envint.2007.12.011 PMID:18234340

Empirical Study of Air Pollution and Sustainable Development of. (2014, September 19). Retrieved April 14, 2018 from shodhganga.inflibnet.ac.in/bitstream/10603/109618/15/15%20 chapter%206.pdf

Encyclopaedia Britannica. (2012). *Greenhouse effect.* Retrieved from https://www.britannica.com/science/electromagnetic-radiation/-continuous-spectra-of-electromagnetic-radiation/media/183228/85040

Encyclopaedia Britannica. (2018). *Meteorology-Science.* Retrieved from https://www.britannica.com/science/meteorology

Environmental Protection Agency (EPA). (2017). *Positive matrix factorization (PMF) 5.0: fundamentals and user guide. Research and Development, 2014.* Available at: http://www.epa.gov/heasd/documents/PMF_5.0_User_Guide.pdf

Epton, M. J., Dawson, R. D., Brooks, W. M., Kingham, S., Aberkane, T., Cavanagh, J. A., ... Brown, L. (2008). The effect of ambient Atmospheric pollutants on respiratory health of school children: A panel study. *Environmental Health*, *7*(1), 16. doi:10.1186/1476-069X-7-16 PMID:18479529

Escalante-Garcia, J. I., Rios-Escobar, M., Gorokhovsky, A., & Fuentes, A. F. (2008). Fluorgypsum binders with OPC and PFA additions, strength and reactivity as a function of component proportioning and temperature. *Cement and Concrete Composites*, *30*(2), 88–96. doi:10.1016/j.cemconcomp.2007.05.015

Escobedo, F., Kroeger, T., & Wagner, J. (2011). Urban forests and pollution mitigation: Analyzing ecosystem services and disservices. *Environ. Pollut*, *159*(8-9), 2078–2087. doi:10.1016/j.envpol.2011.01.010 PMID:21316130

Esmaeeli, Shakouri, & Sedighi. (2006). Investigation Of Pricing İmpact On The Electrical Energy Consumption Behavior Of The Household Sector By A System Dynamics Approach. *International Conference of the System Dynamics Society.*

Espanol. (2016). *Air pollution: Everything you need to know.* New York: Natural Resources Defense Council.

Estratégia de Resiliência de Porto Alegre. (2017). Disponível em: http://100resilientcities.org/wpcontent/uploads/2017/07/PortoAlegre_Strategy_Portuguese_PDF.pdf

EU. (2013). *Commission staff working document impact assessment.* Retrieved April 13, 2018 from https://eurohealthnet.eu/media/news-releases/eurohealthnet-expresses-concern-european-commission-withdrawal-anticipated-eu-%E2%80%98

European Aviation Environmental Report. (2016). Retrieved October 16, 2016 from https://www.easa.europa.eu/eaer/

European Commission. (1996). ExternE — Externalities of Energy. A Research Project of the European Commission: Vol. 1. *Summary Report EUR 16520 EN*. Brussels: Author.

European Environment Agency (EEA). (2015). *Air pollution due to ozone: health impacts and effects of climate change.* Available: https://www.eea.europa.eu/data-and-maps/indicators/air-pollution-by-ozone-2/assessment

European Environmental Agency. (2018). *Air pollution due to ozone: health impacts and effects of climate change*. Retrieved March 15, 2018 from https://www.eea.europa.eu/data-and-maps/indicators/air-pollution-by-ozone-2/assessment

Eyring, V., Butchart, N., Waugh, D. W., Akiyoshi, H., Austin, J., Bekki, S., ... Cordero, E. (2006). Assessment of temperature, trace species, and ozone in chemistry climate model simulations of the recent past. *Journal of Geophysical Research, D, Atmospheres*, *111*(D22).

Faloona, I., Conley, S. A., Blomquist, B., Clarke, A. D., Kapustin, V., Howell, S., ... Bandy, A. R. (2010). Sulfur dioxide in the tropical marine boundary layer: Dry deposition and heterogeneous oxidation observed during the Pacific Atmospheric Sulfur Experiment. *Journal of Atmospheric Chemistry*, *63*(1), 13–32. doi:10.100710874-010-9155-0

Fan, F. M., & Collischonn, W. (2014). Integraçao do Modelo MGB-IPH com Sistema de Informaçao Geografica. *RBRH - Rev. Bras. Recur. Hídricos*, *19*, 243–254.

Fang, Y., Naik, V., Horowitz, L. W., & Mauzerall, D. L. (2013). Atmospheric pollutants and associated human mortality: The role of air pollutant emissions, climate change and methane concentration increases from the preindustrial period to present. *Atmospheric Chemistry and Physics*, *13*(3), 1377–1394. doi:10.5194/acp-13-1377-2013

FAO. (2009). *How to Feed the World in 2050: Background paper for the High-Level Forum on How to feed the World in 2050*. Rome: FAO.

Fears, D. (2016). Air pollution doesn't just make breathing harder for humans, fish are choking too. *Chicago Tribune*

Felzer, B., Kicklighter, D., Melillo, J., Wang, C., Zhuang, Q., & Prinn, R. (2004). Effects of ozone on net primary production and carbon sequestration in the conterminous United States using a biogeochemistry model. *Tellus*, *56B*(3), 230–248. doi:10.1111/j.1600-0889.2004.00097.x

Fernández, C., Bion, N., Gaigneaux, E. M., Duprez, D., & Ruiz, P. (2016). 2016, Kinetics of hydrogen adsorption and mobility on Ru nanoparticles supported on alumina: Effects on the catalytic mechanism of ammonia synthesis. *Journal of Catalysis*, *344*, 16–28. doi:10.1016/j.jcat.2016.09.013

Final Environmental Analysis, Revised Proposed Short-Lived Climate Pollutant Reduction Strategy, Air Resources Board, March. (2017). Retrieved from https://www.arb.ca.gov/cc/shortlived/meetings/03142017/appendixe.pdf

Fioletov, V. E., McLinden, C. A., Krotkov, N., & Li, C. (2015). Lifetimes and emissions of SO2 from point sources estimated from OMI. *Geophysical Research Letters, 42*(6), 1969–1976. doi:10.1002/2015GL063148

Fiore, A. M. (2014). Atmospheric chemistry: No equatorial divide for a cleansing radical. *Nature, 513*(7517), 176–178. doi:10.1038/513176a PMID:25209794

Fiore, A. M., Jacob, D. J., Field, B. D., Streets, D. G., Fernandes, S. D., & Jang, C. (2002). Linking ozone pollution and climate change: The case for controlling methane. *Geophysical Research Letters, 29*(19), 1919. doi:10.1029/2002GL015601

Fiore, A. M., Naik, V., & Leibensperger, E. M. (2015). Air quality and climate connections. *Journal of the Air & Waste Management Association, 65*(6), 645–685. PMID:25976481

Flato, G., Marotzke, J., Abiodun, B., Braconnot, P., Chou, S. C., Collins, W. J., ... Forest, C. (2013). Evaluation of climate models. In: climate change 2013: the physical science basis. Contribution of working group I to the fifth assessment report of the intergovernmental panel on climate change. *Climatic Change, 5*, 741–866.

Fly ash-A resource material for construction sector. (2014). NTPC Limited, Ash Management Group.

Font, O., Moreno, N., Díez, S., Querol, X., López-Soler, A., Coca, P., & Peña, F. G. (2009). Differential behaviour of combustion and gasification fly ash from Puertollano Power Plants (Spain) for the synthesis of zeolites and silica extraction. *Journal of Hazardous Materials, 166*(1), 94–102. doi:10.1016/j.jhazmat.2008.10.120 PMID:19097700

Font, O., Querol, X., Juan, R., Casado, R., Ruiz, C. R., López-Soler, Á., ... Peña, F. G. (2007). Recovery of gallium and vanadium from gasification fly ash. *Journal of Hazardous Materials, 139*(3), 413–423. doi:10.1016/j.jhazmat.2006.02.041 PMID:16600480

Fordham, D. A., & Akçakaya, H. R. (2012). Plant extinction risk under climate change: Are forecast range shifts alone a good indicator of species vulnerability to global warming? *Global Change Biology, 18*(4), 1357–1371. doi:10.1111/j.1365-2486.2011.02614.x

Foresti, A. J. (2015). *Um arranjo interdisciplinar para a gestão de riscos de desastres socionaturais com base na Engenharia da Resiliência, RS* (Tese apresentada, como requisito para obtenção do título de Doutora). ao Programa de Pós-graduação em Meio Ambiente, da Universidade do Estado do Rio Grande do Sul.

Forkel, R., & Knoche, R. (2006). Regional climate change and its impact on photooxidant concentrations in southern Germany: Simulations with a coupled regional climate chemistry model. *Journal of Geophysical Research, D, Atmospheres, 111*(D12).

Forrester, . (1995). *Road Map 1: Counterintuitive Behaviour Of Social Systems*. MIT System Dynamics in Education Project.

Forster, P. M., Andrew, T., Good, P., & Gregory, J. M. (2013). Evaluating adjusted forcing and model spread for historical and future scenarios in the CMIP5 generation of climate models. *Journal of Geophysical Research, D, Atmospheres*, *118*(3), 1139–1150. doi:10.1002/jgrd.50174

Fowler, H. J., Blenkinsop, S., & Tebaldi, C. (2007). Linking Climate Change Modelling to Impacts Studies: Recent Advances in Downscaling Techniques for Hydrological Modelling. *International Journal of Climatology*, *27*(12), 1547–1578. doi:10.1002/joc.1556

Frich, P., Alexander, L. V., Della-Marta, P. M., Gleason, B., Haylock, M., Tank, A. K., & Peterson, T. (2002). Observed coherent changes in climatic extremes during the second half of the twentieth century. *Climate Research*, *19*(3), 193–212.

Fröhlich-Nowoiskya, J., & Kamp, C. J. (2016). Bioaerosols in the Earth system: Climate, health, and ecosystem interactions. *Atmospheric Research*, *182*, 346–376. doi:10.1016/j.atmosres.2016.07.018

Fry, G. (1984). Australia and the South Pacific. *Diplomacy in the marketplace: Australia in world affairs, 90*, 175-77.Filippidou EC, Koukouliata A. Ozone effects on the respiratory system. *Progress in Health Sciences*, *2011*(1), 144–155.

Fuzzi, S., Baltensperger, U., Carslaw, K., Decesari, S., Denier Van Der Gon, H., Facchini, M. C., ... Nemitz, E. (2015). Particulate matter, air quality and climate: Lessons learned and future needs. *Atmospheric Chemistry and Physics*, *15*(14), 8217–8299. doi:10.5194/acp-15-8217-2015

Ganguly, R., Batterman, S., Isakov, V., & Breen, M. (2015). Effect of geocoding errors on traffic-related air pollutant exposure and concentration estimates. *Journal of Exposure Science & Environmental Epidemiology*, *25*(5), 490–498. doi:10.1038/jes.2015.1 PMID:25670023

Ganguly, R., & Broderick, B. (2008). Performance evaluation and sensitivity analysis of the general finite line source model for CO concentrations adjacent to motorways: A note. *Transportation Research Part D, Transport and Environment*, *13*(3), 198–205. doi:10.1016/j.trd.2008.01.006

Ganguly, R., & Thapa, S. (2016). An assessment of ambient air quality in Shimla city. *Current Science*, *111*(3), 509–516. doi:10.18520/cs/v111/i3/509-516

Gao, J., Tian, H., Cheng, K., Lu, L., Wang, Y., Wu, Y., ... Hao, J. (2014). Seasonal and spatial variation of trace elements in multi-size airborne particulate matters of Beijing, China: Mass concentration, enrichment characteristics, source apportionment, chemical speciation and bioavailability. *Atmospheric Environment*, *99*, 257–265. doi:10.1016/j.atmosenv.2014.08.081

Garfinkel, C. I., Waugh, D. W., Oman, L. D., Wang, L., & Hurwitz, M. M. (2013). Temperature trends in the tropical upper troposphere and lower stratosphere: Connections with sea surface temperatures and implications for water vapor and ozone. *Journal of Geophysical Research, D, Atmospheres*, *118*(17), 9658–9672.

Gasana, J., Dillikar, D., Mendy, A., Forno, E., & Ramos Vieira, E. (2012). Motor vehicle air pollution and asthma in children: A meta-analysis. *Environmental Research*, *117*, 36–45. doi:10.1016/j.envres.2012.05.001 PMID:22683007

Gaudel, A., Cooper, O. R., Ancellet, G., Barret, B., Boynard, A., Burrows, J. P., ... Doniki, S. (2018). Tropospheric Ozone Assessment Report: Present-day distribution and trends of tropospheric ozone relevant to climate and global atmospheric chemistry model evaluation. *Elem Sci Anth, 6*(1).

Gauderman, W. J., Vora, H., McConnell, R., Berhane, K., Gilliland, F., Thomas, D., ... Peters, J. (2007). Effect of exposure to traffic on lung development from 10 to 18 years of age: A cohort study. *Lancet, 369*(9561), 571–577. doi:10.1016/S0140-6736(07)60037-3 PMID:17307103

Gauss, M., Myhre, G., Isaksen, I. S. A., Grewe, V., Pitari, G., Wild, O., ... Sanderson, M. G. (2003). Radiative forcing in the 21st century due to ozone changes in the troposphere and the lower stratosphere. *Journal of Geophysical Research, 108*(D9), 4292. doi:10.1029/2002JD002624

Gentler, A. W., Gillies, J. A., & Pierson, W. R. (2000). an assessment of the mobile source contribution in U.S. *Water, Air, and Soil Pollution, 123*(1/4), 203–212. doi:10.1023/A:1005263220659

Ghio, A. J., & Huang, Y. C. (2004). Exposure to concentrated ambient particles (CAPs): A review. *Inhalation Toxicology, 16*, 53.

Ghio, A. J., & Samet, J. (1999). *Atmospheric pollutants and health: metals and Atmospheric pollutants particles.* Academic Press.

Giere, R., Carleton, L. E., & Lumpkin, G. R. (2003). Micro- and nanochemistry of fly ash from a coal-fired power plant. *The American Mineralogist, 88*(11-12), 1853–1865. doi:10.2138/am-2003-11-1228

Gillis, J. (2013). Heat-trapping gas passes milestone, raising fears. *New York Times, 11.*

Giorgi, F. (2006, December). Regional climate modeling: Status and perspectives. Journal de Physique IV (Proceedings), 139, 101-118.

Giorgi, F., & Meleux, F. (2007). Modelling the regional effects of climate change on air quality. *Comptes Rendus Geoscience, 339*(11-12), 721–733.

Giorgio, Enzo, & Stefania, . (2008). Renewable Energy İn Italy: Scenarios To 2030. *International Conference Of The System Dynamics Society.*

Global Carbon Project. (2017). *Carbon Budget 2017.* Retrieved on November 20, 2017, from: http://www.globalcarbonproject.org/carbonbudget/index.htm

Gottfried, M., Pauli, H., Futschik, A., Akhalkatsi, M., Barančok, P., Benito Alonso, J. L., ... Grabherr, G. (2012). Continent-wide response of mountain vegetation to climate change. *Nature Climate Change, 2*(2), 111–115. doi:10.1038/nclimate1329

Greaver, T. L., Sullivan, T. J., Herrick, J. D., Barber, M. C., Baron, J. S., Cosby, B. J., ... Novak, K. J. (2012). Ecological effects of nitrogen and sulfur Atmospheric pollutants in the US: What do we know? *Front. Ecol, 10*(7), 365–372. doi:10.1890/110049

Greben, H., & Oelofse, S. (2009). *Unlocking the Resource Potentail of Organic Waste: A South African Persepctive*. Pretoria: CSIR- Natural Resource and the Environment, Waste Management Research.

Greenhouse Gases Equivalencies Calculator - Calculations and References. (n.d.). US Environmental Protection Agency. Accessed on July 12, 2018 https://www.epa.gov/energy/greenhouse-gases-equivalencies-calculator-calculations-and-references

Gregory, J. M., Andrews, T., Good, P., Mauritsen, T., & Forster, P. M. (2016). Small global-mean cooling due to volcanic radiative forcing. *Climate Dynamics*, *47*(12), 3979–3991.

Griffin, R. J., Dabdub, D., & Seinfeld, J. H. (2005). Development and initial evaluation of a dynamic species resolved model for gas phase chemistry and size resolved gas/particle partitioning associated with secondary organic aerosol formation. *Journal of Geophysical Research, D, Atmospheres*, *110*(D5).

Grulke, N. E., Minnich, R. A., Paine, T. D., Seybold, S. J., Chavez, D. J., Fenn, M. E., . . . Dunn, A. (2009). Atmospheric pollutants increases forest susceptibility to wildfires: A case study in the San Bernardino Mountains in Southern California. Academic Press.

Grulke, N. E., Paoletti, E., & Heath, R. L. (2006). Direct measurements of foliar ozone uptake in crop and tree species. *Environ. Pollut.*, *146*, 640–647. doi:10.1016/j.envpol.2006.04.014 PMID:16777298

Gryparis, A., Forsberg, B., Katsouyanni, K., Analitis, A., Touloumi, G., & Schwartz, J. (2004). al. Acute effects of ozone on mortality from the "Atmospheric pollutants and health: A European approach" project. *American Journal of Respiratory and Critical Care Medicine*, *170*, 1080–1087. doi:10.1164/rccm.200403-333OC PMID:15282198

Guan, W., Zheng, X., Chung, K. F., & Zhong, N. (2016). Impact of air pollution on the burden of chronic respiratory diseases in China: Time for urgent action. *Lancet*, *388*(10054), 1939–1951. doi:10.1016/S0140-6736(16)31597-5 PMID:27751401

Guariso, G., & Volta, M. (Eds.). (2017). *Air Quality Integrated Assessment*. Springer Briefs in Applied Sciences and Technology. Springer. Retrieved April 10, 2018 from https://link.springer.com/chapter/10.1007/978-3-319-33349-6_6#

Guha-Sapir, D., Hoyois, P., & Below, R. (2016). *Annual Disaster Statistical Review 2015: The Numbers and Trends*. Brussels: CRED.

Guilyardi, E., Wittenberg, A., Fedorov, A., Collins, M., Wang, C., Capotondi, A., ... Stockdale, T. (2009). Understanding El Niño in ocean–atmosphere general circulation models: Progress and challenges. *Bulletin of the American Meteorological Society*, *90*(3), 325–340.

Guimarães, E., Domingos, M., Alves, E. S., Caldini, N. Jr, Lobo, D. J. A., Lichtenfels, A. J. F. C., & Saldiva, P. H. N. (2000). Detection of the genotoxicity of air pollutants in and around the city of São Paulo (Brazil) with the *Tradescantia*-micronucleus (Trad-MCN) assay. *Environmental and Experimental Botany*, *44*(1), 1–8. doi:10.1016/S0098-8472(00)00050-2 PMID:10927123

Guo, Y., Li, Y., Cheng, F., Wang, M., & Wang, X. (2013). Role of additives in improved thermal activation of coal fly ash for alumina extraction. *Fuel Processing Technology*, *110*, 114–121. doi:10.1016/j.fuproc.2012.12.003

Gurupia, T., Jones, C. L., Howard, A., Lockert, C., Wandell, T., & Stencel, J. M. (2001). *International Ash Utilization Symposium, Centre for Applied Energy Research*. University of Kentucky, Paper#44.

Guttikunda, S. K., & Jawahar, P. (2014). Atmospheric emissions and pollution from coal-fired thermal power plants in India. Atmospheric environment. *Atmospheric Environment*, *92*, 449–460. doi:10.1016/j.atmosenv.2014.04.057

Haase, D., Larondelle, N., Andersson, E., Artmann, M., Borgström, S., Breuste, J., ... Kabisch, N. (2014). A quantitative review of urban ecosystem service assessments: Concepts, models, and implementation. *Ambio*, *43*(4), 413–433. doi:10.100713280-014-0504-0 PMID:24740614

Haghnegahdar, A., Tolson, B. A., Craig, J. R., & Paya, K. T. (2015). Assessing the performance of a semi-distributed hydrological model under various watershed discretization schemes. *Hydrological Processes*, *29*(18), 4018–4031. doi:10.1002/hyp.10550

Haklay, M. (2013). Citizen science and volunteered geographic information: Overview and typology of participation. In *Crowdsourcing Geographic Knowledge*. Heidelberg, Germany: Springer. doi:10.1007/978-94-007-4587-2_7

Haleem, D., Von Wintzingerode, F., Moter, A., Moawad, H., & Göbel, U. B. (2000). Phylogenetic analysis of rhizosphere-associated β-subclass proteobacterial ammonia oxidizers in a municipal wastewater treatment plant based on rhizoremediation technology. *Letters in Applied Microbiology*, *31*(1), 34–38. doi:10.1046/j.1472-765x.2000.00760.x PMID:10886611

Hamlyn, O. (2015). Sustainability and the Failure of Ambition in European Pesticides Regulation. *Journal of Environmental Law*, *0*, 1–25. Retrieved from http://discovery.ucl.ac.uk/1469006/7/Sustainability%20and%20the%20Failure%20of%20Ambition%20in%20European%20Pesticides%20Regulation.pdf

Hansen, C., Luben, T. J., Sacks, J. D., Olshan, A., Jeffay, S., Strader, L., & Perreault, S. D. (2010). The effect of ambient Atmospheric pollutants on sperm quality. *Environmental Health Perspectives*, *118*(2), 203–209. doi:10.1289/ehp.0901022 PMID:20123611

Han, & Yoshitsugu, . (2008). *A System Dynamics Model Of CO2 Mitigation İn China's İnter-City Passenger Transport*. Elsevier Science Direct.

Haramis, G. M. (1984). Acid alteration of experimental wetlands: Effects on diversity andabundance of macroinvertebrates and growth, behaviours and survival of juvenile black ducks. Draft research summary, US Food and Wildlife Service, Eastern Energy and Land Use Team, Kearneyville, WV.

Hare, J. A., Wuenschel, M. J., & Kimball, M. E. (2012). Projecting range limits with coupled thermal tolerance-climate change models: An example based on gray snapper (Lutjanus griseus) along the US east coast. *PLoS One*, *7*(12), e52294. PMID:23284974

Harry, E. G. (1978). Atmospheric pollutants in farm buildings and methods of control: A review. *Avian Pathology*, *7*(4), 441–454. doi:10.1080/03079457808418301 PMID:18770401

Hasenfratz, D., Saukh, O., Sturzenegger, S., & Thiele, L. (2012). Participatory air pollution monitoring using smartphones. *Proceedings of the 2nd International Workshop on Mobile Sensing*.

Hauglustaine, D. A., Lathiere, J., Szopa, S., & Folberth, G. A. (2005). Future tropospheric ozone simulated with a climate chemistry biosphere model. *Geophysical Research Letters*, *32*(24).

Headwaters Resources Class F fly ash safety data sheet, (2016). Headwaters Resources.

Heal, M. R., Kumar, P., & Harrison, R. M. (2012). Particles, Air Quality, Policy and Health. *Chemical Society Reviews*, *41*(19), 6606–6630. doi:10.1039/c2cs35076a PMID:22660420

Health and Environment Alliance. Climate Change. (2009). Retrieved from www.env-health.org/r/93

Heath, Pollard, & Thomas. (1984). *Atlas of Butterflies of Britain and Ireland*. Viking Press.

He, H., Liang, X. Z., & Wuebbles, D. J. (2018). Effects of emissions change, climate change and long-range transport on regional modeling of future US particulate matter pollution and speciation. *Atmospheric Environment*, *179*, 166–176.

HEI & IHME. (2018). *State of Global Air/2018. A Special Report on Global Exposure to Air Pollution and its Disease Burden.* Retrieved from http://stateofglobalair.org/sites/default/files/soga-2018-report.pdf

Heidorn, K. C. (1970). A Chronology of Important Events in the History of Air Pollution Meteorology to 1970. *Bulletin of the American Meteorological Society*, *59*(12), 1589–1597. doi:10.1175/1520-0477(1978)059<1589:ACOIEI>2.0.CO;2

Held, I. M., & Soden, B. J. (2000). Water vapor feedback and global warming. *Annual Review of Energy and the Environment*, *25*(1), 441–475.

Henze, D., & Lee, H. (2012). GEOS-Chem Adjoint User's Guide (gcadj v32). Academic Press.

Herring, S. C., Hoell, A., Hoerling, M. P., Kossin, J. P., Schreck Iii, C. J., & Stott, P. A. (2016). Explaining Extreme Events of 2015 from a Climate Perspective. *American Meteorological Society*, *9*(12), S1–S145.

Hoffman, D. J., Butler, J. H., Dlugokencky, E. J., Elkins, J. W., Masarie, K., Montzka, S. A., & Tans, P. (2006). The role of carbon dioxide in climate forcing from 1979 to 2004: Introduction of the Annual Greenhouse Gas Index. *Tellus. Series B, Chemical and Physical Meteorology*, *58*(5), 614–619.

Holling, C. S., & Gunderson, L. H. (2002). *Resilience and adaptive cycles. In Panarchy: understanding transformations in human and natural systems* (pp. 25–62). Washington, DC: Island Press.

Hollnagel, E., Woods, D. D., & Leveson, N. (2006). *Resilience engineering: Concepts and precepts*. Aldershot, UK: Ashgate.

Holloway, T., Spak, S. N., Barker, D., Bretl, M., Moberg, C., Hayhoe, K., ... Wuebbles, D. (2008). Change in ozone air pollution over Chicago associated with global climate change. *Journal of Geophysical Research, D, Atmospheres*, *113*(D22).

Holzworth, C. G. (1971). Meteorological Potential for Urban Air Pollution in the Contiguous United States. *Proc. 2nd Intl. Clean Air Congress*, 1076-1080. 10.1016/B978-0-12-239450-8.50192-1

Hong, S., Chung, Y., Kim, J., & Chun, D. (2016). Analysis on the level of contribution to the national greenhouse gas reduction target in Korean transportation sector using LEAP model. *Renewable & Sustainable Energy Reviews*, *60*, 549–559. doi:10.1016/j.rser.2015.12.164

Honnay, O., Verheyen, K., Butaye, J., Jacquemyn, H., Bossuyt, B., & Hermy, M. (2002). Possible effects of habitat fragmentation and climate change on the range of forest plant species. *Ecology Letters*, *5*(4), 525–530. doi:10.1046/j.1461-0248.2002.00346.x

Horne, J. R., & Dabdub, D. (2017). Impact of global climate change on ozone, particulate matter, and secondary organic aerosol concentrations in California: A model perturbation analysis. *Atmospheric Environment*, *153*, 1–17.

Huang, J., McQueen, J., Wilczak, J., Djalalova, I., Stajner, I., Shafran, P., ... Huang, H. C. (2017). Improving NOAA NAQFC PM2. 5 Predictions with a Bias Correction Approach. *Weather and Forecasting*, *32*(2), 407–421.

Huang, W., Fan, H., Qiu, Y., Cheng, Z., Xu, P., & Qian, Y. (2016). Causation mechanism analysis for haze pollution related to vehicle emission in Guangzhou, China by employing the fault tree approach. *Chemosphere*, *151*, 9–16. doi:10.1016/j.chemosphere.2016.02.024 PMID:26923237

Huang, X., Wang, H., Yin, S., Chen, X., Chen, W., & Yang, H. (2009). sterlizatioin system for air purifier by combining ultraviolet light emitted diodes. *Journal of Chemical Technology and Biotechnology (Oxford, Oxfordshire)*, *84*(10), 1437–1440. doi:10.1002/jctb.2180

Huang, Y. C., & Ghio, A. J. (2006). Vascular effects of ambient pollutant particles and metals. *Current Vascular Pharmacology*, *4*(3), 199–203. doi:10.2174/157016106777698351 PMID:16842137

Huanqiunet. (2014). *WHO reported that 7 million people died of air pollution in 2012*. Retrieved April 10, 2018 from http://world.huanqiu.com/exclusive/2014-03/4929416.html

Huanqiunet. (2015). *WHO: Air Pollution has become the World's largest Environmental Health risk*. Retrieved January 22, 2018 from https://www.ncbi.nlm.nih.gov/pmc/articles/PMC4816944/

Hulin, M., Simoni, M., Viegi, G., & Annesi-Maesano, I. (2012). Respiratory health and indoor Atmospheric pollutants based on quantitative exposure assessments. *The European Respiratory Journal*, *40*(4), 1033–1045. doi:10.1183/09031936.00159011 PMID:22790916

Huo, P., Yan, Y., Li, S., Li, H., Huang, W., Chen, S., & Zhang, X. (2010). H2O2 modified surface of TiO2/fly ash cenospheres and enhanced photocatalytic activity on methylene blue.Desalination. *Desalination*, *263*(1-3), 258–263. doi:10.1016/j.desal.2010.06.067

Hurley, J. P., Physick, L. W., & Luhar, K. A. (2005). TAPM: A practical approach to prognostic meteorological and air pollution modelling. *Environmental Modelling & Software*, *20*(6), 737–752. doi:10.1016/j.envsoft.2004.04.006

Hurrell, J. W., Hack, J. J., Shea, D., Caron, J. M., & Rosinski, J. (2008). A new sea surface temperature and sea ice boundary dataset for the Community Atmosphere Model. *Journal of Climate*, *21*(19), 5145–5153.

IEA. (2011). *Energy For All: Financing Access for the Poor, Special Early Except of the World Energy Outlook*. Oslo, Norway: IEA.

Illi, J. C., Vancetta, T., Alves, D. D., Osório, D. M. M., Bianchin, L., Quevedo, D. M., & Juchem, F. (2017). Integrated assessment of air pollution by metals and source apportionment using ryegrass (*Lolium multiflorum Lam.*) in southern Brazil. *Environmental Science and Pollution Research International*, *24*(3), 2790–2803. doi:10.100711356-016-8041-y PMID:27837473

Inouye, D. W., Barr, B., Armitage, K. B., & Inouye, B. D. (2000). Climate change is affecting altitudinal migrants and hibernating species. *Proceedings of the National Academy of Sciences of the United States of America*, *97*(4), 1630–1633. doi:10.1073/pnas.97.4.1630 PMID:10677510

Intergovernmental Panel on Climate Change (CH). Climate change. (2007). *The physical science basis. Geneva: Intergovernmental Panel on Climate Change; 2007 (cited 2007 Jan 01)*. Available from: http://www.ipcc.ch/pdf/assessment-report/ar4/wg1/ar4_wg1_full_report.pdf

IPCC. (2005). *2005: Carbon Dioxide Capture and Storage Summary for Policymakers*. Intergovernmental Panel on Climate Change. Available.

IPCC. (2013). Summary for Policymakers. In: Climate Change 2013: The Physical Science Basis. Contribution of Working Group I to the Fifth Assessment Report of the Intergovernmental Panel on Climate Change. Cambridge University Press.

IPCC. (2014). *Climate Change 2014: Synthesis Report. Contribution of Working Groups I, II and III to the Fifth Assessment Report of the Intergovernmental Panel on Climate Change* (R. K. Pachauri & L. A. Meyer, Eds.). Geneva, Switzerland: IPCC.

IPCC. (2014). *Fifth Assessment Report – IPCC-AR5. Climate Change – Synthesis Report*. Disponível em: http://www.ipcc.ch/report/ar5/syr/

IPCC. (2014). Summary for Policymakers. In Climate Change 2014: Mitigation of Climate Change. Contribution of Working Group III to the Fifth Assessment. IPCC.

Iverson, L. R., & Prasad, A. M. (1998). Predicting abundance of 80 tree species following climate change in the eastern United States. *Ecological Monographs*, *68*(4), 465–485. doi:10.1890/0012-9615(1998)068[0465:PAOTSF]2.0.CO;2

Jacob, D. J., & Winner, D. A. (2009). Effect of climate change on air quality. *Atmospheric Environment, 43*(1), 51–63.

Jacobsen, C. S. (1997). Plant protection and rhizosphere colonization of barley by seed inoculated herbicide degrading Burkholderia (*Pseudomonas*) cepacia DBO1 (pRO101) in 2, 4-D contaminated soil. *Plant and Soil, 189*(1), 139–144. doi:10.1023/A:1004296615446

Jacobson, M. Z. (2007). Effects of ethanol (E85) versus gasoline vehicles on cancer and mortality in the United States. *Environmental Science & Technology, 41*(11), 4150–4157. doi:10.1021/es062085v PMID:17612204

Jäger, T., Schmídt, S., & Karl, U. (2009). A System Dynamics Model For The German Electricity Market – Model Development And Application. *International Conference of the System Dynamics Society.*

Jalkanen, L. (2011). WMO addressing climate and air quality. *Technical Workshop on Science and Policy of Short-lived Climate Forcers.* Retrieved March 31, 2018 from http://mce2.org/SLCFWorkshop/docs/%28Jalkanen%29%20SLCF%20Mexico% 20WMO.pdf

Jalota, S. K., Kaur, H., Kaur, S., & Vashisht, B. B. (2013). Impact of climate change scenarios on yield, water and nitrogen-balance and use efficiency of rice-wheat cropping system. *Agricultural Water Management, 116*, 29–38. doi:10.1016/j.agwat.2012.10.010

Jerrett, M., Burnett, R. T., Pope, C. A., Thurston, G., & Krewski, D. (2009). Long-term ozone exposure and mortality. *The New England Journal of Medicine, 360*(11), 1085–1095. doi:10.1056/NEJMoa0803894 PMID:19279340

Jiang, X., Sommer, S., & Chritensen, K. (2011). A Reveiw of the Biogas Industry in China. *Energy Policy, 39*, 6073–6081.

Johnson, B. D. (2003). High –power short –wave LED purifies air. *Photonics Spectra, 37*, 111.

Joyce, C. (2012). *Pollution playing a major role in sea temperatures.* National Public Radio.

Jurewicz, J., Radwan, M., Sobala, W., Polańska, K., Radwan, P., Jakubowski, L., ... Hanke, W. (2015). The relationship between exposure to air pollution and sperm disomy. *Environmental and Molecular Mutagenesis, 56*(1), 50–59. doi:10.1002/em.21883 PMID:24989325

Kagawa, J. (1985). Evaluation of biological significance of nitrogen oxides exposure. *The Tokai Journal of Experimental and Clinical Medicine, 10*, 348.

Kahn Ribeiro, S., Kobayashi, S., Beuthe, M., Gasca, J., Greene, D., Lee, D. S., ... Wit, R. (2007). *Transport and its infrastructure. Climate change 2007: mitigation. contribution of working group III to the fourth assessment report of the intergovernmental panel on climate change.* Cambridge, UK: Cambridge University Press.

Kala. (2014). Rhizoremediation: A Promising Rhizosphere Technology. *IOSR-JESTFT, 8*(8), 23-27.

Kars, C., & Altunbay, S. (2010). *Petrol Kullanımının Küresel Isınmaya Etkisi Ve Yenilebilir Enerji Yatırımları- Bir Sistem Dinamiği Modeli.* YAEM.

Kassiola, J. (2002). Why Environmental Thought and Action Must Include Considerations of Social Justice. In M. Gillroy & J. Bowersox (Eds.), *The Moral Austerity of Environmental Decision Making Sustainability, Democracy, and Normative Argument in Policy and Law*. Duke University Press.

Kassomenos, P., Skouloudis, A. N., Lykoudis, S., & Flocas, H. A. (1999). Air-Quality Indicators for Uniform Indexing of Atmospheric Pollution Over Large Metropolitan Areas. *Atmospheric Environment, 33*(12), 1861–1879. doi:10.1016/S1352-2310(98)00355-0

Kawazoe, S., & Gutowski, W. J. Jr. (2013). Regional, Extreme Daily Precipitation in CMIP5 Simulations. *Journal of Hydrometeorology, 14*(4), 1228–1242. doi:10.1175/JHM-D-12-0112.1

Kelley, S. L., Aitchison, E. W., Deshpande, M., Schnoor, J. L., & Alvarez, P. J. (2001). Biodegradation of 1, 4-dioxane in planted and unplanted soil: Effect of bioaugmentation with *Amycolata* sp. CB1190. *Water Research, 35*(16), 3791–3800. doi:10.1016/S0043-1354(01)00129-4 PMID:12230161

Kenkel, J. A., Sisk, T. D., Hultine, K. R., Sesnie, S. E., Bowker, M. A., & Johnson, N. C. (2016). Nancy Collins Johnson; Indicators of vehicular emission inputs into semi-arid roadside Ecosystems. *Journal of Arid Environments, 134*, 150–159. doi:10.1016/j.jaridenv.2016.06.007

Khoiyangbam, R. (2011). Environmental Implication of Biomethanation in Conentional Plants. *Energy & Environment, 2*, 181–187.

Kim, H., Cho, Y., Seo, J., Kim, T., Song, K., & Kim, K. (2017). *The increased prevalence of chronic actinic dermatitis in accordance with climate change and air pollution during the recent 15 years*. Academic Press.

Kimbrough, R. D., Carter, C. D., Liddle, J. A., Cline, R. E., & Phillips, P. E. (1977). Epidemiology and pathology of a tetrachlorodibenzodioxin poisoning episode. *Archives of Environmental Health, 32*(2), 77–86. doi:10.1080/00039896.1977.10667259 PMID:557961

Kim, E., & Hopke, P. K. (2007). Source Identifications of airborne fine particles using positive matrix factorization and U.S. Environmental Protection Agency Positive Matrix Factorization. *Journal of the Air & Waste Management Association, 57*(7), 811–819. doi:10.3155/1047-3289.57.7.811 PMID:17687996

Kingsley, M. T., Fredrickson, J. K., Metting, F. B., & Seidler, R. J. (1994). *Environmental restoration using plant-microbe bioaugmentation. In Bioremediation of chlorinated and polyaromatic hydrocarbon compounds* (pp. 287–292). Boca Raton, Fla: Lewis Publishers.

Kinney, P. L. (2008). Climate change, air quality, and human health. *American Journal of Preventive Medicine, 35*(5), 459–467. doi:10.1016/j.amepre.2008.08.025 PMID:18929972

Kleeman, M. J. (2008). A preliminary assessment of the sensitivity of air quality in California to global change. *Climatic Change, 87*(1), 273–292.

Klingmüller, K., Pozzer, A., Metzger, S., Stenchikov, G. L., & Lelieveld, J. (2016). Aerosol optical depth trend over the Middle East. *Atmospheric Chemistry and Physics*, *16*(8), 5063–5073. doi:10.5194/acp-16-5063-2016

Klumpp, A., Ansel, W., & Klumpp, G. (2004). *Urban Air Pollution, bioindication and environmental awareness*. Göttingen: Cuvillier Verlang.

Knutti, R., & Sedlácek, J. (2012). Robustness and uncertainties in the new CMIP5 climate model projections. *Nature Climate Change*. doi:10.1038/nclimate1716

Kopacz, M., Jacob, D. J., Fisher, J. A., Logan, J. A., Zhang, L., Megretskaia, I. A., ... Buchwitz, M. (2010). Global estimates of CO sources with high resolution by adjoint inversion of multiple satellite datasets (MOPITT, AIRS, SCIAMACHY, TES). *Atmospheric Chemistry and Physics*, *10*(3), 855–876.

Koposzki, H. (1992). Effects of acid and nitrogen deposition on the mesofauna, especially the collembolan. In *Proceedings of Atmospheric pollutants and Interactions between Organisms in Forest Ecosystems, 15th IUFRO International Meeting of Specialists on Atmospheric pollutants Effects on Forest Ecosystems* (p. 46). Academic Press.

Korade, D. L., & Fulekar, M. H. (2009). Rhizosphere remediation of chlorpyrifos in mycorrhizospheric soil using ryegrass. *Journal of Hazardous Materials*, *172*(2), 1344–1350. doi:10.1016/j.jhazmat.2009.08.002 PMID:19720454

Kousoulidou, M., & Lonza, L. (2016). Biofuels in aviation: Fuel demand and CO2 emissions evolution in Europe toward 2030. *Transp. Res. D.*, *46*, 166–181. doi:10.1016/j.trd.2016.03.018

Kovats, R. S., Johnson, H., & Griffiths, C. (2006). Mortality in southern England during the 2003 heat wave by place of death. *Health Statistics Quarterly*, *29*, 6–8. PMID:16523675

Kowalska, M., Osrodka, L., Klejnowski, K., Zejda, J. E., Krajny, E., & Wojtylak, M. (2009). Air quality index and its significance in environmental health riskcommunication. *Archives of Environmental Protection*, *35*(1), 13–21.

Krol, M. S. B. M. A. P. W. F. P., Houweling, S., Bregman, B., Broek, M. V., Segers, A., Velthoven, P. V., ... Bergamaschi, P. (2005). The two-way nested global chemistry-transport zoom model TM5: Algorithm and applications. *Atmospheric Chemistry and Physics*, *5*(2), 417–432.

Krupnick, A., McConnell, V., Austin, D., Cannon, M., Stoessell, T., & Morton, B. (1998). The Chesapeake Bay and the Control of NOX Emissions: Policy Analysis. *Resources for the Future, 1998*. Discussion Paper 98-46. Retrieved February 28, 2018 from http://ageconsearch.umn.edu/bitstream/10576/1/dp980046.pdf

Kumar, A., & Omidvarborna, H. (2018a). Introduction to Air Pollution. In Handbook of Environmental Engineering. Academic Press.

Kumar, A., & Omidvarborna, H. (2018b). Technology for Air Pollution Control. In Handbook of Environmental Engineering. Academic Press.

Kumar, A., & Omidvarborna, H. (2018c). Role of Air Pollution in Climate Change and Global Warming: Present Global Scenario. In Handbook of Environmental Engineering. Academic Press.

Kumar, A., & Goyal, P. (2011). Forecasting of air quality in Delhi using principal component regression technique. *Atmospheric Pollution Research*, *2*(4), 436–444. doi:10.5094/APR.2011.050

Kumar, A., & Goyal, P. (2011). Forecasting of daily air quality index in in Delhi. *The Science of the Total Environment*, *409*(24), 5517–5523. doi:10.1016/j.scitotenv.2011.08.069 PMID:21962560

Kumar, P., Andrade, M. F., Ynoue, R. Y., Fornaro, A., & de Freitas, E. D. (2016). New Directions: From biofuels to wood stoves: the modern and ancient air quality challenges in the megacity of São Paulo. *Atmospheric Environment*, *140*, 364–369. doi:10.1016/j.atmosenv.2016.05.059

Kumar, P., Jain, S., Gurjar, B. R., Sharma, P., Khare, M., Morawska, L., & Britter, R. (2013). Can a "Blue Sky" return to Indian megacities? *Atmospheric Environment*, *71*, 198–201. doi:10.1016/j.atmosenv.2013.01.055

Kumar, P., Khare, M., Harrison, R. M., Bloss, W. J., Lewis, A., Coe, H., & Morawska, L. (2015). New Directions: Air pollution challenges for developing megacities like Delhi. *Atmospheric Environment*, *122*, 657–661. doi:10.1016/j.atmosenv.2015.10.032

Kumar, P., Robins, A., Vardoulakis, S., & Britter, R. (2010). A review of the characteristics of nanoparticles in the urban atmosphere and the prospects for developing regulatory controls mode particles in an urban street canyon. *Atmospheric Environment*, *44*(39), 5035–5052. doi:10.1016/j.atmosenv.2010.08.016

Kunkel, K. E., Andsager, K., & Easterling, D. R. (1999). Long-term trends in extreme precipitation events over the conterminous United States and Canada. *Journal of Climate*, *12*(8), 2515–2527.

Kuo, C. Y., Wong, R. H., Lin, J. Y., Lai, J. C., & Lee, H. (2006). Accumulation ofchromium and nickel metals in lung tumors from lung cancer patients in Taiwan. *Journal of Toxicology and Environmental Health*, *69*(14), 1337–1344. doi:10.1080/15287390500360398 PMID:16760140

Kurpius, M. R., & Goldstein, A. H. (2003). Gas-phase chemistry dominates O3 loss to a forest, implying a source of aerosols and hydroxyl radicals to the atmosphere. *Geophysical Research Letters*, *30*(7), 1371. doi:10.1029/2002GL016785

Lacressonnière, G., Watson, L., Gauss, M., Engardt, M., Andersson, C., Beekmann, M., ... Nyiri, A. (2017). Particulate matter air pollution in Europe in a+ 2° C warming world. *Atmospheric Environment*, *154*, 129–140.

Lago, A. A. C. (2013). *Conferências de desenvolvimento sustentável*. Brasília: FUNAG.

Lamarque, J. F., Shindell, D. T., Josse, B., Young, P. J., Cionni, I., Eyring, V., ... Dalsoren, S. (2013). The Atmospheric Chemistry and Climate Model Intercomparison Project (ACCMIP): Overview and description of models, simulations and climate diagnostics. *Geoscientific Model Development*, *6*, 179–206.

Landulfo, E., Matos, C. A., Torres, A. S., Sawamura, P., & Uehara, S. T. (2007). Air Quality Assessment Using a Multi-Instrument Approach and Air Quality Indexing in an Urban Area. *Atmospheric Research, 85*(1), 98–111. doi:10.1016/j.atmosres.2006.11.008

Lane & Sterman. (2011). Profiles in Operations Research: Jay Wright Forrester. In *Profiles in Operations Research: Pioneers and Innovators*. New York: Springer. Retrieved from http://jsterman.scripts.mit.edu/docs/Lane2011%20Profiles%20in%20Operations%20Research.pdf

Langholz, B., Ebi, K. L., Thomas, D. C., Peters, J. M., & London, S. J. (2002). Traffic Density and the Risk of Childhood Leukemia in a Los Angeles Case-Control Study. *Annals of Epidemiology, 12*(7), 482–487. doi:10.1016/S1047-2797(01)00317-9 PMID:12377426

Law, K. (2010). More ozone over North America. *Nature, 463*, 307–308.

Lee, T. M., Markowitz, E. M., Howe, P. D., Ko, C., & Leiserowitz, A. A. (2015). Predictors of public climate change awareness and risk perception around the world. *Nature Climate Change, 5*(11), 1014–1020. doi:10.1038/nclimate2728

Lehohla, P. (2014). *Poverty Trends in South Africa: An examination of absolute poverty between 2006 and 2011*. Pretoria: Statistics South Africa.

Leung, D. Y. C. (2015). Outdoor-indoor air pollution in urban environment: Challenges and opportunity. *Frontiers in Environmental Science*, 2.

Li, H., Hopke, P. K., Liu, X., Du, X., & Li, F. (2015). Application of positive matrix factorization to source apportionment of surface water quality of the Daliao River basin, northeast China. *Environmental Monitoring and Assessment, 187*(3), 80. doi:10.100710661-014-4154-2 PMID:25655122

Lilleskov, E. A., Liu, L., Lynch, J., Nadelhoffer, K., Perakis, S. S., Robin-Abbott, M. J., ... Dennis, R. L. (2011). Effects of nitrogen deposition and empirical nitrogen critical loads for ecoregions of the United States. *Ecological Applications, 21*(8), 3049–3082. doi:10.1890/10-2341.1

Lindgren, A., Bjork, J., Stroh, E., & Jakobsson, K. (2010). Adult asthma and traffic exposure at residential address, workplace address, and self-reported daily time outdoor in traffic: A two-stage case-control study. *BMC Public Health, 10*(1), 716. doi:10.1186/1471-2458-10-716 PMID:21092159

Lin, J. T., Patten, K. O., Hayhoe, K., Liang, X. Z., & Wuebbles, D. J. (2008). Effects of future climate and biogenic emissions changes on surface ozone over the United States and China. *Journal of Applied Meteorology and Climatology, 47*(7), 1888–1909.

Lin, J.-T., Patten, K. O., Liang, X.-Z., & Wuebbles, D. J. (2008). Effects of intercontinental transport on surface ozone over the United States at the present and future. *Geophysical Research Letters, 35*(2), L02805. doi:10.1029/2007GL031415

Lin, Y., Liu, Y., Sun, Y., Zhu, X., Lai, J., & Heynderickz, I. (2014). Model predicting discomfort glare caused by LED road light. *Optics Express, 22*(15), 18056–18071. doi:10.1364/OE.22.018056 PMID:25089425

Lipsett, M. J., Ostro, B. D., Reynolds, P., Goldberg, D., Hertz, A., Jerrett, M., ... Bernstein, L. (2011). Long-term exposure to air pollution and cardiorespiratory disease in the California teachers study cohort. *American Journal of Respiratory and Critical Care Medicine*, *184*(7), 828–835. doi:10.1164/rccm.201012-2082OC PMID:21700913

Li, S., Williams, G., Jalaludin, B., & Baker, P. (2012). Panel studies of Atmospheric pollutants on children's lung function and respiratory symptoms: A literature review. *The Journal of Asthma*, *49*(9), 895–910. doi:10.3109/02770903.2012.724129 PMID:23016510

Liu, J. C., & Peng, R. D. (2018). Health effect of mixtures of ozone, nitrogen dioxide, and fine particulates in 85 US counties. *Air Quality, Atmosphere & Health*, 1–14.

Liu, Mao, Ren, Li, Guo, & Zhang, . (2015). How Might China Achieve Its 2020 Emissions Target? A Scenario Analysis Of Energy Consumption And CO2 Emissions Using The System Dynamics Model. *Journal of Cleaner Production*, *103*, 401–410. doi:10.1016/j.jclepro.2014.12.080

Liu, S., Bond-Lamberty, B., Boysen, L. R., Ford, J. D., Fox, A., Gallo, K., ... Loveland, T. R. (2017). Grand challenges in understanding the interplay of climate and land changes. *Earth Interactions*, *21*(2), 1–43.

Liu, T., Gong, S., He, J., Yu, M., Wang, Q., Li, H., ... Zhao, Q. (2017). Attributions of meteorological and emission factors to the 2015 winter severe haze pollution episodes in China's Jing-Jin-Ji area. *Atmospheric Chemistry and Physics*, *17*(4), 2971–2980. doi:10.5194/acp-17-2971-2017

Liu, Y., Stanturf, J., & Goodrick, S. (2010). Trends in global wildfire potential in a changing climate. *Forest Ecology and Management*, *259*(4), 685–697. doi:10.1016/j.foreco.2009.09.002

Li, Z., Liu, Y., Yang, X., Xing, Y., Tsai, C.-J., Meng, M., & Yang, R. T. (2017). Performance of mesoporous silicas and carbon in adsorptive removal of phenanthrene as a typical gaseous polycyclic aromatic hydrocarbon. *Microporous and Mesoporous Materials*, *239*, 9–18. doi:10.1016/j.micromeso.2016.09.027

Localized Pollution Potentially Plays Large Role in Future Climate Change. (2008). National Oceanic and Atmospheric Administration (NOAA). Retrieved from http://www.noaanews.noaa.gov/stories2008/20080904_pollution.html

Loghman-Adham, M. (1997). Renal effects of environmental and occupational lead exposure. Environ. Health Perspect. 105, 928.Jarup, L., 2003. Hazards of heavy metal contamination. *British Medical Bulletin*, *68*, 167.

Longhurst, J. (2005). 1 to 100: Creating an Air Quality Index in Pittsburg. *Environmental Monitoring and Assessment*, *106*(1-3), 27–42. doi:10.100710661-005-0758-x PMID:16001710

Luhar, A. K., Galbally, I. E., Woodhouse, M. T., & Thatcher, M. (2017). An improved parameterisation of ozone dry deposition to the ocean and its impact in a global climate-chemistry model. *Atmospheric Chemistry and Physics*, *17*(5), 3749.

Maani & Cavana. (2007). Systems Thinking And Modelling Understanding Change and Complexity. Pearson Education.

Macal, C. M. (2004). To Agent-Based Simulation From System Dynamics. *International Conference of the System Dynamics Society.*

Madden, E. F., & Fowler, B. A. (2000). Mechanisms of nephrotoxicity from metal combinations: A review. *Drug and Chemical Toxicology, 23*(1), 1–12. doi:10.1081/DCT-100100098 PMID:10711385

Madronich, S., Shao, M., Wilson, S. R., Solomon, K. R., Longstreth, J. D., & Tang, X. Y. (2015). Changes in air quality and tropospheric composition due to depletion of stratospheric ozone and interactions with changing climate: Implications for human and environmental health. *Photochemical & Photobiological Sciences, 14*(1), 149–169. doi:10.1039/C4PP90037E PMID:25380416

Manahan, S. E. (2011). *Environmental Science and Technology – A Sustainable Approach to Green Science and Technology*. Boca Raton, FL: CRC Press, Taylor & Francis.

Mandal, P. K. (2005). Dioxin: A review of its environmental effects and its aryl hydrocarbon receptor biology. *Journal of Comparative Physiology, 175*(4), 221–230. doi:10.100700360-005-0483-3 PMID:15900503

Manzo, N. D., Slade, R., & Dye, J. A. (2010). Interrelationship of lung inflammation and air pollutant exposure on cellular oxidative stress and epithelial injury. *Proceedings 27th Symposium of the Veterinary Comparative Respiratory Society.*

Marcott, S. A., Shakun, J. D., Clark, P. U., & Mix, A. C. (2013). A Reconstruction of Regional and Global Temperature for the Past 11,300 Years. *Science, 339*(6124), 1198–1201. doi:10.1126cience.1228026 PMID:23471405

Marengo, J. A. (2011). *Riscos das mudanças climáticas no Brasil: Análise conjunta Brasil-Reino Unido sobre os impactos das mudanças climáticas e do desmatamento na Amazônia*. Projeto colaborativo realizado pelo Centro de Ciência do Sistema Terrestre (CCST) do Instituto Nacional de Pesquisas Espaciais (INPE) do Brasil e o Met Office Hadley Centre (MOHC) do Reino Unido.

Marland, G., & Rotty, R. M. (1984). Carbon dioxide emissions from fossil fuels: A procedure for estimation and results for 1950–1982. *Tellus. Series B, Chemical and Physical Meteorology, 36*(4), 232–261.

Marroccoli, M., Pace, M. L., Telesca, A., Valenti, G. L., & Montagnaro, F. (2010). Utilization of coal combustion ashes for the synthesis of ordinary and special cements. *Combustion Science and Technology, 182*(4-6), 588–599. doi:10.1080/00102200903466210

Mason, C. F., & MacDonald, S. M. (1987). Acidification and otter (Lutra lutra) distribution on a British river. *Mammalia, 51*(1), 81–87.

Mathias, J. M., & Thomas, R. B. (2018). Disentangling the effects of acidic air pollution, atmospheric CO2, and climate change on recent growth of red spruce trees in the Central Appalachian Mountains. *Global Change Biology, 24*(9), 3938–3953. doi:10.1111/gcb.14273 PMID:29781219

Matthews, H. D., Gillett, N. P., Stott, P. A., & Zickfeld, K. (2009). The proportionality of global warming to cumulative carbon emissions. *Nature*, *459*(7248), 829. PMID:19516338

Matyssek, R., Clarke, N., Cudlin, P., Mikkelsen, T. N., Tuovinen, J. P., Wieser, G., & Paoletti, E. (2013). Climate Change, Air Pollution and Global Challenges: Chapter 1. In Climate Change, Air Pollution and Global Challenges: Understanding and Perspectives from Forest Research (Vol. 13). Elsevier Inc.

Matyssek, R., Kozovits, A.R., Schnitzler, J. P., Pretzsch, H., Dieler, J., & Wieser G. (2014). Trees in a Changing Environment: Forest Trees Under Air Pollution as a Factor of Climate Change. *Plant Ecophysiology*, 9.

Mayer, H., & Kalberlah, F. (2008). Two Impact Related Air Quality Indices as Tools to Assess the Daily and Long-Term Air Pollution. *International Journal of Environment and Pollution*, *32*, 232–243.

McDonough, C., Jaffe, D., Watzin, M., & McGinley, M. (2018). Indicator species. In C. J. Cleveland (Ed.), *Encyclopedia of Earth*. Washington, DC: Environmental Information Coalition, National Council for Science and the Environment.

McGrath, J. M., Betzelberger, A. M., Wang, S., Shook, E., Zhu, X. G., Long, S. P., & Ainsworth, E. A. (2015). An analysis of ozone damage to historical maize and soybean yields in the United States. *Proceedings of the National Academy of Sciences of the United States of America*, *112*(46), 14390–14395. doi:10.1073/pnas.1509777112 PMID:26578785

McHugh, A. J., Griffiths, G. W., & Schiesser, W. E. (2015). *An Introductory Global CO2 Model:(with Companion Media Pack)*. World Scientific Publishing Company.

McLachlan, J. S., Clark, J. S., & Manos, P. S. (2005). Molecular indicators of tree migration capacity under rapid climate change. *Ecology*, *86*(8), 2088–2098. doi:10.1890/04-1036

McLinden, C. A., Fioletov, V., Boersma, K. F., Kharol, S. K., Krotkov, N., Lamsal, L., ... Yang, K. (2014). Improved satellite retrievals of NO2 and SO2 over the Canadian oil sands and comparisons with surface measurements. *Atmospheric Chemistry and Physics*, *14*(7), 3637–3656. doi:10.5194/acp-14-3637-2014

Meadows, D. H., Randers, J., Meadows, D. L., & Behrens, W. W. (1972). *The Limits to Growth: Report for the Club of Rome's Project on the Predicament of Mankind*. New York: Universe Books.

Meireles, J. R. C., & Cerqueira, E. M. M. (2011). *Use of the Micronucleus Test on Tradescantia (Trad-MCN) to Evaluate the Genotoxic Effects of Air Pollution*. Air Pollution – New Developments.

Meleux, F., Solmon, F., & Giorgi, F. (2007). Increase in summer European ozone amounts due to climate change. *Atmospheric Environment*, *41*(35), 7577–7587.

Melillo, J. M., Richmond, T. C., & Yohe, G. W. (Eds.). (2014). *Climate Change Impacts in the United States: The Third National Climate Assessment*. Washington, DC: U.S. Global Change Research Program; doi:10.7930/J0Z31WJ2

Menéndez, R., Megías, A. G., Hill, J. K., Braschler, B., Willis, S. G., Collingham, Y., ... Thomas, C. D. (2006). Species richness changes lag behind climate change. *Proceedings. Biological Sciences*, *273*(1593), 1465–1470. doi:10.1098/rspb.2006.3484 PMID:16777739

Menzel, A., & Sparks, T. H. (2006). European phenological response to climate change matches the warming pattern. *Global Climate Change*, *12*(10), 1969-1976.

Mieszkowska, N. (2006). Changes in the range of some common rocky shore species in Britain - A response to climate change? Marine Biodiversity. *Developments in Hydrobiology*, 241-251.

Miller, K. A., Siscovick, D. S., Sheppard, L., Shepherd, K., Sullivan, J. H., Anderson, G. L., & Kaufman, J. D. (2007). Long-term exposure to air pollution and incidence of cardiovascular events in women. *The New England Journal of Medicine*, *356*(5), 447–458. doi:10.1056/NEJMoa054409 PMID:17267905

Miller, S. G. (2017). *Here's how climate change can cause more air pollution (All by itself)*. LiveScience.

Millstein, D. E., & Harley, R. A. (2009). Impact of climate change on photochemical air pollution in Southern California. *Atmospheric Chemistry and Physics*, *9*(11), 3745–3754.

Mintzberg, H., Ahlstrand, B., & Lampel, J. (2009). *Strategy Safari: Your complete guide through the wilds of strategic management*. London: Prentice Hall.

Mishchenko, M. I., Geogdzhayev, I. V., Rossow, W. B., Cairns, B., Carlson, B. E., Lacis, A. A., ... Travis, L. D. (2007). Long-term satellite record reveals likely recent aerosol trend. *Science*, *315*(5818), 1543. doi:10.1126cience.1136709 PMID:17363666

Mohan, D., Singh, K. P., Singh, G., & Kumar, K. (2002). Removal of dyes from wastewater using flyash, a low-cost adsorbent. *Industrial & Engineering Chemistry Research*, *41*(15), 3688–3695. doi:10.1021/ie010667+

Molod, A., Takacs, L., Suarez, M., & Bacmeister, J. (2015). Development of the GEOS-5 atmospheric general circulation model: Evolution from MERRA to MERRA2. *Geoscientific Model Development*, *8*(5), 1339.

Monks, P. S., Archibald, A. T., Colette, A., Cooper, O., Coyle, M., Derwent, R., ... Stevenson, D. S. (2015). Tropospheric ozone and its precursors from the urban to the global scale from air quality to short-lived climate forcer. *Atmospheric Chemistry and Physics*, *15*(15), 8889–8973.

Monteiro, A., Viera, M., Gama, C., & Miranda, A. I. (2016). Towards an improved air quality index. *Air Quality, Atmosphere & Health*. doi:10.100711869-016-0435-y

Moore, J., & Ouellet, M. (2015). Questioning the use of an amphibian colour morph as an indicator of climate change. *Global Change Biology*, *21*(2), 566–571. doi:10.1111/gcb.12744 PMID:25230362

Mora, C., Frazier, A. G., Longman, R. J., Dacks, R. S., Walton, M. M., Tong, E. J., ... Giambelluca, T. W. (2013). The projected timing of climate departure from recent variability. *Nature*, *502*(7470), 183–187. doi:10.1038/nature12540 PMID:24108050

Morawska, L., Ristovski, Z., Jayaratne, E. R., Keogh, D. U., & Ling, X. (2008). Ambient nano and ultrafine particles from motor vehicle emissions: Characteristics, ambiente processing and implications on human exposure. *Atmospheric Environment*, *42*(35), 8113–8138. doi:10.1016/j.atmosenv.2008.07.050

Morecroft. (2015). *Strategic Modelling And Business Dynamics: A Feedback Systems Approach*. Chichester, UK: John Wiley & Sons.

Moreira, H. M., & Giometti, A. B. R. (2008). O Protocolo de Quioto e as Possibilidades de Inserção do Brasil no Mecanismo de Desenvolvimento Limpo por meio de Projetos em Energia Limpa. Contexto Internacional, 30, 9-47. doi:10.1590/S0102-85292008000100001

Morelli, F., Møller, A., Nelson, E., Benedetti, Y., Liang, W., Šímová, P., ... Tryjanowski, P. (2017). The common cuckoo is an effective indicator of high bird species richness in Asia and Europe. *Scientific Reports*, *7*(1), 4376. doi:10.103841598-017-04794-3 PMID:28663558

Morgenstern, O., Hegglin, M. I., Rozanov, E., O'Connor, F. M., Abraham, N. L., Akiyoshi, H., ... Deushi, M. (2017). Review of the global models used within phase 1 of the Chemistry-Climate Model Initiative (CCMI). *Geoscientific Model Development*, *10*(2), 639.

Moss, R. H., Richard, H. M., Jae, A. E., Kathy, A. H., Martin, R. M., Steven, K. R., ... Thomas, J. W. (2010). The next generation of scenarios for climate change research and assessment. *Nature*, *463*(7282), 747–756. doi:10.1038/nature08823 PMID:20148028

Mujumdar, P. P., & Kumar, N. D. (2012). *Floods in a Changing Climate: Hydrologic Modeling*. Cambridge University Press. doi:10.1017/CBO9781139088428

Murakami, M., Abe, M., Kakumoto, Y., Kawano, H., Fukasawa, H., Saha, M., & Takada, H. (2012). Evaluation of ginkgo as a biomonitor of airborne polycyclic aromatic. *Atmospheric Environment*, *54*, 9–17. doi:10.1016/j.atmosenv.2012.02.014

Murena, F. (2004). Measuring Air Quality over Large Urban Areas: Development and Application of an Air Pollution Index at the Urban Area of Naples. *Atmospheric Environment*, *38*(36), 6195–6202. doi:10.1016/j.atmosenv.2004.07.023

Musango, J. K., Brent, A. C., & Bassí, A. (2009). South African Energy Model: A System Dynamics Approach. *International Conference of the System Dynamics Society.*

Nabat, P., Somot, S., Mallet, M., Sevault, F., Chiacchio, M., & Wild, M. (2015). Direct and semi-direct aerosol radiative effect on the Mediterranean climate variability using a coupled regional climate system model. *Climate Dynamics*, *44*(3-4), 1127–1155.

Nagendra, S. M. S., Venugopal, K., & Jones, S. L. (2007). Assessment of air quality near traffic intersectionsin Bangalore city using air quality index. *Transportation Research Part D, Transport and Environment*, *12*(3), 167–176. doi:10.1016/j.trd.2007.01.005

Nakicenovic, N., & Swart, R. (2000). *Special Report on Emissions Scenarios: A Special Report of Working Group III of the Intergovernmental Panel on Climate Change*. Cambridge, UK: Cambridge University Press.

National climate adaptation plan – 1° Nap Monitoring and Evaluation Report - 2016/2017 – Brazil. (n.d.). Disponível em: http://www.mma.gov.br/clima/adaptacao/plano-nacional-de-adaptacao?tmpl=component&print=1

National Drought Mitigation Center. (2016). *Maps and data*. Retrieved from http://droughtmonitor.unl.edu/MapsAndData.aspx

Navarro, R., Guzman, J., Saucedo, I., Revilla, J., & Guibal, E. (2007). Vanadium recovery from oil fly ash by leaching, precipitation and solvent extraction processes. *Waste Management (New York, N.Y.)*, *27*(3), 425–438. doi:10.1016/j.wasman.2006.02.002 PMID:16563726

Nawrot, T., Plusquin, M., Hogervorst, J., Roels, H. A., Celis, H., Thijs, L., ... Staessen, J. A. (2006). Environmental exposure to cadmium and risk of cancer: A prospective population-based study. *The Lancet Oncology*, *7*(2), 119–126. doi:10.1016/S1470-2045(06)70545-9 PMID:16455475

Neas, L., Williams, R. W., & Devlin, R. B. (2004). Particulate matter exposure in cars is associated with cardiovascular effects in healthy young men. *American Journal of Respiratory and Critical Care Medicine*, *169*(8), 934–940. doi:10.1164/rccm.200310-1463OC PMID:14962820

Neeling, J. D. (2011). *Climate Change and Climate Modeling*. Cambridge university press.

Nichols, T. D., Wolf, D. C., Rogers, H. B., Beyrouty, C. A., & Reynolds, C. M. (1997). Rhizosphere microbial populations in contaminated soils. *Water, Air, and Soil Pollution*, *95*(1-4), 165–178. doi:10.1007/BF02406163

Nijaguma, B. (2002). Biogas technology. New Technology International, 46-48.

Nowak, D. J. (2000). Tree species selection, design, and management to improve air quality. *Construction Technology: Annual Meeting Proceedings of the American Society of Landscape Architects*.

Nowak, D. J., Crane, D. E., & Stevens, J. C. (2006). Atmospheric pollutants removal by urban trees and shrubs in the United States. *Urban Forestry & Urban Greening*, *4*(3-4), 115–123. doi:10.1016/j.ufug.2006.01.007

NPTEL. (2018). *Air Quality Modelling*. Retrieved from https://nptel.ac.in/courses/105102089/air%20pollution%20(Civil)/Module-5/1.htm

NRDC. (2007). *Medium and Long-term Develoment Plan for Renewable Energy in China*. Beijing: National Development and Reform Commission.

Nurul, M., & Ruma, R. (2015). An efficient solid base catalyst from coal combustion fly ash for green synthesis of dibenzylideneacetone. *Journal of Industrial and Engineering Chemistry*, *29*, 401–414.

Obbard. (2017). *APCEL Climate Change Adaptation Platform COP23 Bonn, Germany: The U.S. Stands Alone*. Asia Pacific Centre for Environmental Law. National University of Singapore.

Ochoa-Hueso, R., Munzi, S., Alonso, R., Arróniz-Crespo, M., Avila, A., Bermejo, V., ... Theobald, M. R. (2017). Ecological impacts of atmospheric pollution and interactions with climate change in terrestrial ecosystems of the Mediterranean Basin: Current research and future directions. *Environmental Pollution*, *227*, 194–206. doi:10.1016/j.envpol.2017.04.062 PMID:28460237

Öğüt & Şahin. (2012). *Türkiye'de Finansal İstikrar Sorununa Sistem Dinamiği Yaklaşımı*. Türkiye Ekonomi Kurumu.

Oliva, M. A., & Figueiredo, J. G. (2005). Gramíneas biondicadoras da presença de flúor em regiões tropicais. *Revista Brasileira de Botanica. Brazilian Journal of Botany*, *28*(2), 389–397. doi:10.1590/S0100-84042005000200017

Ollinger, S. V., Aber, J. D., & Reich, P. B. (1997). Simulating ozone effects on forest productivity: Interactions among leaf-, canopy-, and stand-level processes. *Ecological Applications*, *7*(4), 1237–1251. doi:10.1890/1051-0761(1997)007[1237:SOEOFP]2.0.CO;2

Olsson, D., Mogren, I., & Forsberg, B. (2013). Atmospheric pollutants exposure in early pregnancy and adverse pregnancy outcomes: A register-based cohort study. *BMJ Open*, *3*(2), e001955. doi:10.1136/bmjopen-2012-001955 PMID:23386578

Omidvarborna, H. (2017). *Combustion chemistry of biodiesel for the use in urban transport buses: experiment and modeling* (PhD Thesis). University of Toledo.

Omidvarborna, H., Kumar, A., Kim, D. S., & Shandilya, K. K. (2014). Analysis of particulate matter from the exhaust of biodiesel transit buses under idling conditions. In Air Waste Manage. Assoc. Annu. Meeting Conf. Exhibition (Vol. 106, No. 4, pp. 2712-2725). Academic Press.

Omidvarborna, H., Kumar, A., & Kim, D. S. (2015). Recent studies on soot modeling for diesel combustion. *Renewable & Sustainable Energy Reviews*, *48*, 635–647.

Omidvarborna, H., Kumar, A., & Kim, D. S. (2016). Variation of diesel soot characteristics by different types and blends of biodiesel in a laboratory combustion chamber. *The Science of the Total Environment*, *544*, 450–459. PMID:26657390

Orru, H., Andersson, C., Ebi, K. L., Langner, J., Astrom, C., & Forsberg, B. (2013). Impact of climate change on ozone-related mortality and morbidity in Europe. *The European Respiratory Journal*, *41*(2), 285–294. doi:10.1183/09031936.00210411 PMID:22743679

Orru, H., Ebi, K. L., & Forsberg, B. (2017). The interplay of climate change and air pollution on health. *Current Environmental Health Reports*, *4*(4), 504–513. doi:10.100740572-017-0168-6 PMID:29080073

Pachauri, R. K., & Reisinger, A. (2007). *IPCC fourth assessment report*. Geneva: IPCC.

Padmanabha Murty, B. (2004). *Environmental Meteorology*. New Delhi: I.K. International Pvt. Ltd.

Pai, S. G., Riley, M. B., & Camper, N. D. (2001). Microbial degradation of mefenoxam in rhizosphere of Zinnia angustifolia. *Chemosphere*, *44*(4), 577–582. doi:10.1016/S0045-6535(00)00368-4 PMID:11482644

Palacio, I. C., Oliveira, I. F., Franklin, R. L., Barros, S. B. M., & Roubicek, D. A. (2016). Evaluating the mutagenicity of the water-soluble fraction of air particulate matter: A comparison of two extraction strategies. *Chemosphere*, *158*, 124–130. doi:10.1016/j.chemosphere.2016.05.058 PMID:27258903

Pandey, B., Subedi, P., Sengendo, M., & Monroe, I. (2007). *Report on the Feasibility for a national Household Commercialization program in Uganda*. Brussels: Winrock International.

Pan, L. L., Bowman, K. P., Atlas, E. L., Wofsy, S. C., Zhang, F., Bresch, J. F., ... Cooper, W. A. (2010). The stratosphere–troposphere analyses of regional transport 2008 experiment. *Bulletin of the American Meteorological Society*, *91*(3), 327–342.

Pardo, L. H., Fenn, M. E., Goodale, C. L., Geiser, L. H., Driscoll, C. T., Allen, E. B., ... Haines, A. (2014). Climate change: Challenges and opportunities for global health. *Journal of the American Medical Association*, *312*(15), 1565–1580. doi:10.1001/jama.2014.13186 PMID:25244362

Pasquill, F. (1961). The Estimation of the Dispersion of Windborne Material. *The Meteorological Magazine*, *90*(1063), 33–49.

Pauli, H., Gottfried, M., & Grabherr, G. (2003). Effects of climate change on the alpine and nival vegetation of the Alps. *The Journal of Mountain Ecology*, 7.

Pawson, S., Stolarski, R. S., Douglass, A. R., Newman, P. A., Nielsen, J. E., Frith, S. M., & Gupta, M. L. (2008). Goddard Earth Observing System chemistry climate model simulations of stratospheric ozone temperature coupling between 1950 and 2005. *Journal of Geophysical Research, D, Atmospheres*, *113*(D12).

Pearce, D., & Ozdemiroglu, E. (2002). *Economic Valuation with Stated preference Techniques*. London: Renwick, Department of Tranport, Local Government and the Regions.

Pearson, R. G., & Dawson, T. P. (2003). Predicting the impacts of climate change on the distribution of species: Are bio-climate envelope models useful? *Global Ecology and Biogeography*, *12*(5), 361–371. doi:10.1046/j.1466-822X.2003.00042.x

Pecl, G. T., Araujo, M. B., Bell, J. D., Blanchard, J., Bonebrake, T. C., Chen, I., ... Williams, S. E. (2017). Biodiversity redistribution under climate change: Impacts on ecosystems and human well-being. *Science*, *355*(6332), 1–10. doi:10.1126cience.aai9214 PMID:28360268

Perry, C. T. (2010). *Psidium guajava* as a bioaccumulator of nickel around an oil refinery, southern Brazil. *Ecotoxicology and Environmental Safety*, *73*(4), 647–654. doi:10.1016/j.ecoenv.2009.10.001 PMID:20036423

Peters, A., von Klot, S., Heier, M., Trentinaglia, I., Hörmann, A., Wichmann, H. E., & Löwel, H. (2004). Exposure to traffic and the onset of myocardial infarction. *The New England Journal of Medicine*, *351*(17), 1721–1730. doi:10.1056/NEJMoa040203 PMID:15496621

Pickrell, J. (1991). Hazards in confinement housing - gases and dusts in confined animal houses for swine, poultry, horses and humans. *Veterinary and Human Toxicology*, *33*, 32–39. PMID:2017863

Pidwirny, M., & Jones, S. (2009). *Atmospheric Effects on Incoming Solar Radiation*. Retrieved from http://www.physicalgeography.net/fundamentals/

Pielke, R. A., & Wilby, R. L. (2012). Regional climate downscaling: What's the point? *Eos (Washington, D.C.)*, *93*(5), 52–53.

Pinder, R.W., Bettez, N.D., Bonan, G.B., Greaver, T. L., Widier, W.R., Schlesinger, W.H., & Davidson, E. A. (2013). Impacts of human alteration of the nitrogen cycle in the US on radiative forcing. *Biogeochemistry*, *114*, 25–40.

Pleijel, H. (2009). Air pollution & climate change, two sides of the same coin? In Common roots of air pollution and climate change. Swedish Environmental protection Agency.

Pope, C. A. III, Ezzati, M., & Dockery, D. W. (2009). Fine-particulate air pollution and life expectancy in the United States. *The New England Journal of Medicine*, *360*(4), 376–386. doi:10.1056/NEJMsa0805646 PMID:19164188

Pye, H. O. T., Liao, H., Wu, S., Mickley, L. J., Jacob, D. J., Henze, D. K., & Seinfeld, J. H. (2009). Effect of changes in climate and emissions on future sulfate nitrate ammonium aerosol levels in the United States. *Journal of Geophysical Research, D, Atmospheres*, *114*(D1).

Qian, . (2008). An Experimental Study On The Construction Of Electricity Power Stations İn China. *International Conference of the System Dynamics Society*.

Qin, L., Han, J., Zhan, Y., Chen, W., & Kim, H. (2016). 2016, Reduction of Polycyclic Aromatic Hydrocarbon Emission by Porous Alumina Bed Material during Sewage Sludge Incineration. *Energy & Fuels*, *30*(1), 544–550. doi:10.1021/acs.energyfuels.5b01942

Racherla, P. N., & Adams, P. J. (2006). Sensitivity of global tropospheric ozone and fine particulate matter concentrations to climate change. *Journal of Geophysical Research, D, Atmospheres*, *111*(D24).

Radzicki, M. J. (1990). Methodologia Oeconomiae Et Systematis Dynamis. *System Dynamics Review*, *6*(2), 123–147. doi:10.1002dr.4260060202

Rafieisakhaei, Barazandeh, Moosavi, Fekri, & Bastani. (2016). *Modeling Dynamics of the Carbon Market: A System Dynamics Approach on the CO2 Emissions and its Connections to the Oil Market*. International Conference of the System Dynamics Society, Delft, The Netherlands.

Rajdurai, Rao, & Kamalakkannan. (2016). CO_2 Capture Using Activated Alumina in Gasoline Passenger Vehicles. *Int. Journal of Engineering Research and Applications*, *6*(5-4), 73-77.

Ramage. (2009). *Sytem Thinkers, The Open University Walton Hall*. Milton Keynes, UK: Springer London.

Ramanathan, V. (2014). The Two Worlds Approach for Mitigating Air Pollution and Climate Change. *Sustainable Humanity, Sustainable Nature: Our Responsibility*. Extra Series: 41, Pontifical Academy of Sciences. Retrieved April 13, 2018 from http://www.pas.va/content/dam/accademia/pdf/es41/es41-ramanathan.pdf

Ramanathan, V. (1977). Interactions between ice-albedo, lapse-rate and cloud-top feedbacks: An analysis of the nonlinear response of a GCM climate model. *Journal of the Atmospheric Sciences*, *34*(12), 1885–1897.

Ramanathan, V., & Carmichael, G. (2008). Global and regional climate changes due to black carbon. *Nature Geoscience*, *1*(4), 221.

Ramaswamy, V., Boucher, O., Haigh, J., Hauglustine, D., Haywood, J., Myhre, G., ... Solomon, S. (2001). Radiative forcing of climate. *Climatic Change*, 349.

Rano, R. (2005). *Evaluation of the Physico-Chemical and Morphological Properties of Flyash obtained from Different Thermal Power Plants for their Potential Utilisation* (PhD thesis).

Rao, M. N., & Rao, H. V. N. (1986). Air pollution. TATA McGraw-Hill Publishing Company.

Rasmussen, D. J., Fiore, A. M., Naik, V., Horowitz, L. W., McGinnis, S. J., & Schultz, M. G. (2012). Surface ozone-temperature relationships in the eastern US: A monthly climatology for evaluating chemistry-climate models. *Atmospheric Environment*, *47*, 142–153.

Rastogi, S. K., Gupta, B. N., Husain, T., Chandra, H., Mathur, N., Pangtey, B. S., ... Ravenscroft, J. (2006). Effects of pollution on human growth and development: An introduction. *Journal of Physiological Anthropology*, *25*(1), 103–112. doi:10.2114/jpa2.25.103 PMID:16617215

Reducing Short-Lived Climate Pollutants in California. (n.d.). Air Resources Board, California Environmental Agency. Retrieved from https://www.arb.ca.gov/cc/shortlived/shortlived.htm

Reff, A., Eberly, S. I., & Bhav, E. P. V. (2007). Receptor modeling of ambient particulate matter data using positive matrix factorization: Review of existing methods. *Air & Waste Management Association*, *57*(2), 146–154. doi:10.1080/10473289.2007.10465319 PMID:17355075

Reich, P.B., & Amundson, R.G. (n.d.). Ambient levels of ozone reduce net photosynthesis in tree and crop species. *Science, 230*, 566–570.

Renwick, M., Suebi, S. & Hutton, G. (2007). *A Cost-Benefit Analysis of National and Regional Integrated Biogas and Sanitation Programmes in sub-Saharan Africa.* Winrock International.

Reyes-García, V., Fernández-Llamazares, A., Guèze, M., Garcés, A., Mallo, M., Vila-Gómez, M., & Vilaseca, M. (2016). Local indicators of climate change: The potential contribution of local knowledge to climate research. *Wiley Interdisciplinary Reviews: Climate Change*, *7*(1), 109–124. doi:10.1002/wcc.374 PMID:27642368

Richardson. (1999). *Feedback Throught in Social Science And Systems Theory*. Waltham: Pegasus Communications, Inc.

Riebau, A. R., & Andersen, C. (1985). Wildland Fires and Atmospheric pollutants. In *Developments in Environmental Science* (Vol. 8, pp. 365–403). Elsevier.

Ritchie, H, & Roser, M. (2018). *Air Pollution.* OurWorldInData.org.

Ronchel, M. C., & Ramos, J. L. (2001). Dual system to reinforce biological containment of recombinant bacteria designed for rhizoremediation. *Applied and Environmental Microbiology, 67*(6), 2649–2656. doi:10.1128/AEM.67.6.2649-2656.2001 PMID:11375176

Root, H. T., Geiser, L. H., Jovan, S., & Neitlich, P. (2015). Epiphytic macro lichen indication of air quality and climate in interior forested mountains of the Pacific Northwest, USA. *Ecological Indicators, 53*, 95–105. doi:10.1016/j.ecolind.2015.01.029

Rosenzweig, C., Iglesias, A., Yang, X. B., Epstein, P. R., & Chivian, E. (2001). Climate change and extreme weather events; implications for food production, plant diseases, and pests. *Global Change and Human Health, 2*(2), 90–104.

Rotterdam Resilience Strategy - Consultation Document – 2016. (n.d.). Disponível em: http://100resilientcities.org/wp-content/uploads/2017/06/strategy-resilient-rotterdam.pdf

Roupsard, P. (2013). *Etude Phénoménologique Du dépôt Sec d'aérosols En Milieu Urbain* (Ph.D Thesis). University of Rouen.

Roy, A., Chatterjee, A., Tiwari, S., Sarkar, C., Das, S. K., Ghosh, S. K., & Raha, S. (2016). Precipitation chemistry over urban, rural and high altitude Himalayan stations in eastern India. *Atmospheric Research, 181*, 44–53. doi:10.1016/j.atmosres.2016.06.005

Rubin, J. I., Kean, A. J., Harley, R. A., Millet, D. B., & Goldstein, A. H. (2006). Temperature dependence of volatile organic compound evaporative emissions from motor vehicles. *Journal of Geophysical Research, 111*(D3), D03305. doi:10.1029/2005JD006458

Rummukainen, M. (2010). State of the art with Regional Climate Models. *Wiley Interdisciplinary Reviews: Climate Change, 1*(1), 82–96.

Sajjad, F., Noreen, U., & Zaman, K. (2014). Climate change and air pollution jointly creating nightmare for tourism industry. *Environmental Science and Pollution Research International, 21*(21), 12403–12418. doi:10.100711356-014-3146-7 PMID:24938808

Sandmann, E. R. I. C., & Loos, M. A. (1984). Enumeration of 2, 4-D-degrading microorganisms in soils and crop plant rhizospheres using indicator media; high populations associated with sugarcane (*Saccharum officinarum). Chemosphere, 13*(9), 1073–1084. doi:10.1016/0045-6535(84)90066-3

Santini, L., Cornulier, T., Bullock, J. M., Palmer, S. C. F., White, S. M., Hodgson, J. A., ... Travis, J. M. J. (2016). A trait-based approach for predicting species responses to environmental change from sparse data: How well might terrestrial mammals track climate change? *Global Change Biology, 22*(7), 2415–2424. doi:10.1111/gcb.13271 PMID:27073017

Saravanakumar, R., Sivalingam, S., & Elangovan, S. (2016). Assessment of Air Quality Index of Coimbatore City in Tamil Nadu. *Indian Journal of Science and Technology*, *9*(41). doi:10.17485/ijst/2016/v9i41/99185

Sarkar, A., Rano, R.& Kumar A (2010). Characterization of high carbon ash from thermal power station using pulverized fuel combustor. *Energy. Source Part- A*, *32*, 607-609.

Sarkar, A., Rano, R., Mishra, K. K., & Sinha, I. N. (2005). Particle size distribution profiles of some Indian fly ash-A comparative study to assess their possible uses. *Fuel Processing Technology*, *86*(11), 1221–1238. doi:10.1016/j.fuproc.2004.12.002

Sarkar, A., Rano, R., Udaybhanu, G., & Basu, A. K. (2006). A comprehensive characterization of fly ash from a thermal power plant in Eastern India. *Fuel Processing Technology*, *87*(3), 259–277. doi:10.1016/j.fuproc.2005.09.005

Scherer, M., & Unshelm, J. (1995). Untersuchungen zur Belastung von Mastschweinen sowie der Umgebung von Mastschweinstallen durch atembaren Feinstaub, stallspezifische Bakterien und Ammoniak. *Zentralblatt fur Hygiene und Umweltmedizin*, *196*, 399–415. PMID:7727021

Schlenker, W., & Walker, W. R. (2016). Airports, Air Pollution, and Contemporaneous Health. *The Review of Economic Studies*, *83*(2), 768–809. doi:10.1093/restud/rdv043

Schlesinger, R. B. (1979). Natural removal mechanisms for chemical pollutants in the environment. *Bioscience*, *29*(2), 95–101. doi:10.2307/1307745

Schmale, J., Shindell, D., von Schneidemesser, E., Chabay, I., & Lawrence, M. (2014a). Air pollution: clean up our skies. *Nature, 515*, 335–337. Retrieved April 14, 2018 from https://www.nature.com/news/air-pollution-clean-up-our-skies-1.16352

Schmale, J., van Aardenne, J., & von Schneidemesser, E. (2014). New Directions: Support for integrated decision-making in air and climate policies – Development of a metrics-based information portal. E. *Atmospheric Environment*, *90*, 146–148. doi:10.1016/j.atmosenv.2014.03.016

Schneider, T., Lan, S., Stuart, A., & Teixeira, J. (2017). Earth System Modeling 2.0: A Blueprint for Models That Learn From Observations and Targeted High-Resolution Simulations. *Geophysical Research Letters*, *44*(24), 12,396–12,417. doi:10.1002/2017GL076101

Schnoor, J. L., Aitchison, E. W., Kelley, S. L., & Alvarez, P. J. J. (1998). Phytoremediation and bioaugmentation of 1, 4-dioxane. In*Abstr. Adv. Biol. Syst. Rem. Conf* (pp. 87-91). Academic Press.

Schreiner, C., Gonyo, E., Shandilya, K. K., Buonocore, J., Zhou, X., Salamero, G., ... Ferro, A. (2006). Children's Exposure to Diesel Exhaust at a Rural Elementary School. *Epidemiology (Cambridge, Mass.)*, *17*, S146–S147.

Schroder, M., Linkert, K. H., Pedersen, S., Takai, H., Johnsen, J. O., Groot Koerkamp, P. W. G., ... Linke, S. (1998). Emissions of Aerial Pollutants in Livestock Buildings in Northern Europe: Overview of a Multinational Project. *Journal of Agricultural Engineering Research*, *70*(1), 3–9. doi:10.1006/jaer.1998.0278

Schulz, M., Chin, M., & Kinne, S. (2009). The Aerosol Model Comparison Project, Aerocom Phase II: Clearing up Diversity. *IGAC Newslett., 41*. Retrieved from http://aerocom.met.no/pdfs/May_2009_IGAC_41.pdf

Schumann, A.W. & Sumner, M.E. (n.d.). *Chemical Evaluation of Nutrient Supply From Fly Ash- Substances listed in the thirteenth report on Carcinogens.* NTPC.

Schumann, A. W., & Sumner, M. E. (2000). Chemical evaluation of nutrient supply from fly ash-biosolids mixtures. *Soil Science Society of America Journal, 64*(1), 419–426. doi:10.2136ssaj2000.641419x

Scinocca, J. F., Kharin, V. V., Jiao, Y., Qian, M. W., Lazare, M., Solheim, L., ... Dugas, B. (2016). Coordinated global and regional climate modeling. *Journal of Climate, 29*(1), 17–35.

Secretariat, C. C. A. C. (2016, May 30). *Two New UNEP Reports On Air Quality Show More Can Be Done To Tackle Air Pollution.* Retrieved February 25, 2018 from http://www.ccacoalition.org/en/news/two-new-unep-reports-air-quality-show-more-can-be-done-tackle-air-pollution

Seinfeld, J. H., & Pandis, S. N. (2006). *Atmospheric chemistry and physics: from air pollution to climatechange* (2nd ed.). New York: John Wiley & Sons.

Seinfeld, J. H., & Pandis, S. N. (2016). *Atmospheric chemistry and physics: from air pollution to climate change.* John Wiley & Sons.

Sezen, H. (2009). *Günal Murat.* Bursa: Yöneylem Araştırmasında Benzetim, Ekin Yayınevi.

Shandilya, K. K. (2012). *Characterization, speciation, and source apportionment of particles inside and from the exhaust of public transit buses fueled with alternative fuels* (PhD Thesis). University of Toledo.

Shandilya, K. K., & Sitaram, P. (2018). On the Development of a Sustainable Stormwater Management System. *Proceedings for American Society of Engineering Education - North Central Section Spring Conference 2018.*

Shandilya, K. K., Khare, M., & Gupta, A. B. (2009). Defining aerosols by physical and chemical characteristics. *Indian Journal of Air Pollution Control, 9*(1), 107-126.

Shandilya, K. K., & Khare, M. (2012). Particulate matter: Sources, emission rates and health effects. *Advances in Environmental Research, 23*, 265–322.

Shandilya, K. K., Khare, M., & Gupta, A. B. (2006). Suspended particulate matter concentrations in commercial east Delhi and south Delhi, India. *Indian Journal of Environmental Protection, 26*(8), 705–717.

Shandilya, K. K., & Kumar, A. (2010). Morphology of single inhalable particle inside public transit biodiesel fueled bus. *Journal of Environmental Sciences (China), 22*(2), 263–270. PMID:20397416

Shandilya, K. K., & Kumar, A. (2011). Physical characterization of fine particulate matter inside the public transit buses fueled by biodiesel in Toledo, Ohio. *Journal of Hazardous Materials*, *190*(1-3), 508–514. PMID:21497439

Shandilya, K., Gupta, A. B., & Khare, M. (2011). Formation of atmospheric nitrate under high particulate matter concentration. *World Review of Science, Technology and Sustainable Development*, *8*(2-4), 148–165.

Shandilya, K., & Kumar, A. (2014). Carbon speciation of exhaust particulate matter of public transit buses running on alternative fuels. *Fuel*, *115*, 678–684.

Sharmaa, A., Masseya, D. D., & Tanejab, A. (2018). A study of horizontal distribution pattern of particulate and gaseous pollutants based on ambient monitoring near a busy highway. *Urban Climate*, *24*, 643–656. doi:10.1016/j.uclim.2017.08.003

Sharma, M., Kiran, Y. N. V. M., & Shandilya, K. K. (2003). Investigations into formation of atmospheric sulfate under high PM10 concentration. *Atmospheric Environment*, *37*(14), 2005–2013.

Sharma, M., Kiran, Y. N. V. M., & Shandilya, K. K. (2004). Subset statistical modelling of atmospheric sulfate. *Clean Air and Environmental Quality*, *38*(4), 37.

Sharma, M., Pandey, R., Maheshwari, M., Sengupta, B., Shukla, B. P., Gupta, N. K., & Johri, S. (2003). Interpretation of Air Quality Data Using an Air Quality Index for the City of Kanpur, India. *Journal of Environmental Engineering and Science*, *2*, 453–462. doi:10.113903-047

Sharma, R. K., & Agrawal, M. (2005). Biological effects of heavy metals: An overview. *Journal of Environmental Biology*, *26*, 301. PMID:16334259

Sharma, V. K. (1999). Development of Air Quality Indices for Mumbai, India. *International Journal of Environment and Pollution*, *11*(2), 141–146. doi:10.1504/IJEP.1999.002254

Sheng, J. X., Weisenstein, D. K., Luo, B. P., Rozanov, E., Stenke, A., Anet, J., ... Peter, T. (2015). Global atmospheric sulfur budget under volcanically quiescent conditions: Aerosol chemistry climate model predictions and validation. *Journal of Geophysical Research, D, Atmospheres*, *120*(1), 256–276.

Shimizu, Y. (n.d.). *A global health guardian: climate change, air pollution, and antimicrobial resistance*. World Health Organization.

Shine, K. P., Stevenson, D. S., Sudo, K., Szopa, S., & Zeng, G. (2006). Radiative forcing since preindustrial times due to ozone change in the troposphere and the lower stratosphere. *Atmospheric Chemistry and Physics*, *6*(3), 575–599. doi:10.5194/acp-6-575-2006

Shon, Z. H., Kim, K. H., & Song, S. K. (2011). Long-term trend in NO2 and NOx levels and their emission ratio in relation to road traffic activities in East Asia. *Atmospheric Environment*, *45*(18), 3120–3131. doi:10.1016/j.atmosenv.2011.03.009

Signh, J., & Sooch, S. (2004). Comparative Studies of Economies of Different Models of Family Size Biogas Plants for State of Punjab. *Energy Conversion and Management*, *45*, 1329–1341.

Sillman, S. (1999). The relation between ozone, NOx and hydrocarbons in urban and polluted rural environments. *Atmospheric Environment*, *33*(12), 1821–1845.

Sillman, S., & Samson, P. J. (1995). Impact of temperature on oxidant photochemistry in urban, polluted rural and remote environments. *Journal of Geophysical Research, D, Atmospheres*, *100*(D6), 11497–11508. doi:10.1029/94JD02146

Silva, R. A., West, J. J., Zhang, Y., Anenberg, S. C., Lamarque, J. F., Shindell, D. T., & Horowitz, L. W. (2013). Global premature mortality due to anthropogenic outdoor Atmospheric pollutants and the contribution of past climate change. *Environmental Research Letters*, *8*(3), 034005. doi:10.1088/1748-9326/8/3/034005

Silveira, C. S., Filho, F. A. S., Cabral, L. S., Lemos, W. E. D., & Porto, V. C. (2013). Avaliação de desempenho dos modelos de mudança climática do CMIP5 quanto à representação da precipitação sobre as bacias do setor elétrico brasileiro no século XX e análise das projeções para o cenário rcp8.5. Em: XX Simpósio de Brasileiro de Recursos Hídricos.

Singh, R., Sharma, C., & Agrawal, M. (2017). Emission inventory of trace gases from road transport in India. *Transportation Research Part D, Transport and Environment*, *52*, 64–72. doi:10.1016/j.trd.2017.02.011

Sitch, S., Cox, P. M., Collins, W. J., & Huntingford, C. (2007). Indirect radiative forcing of climatechange through ozone effects on the land-carbon sink. *Nature*, *448*(7155), 791–794. doi:10.1038/nature06059 PMID:17653194

Slater, L. (2016). update and expansion to data originally published in: Mallakpour, I., G. Villarini. 2015. The changing nature of flooding across the central United States. *Nature Climate Change*, *5*, 250–254.

Smil, V. (2002). Nitrogen and food production: Proteins for human diets. *Ambio*, *31*(2), 126–131. doi:10.1579/0044-7447-31.2.126 PMID:12078001

Smith, M., Goebel, J., & Blignaut, J. (2013). *The Economic and Fiancial Feasibility of Rural Household Biodigesters for Poor Communities in South Africa*. Waste Management.

Solomon, S., Qin, D., Manning, M., Averyt, K., & Marquis, M. (Eds.). (2007). *Climate change 2007-the physical science basis: Working group I contribution to the fourth assessment report of the IPCC* (Vol. 4). Cambridge University Press.

Solomon, S., Qin, D., Manning, M., Marquis, M., Averyt, K., & Tignor, M. l. (2007). Technical summary. In *Climate change. The physical science basis* (pp. 19–940). New York, NY: Cambridge University Press.

Somoano, D. M. (2003). Phase-minerql and chemical composition of fly ashes as a basis for their multicomponent utilisation.1. Characterization of feed coal and fly ashes. *Fuel*, *82*(14), 1793–1811. doi:10.1016/S0016-2361(03)00123-6

Sriprang, R., Hayashi, M., Yamashita, M., Ono, H., Saeki, K., & Murooka, Y. (2002). A novel bioremediation system for heavy metals using the symbiosis between leguminous plant and genetically engineered rhizobia. *Journal of Biotechnology*, *99*(3), 279–293. doi:10.1016/S0168-1656(02)00219-5 PMID:12385715

Stein, A. F., Draxler, R. R., Rolph, G. D., Stunder, B. J. B., & Cohen, M. D. (2015). NOAA's HYSPLIT Atmospheric Transport and Dispersion Modeling System. American Meteorological Society, 2059-2077.

Steiner, A. (2014, June 23). *UNEP*. Retrieved March 30, 2018 from http://drustage.unep.org/transport/astf/events/air-pollution-priority-united-nations-environment-assembly

Steiner, A. (2016, May 30). Two New UNEP Reports On Air Quality Show More Can Be Done To Tackle Air Pollution. *CCAC Secretariat*. Retrieved February 25, 2018 from http://www.ccacoalition.org/en/news/two-new-unep-reports-air-quality-show-more-can-be-done-tackle-air-pollution

Steiner, A. L., Tonse, S., Cohen, R. C., Goldstein, A. H., & Harley, R. A. (2006). Influence of future climate and emissions on regional air quality in California. *Journal of Geophysical Research, D, Atmospheres*, *111*(D18).

Sterman. (2000). Business Dynamics Systems Thinking And Modelling In A Complex World. McGraw-Hill.

Stern, N. H. (2007). *The Economics of Climate Change: The Stern Review*. Cambridge, UK: Cambridge University Press. doi:10.1017/CBO9780511817434

Stevenson, D. S., Young, P. J., Naik, V., Lamarque, J. F., Shindell, D. T., Voulgarakis, A., ... Folberth, G. A. (2013). Tropospheric ozone changes, radiative forcing and attribution to emissions in the Atmospheric Chemistry and Climate Model Intercomparison Project (ACCMIP). *Atmospheric Chemistry and Physics*, *13*(6), 3063–3085.

Steven, V., Kokelj, T. C., Lantz, J. T., Rebecca, S., & Denis, L. (2017). Climate-driven thaw of permafrost preserved glacial landscapes, northwestern Canada. *Geology*.

Stone, D. A. S., & Knutti, R. (2010). Weather and Climate. In C. F. Fung, A. Lopez, & M. New (Eds.), *Modelling the Impact of Climate Change and Water Resources*. John Willey &Sons. doi:10.1002/9781444324921.ch2

Sun, G., McLaughlin, S. B., Porter, J. H., Uddling, J., Mulholland, P. J., Adams, M. B., & Pederson, N. (2012). Interactive influences of ozone and climate on streamflow of forested watersheds. *Global Change Biology*, *18*(11), 3395–3409. doi:10.1111/j.1365-2486.2012.02787.x

Tager, I. B., Balmes, J., Lurmann, F., Ngo, L., Alcorn, S., & Kunzli, N. (2005). Chronic exposure to ambient ozone and lung function in young adults. *Epidemiology (Cambridge, Mass.)*, *16*, 751. doi:10.1097/01.ede.0000183166.68809.b0 PMID:16222164

Tai, A. P. K., Martin, M. V., & Heald, C. L. (2014). Threat to future global food security from climate change and ozone air pollution. *Nature Climate Change*, *4*(9), 817–821. doi:10.1038/nclimate2317

Tai, A. P., Mickley, L. J., & Jacob, D. J. (2010). Correlations between fine particulate matter (PM2. 5) and meteorological variables in the United States: Implications for the sensitivity of PM2. 5 to climate change. *Atmospheric Environment*, *44*(32), 3976–3984.

Tai, A. P., Mickley, L. J., Jacob, D. J., Leibensperger, E. M., Zhang, L., Fisher, J. A., & Pye, H. O. T. (2012). Meteorological modes of variability for fine particulate matter (PM 2.5) air quality in the United States: Implications for PM 2.5 sensitivity to climate change. *Atmospheric Chemistry and Physics*, *12*(6), 3131–3145.

Tainio, M., Juda-Rezler, K., Reizer, M., Warchałowski, A., Trapp, W., & Skotak, K. (2013). Future climate and adverse health effects caused by fine particulate matter air pollution: Case study for Poland. *Regional Environmental Change*, *13*(3), 5–15.

Tallis, J. H. (1985). Mass movement and erosion of a Southern Pennine blanket peat. *Journal of Ecology*, *73*(1), 283–315. doi:10.2307/2259784

Tan, Anderson, & Parker. (2007). Managing Risk in Alternative Energy Product Development. *International Conference of the System Dynamics Society*.

TangVijay, . (2001). *System Dynamics Origins, Development And Future Prospects of a Method.* Research Seminar in Engineering Systems.

Tasic, M., Mijic, Z., Rajsic, S., Zekic, A., Perisic, M., & Stojic, A. (2010). Characteristics and application of receptor models to the atmospheric aerosols research. In *Air Quality*. Rijeka: SciyoIntech. Disponível em: http://www.intechopen.com/books/airquality/characteristics-and-application-of-receptor-models-to-the-atmospheric-aerosols-research

Teixeira, E. C., Feltes, S., & Santana, E. R. R. (2008). Estudo das emissões de fontes móveis na Região Metropolitana de Porto Alegre, Rio Grande do Sul. *Quimica Nova*, *31*(2), 244–248. doi:10.1590/S0100-40422008000200010

Teixeira, E. C., & Mattiuz, I. (2012). Estimated atmospheric emissions from bio- diesel and characterization of pollutants in the metropolitan area of Porto Alegre-RS. *Anais da Academia Brasileira de Ciências*, *84*(3), 655–667. doi:10.1590/S0001-37652012000300008 PMID:22886159

Teutschbein, C., Wetterhall, F., & Seibert, J. (2011). Evaluation of different downscaling techniques for hydrological climate-change impact studies at the catchment scale. *Climate Dynamics*, 1–19.

The Effect of Climate Change on Water Resources and Programs, US Environmental Protection Agency. (n.d.). Retrieved from https://cfpub.epa.gov/watertrain/moduleFrame.cfm?parent_object_id=2414#

The Guardian. (2016, October 20). *Air pollution more deadly in Africa than malnutrition or dirty water, study warns.* Retrieved January 20, 2018 from https://www.theguardian.com/global-development/2016/oct/20/air-pollution-deadlier-africa-than-dirty-water-or-malnutrition-oecd

The Lancet. (2017). *The Lancet Commission on Pollution and Health.* Retrieved March 15, 2018 from http://www.thelancet.com/commissions/pollution-and-health

The Very, Very Simple Climate Model, University Corporation for Atmospheric Research (UCAR). (n.d.). Retrieved from https://scied.ucar.edu/simple-climate-model

The World Bank. (2016). *World Development Indicators*. Author.

The World Bank. (2018, April 5). *Reducing Pollution*. Retrieved April 14, 2018 from http://www.worldbank.org/en/topic/environment/brief/pollution

Thiemens, M. H., & Geophys, M. H. (1999). D10307.D.H. Ehhalt. *Physical Chemistry Chemical Physics*, *1*(24), 5401.

Thiering, E., & Heinrich, J. (2015). Epidemiology of Atmospheric pollutants and diabetes. *Trends in Endocrinology and Metabolism*, *26*(7), 384–394. doi:10.1016/j.tem.2015.05.002 PMID:26068457

Thron, R. W. (1996). Direct and indirect exposure to Atmospheric pollutants. *Otolaryngology - Head and Neck Surgery*, *114*(2), 281–285. doi:10.1016/S0194-5998(96)70184-5 PMID:8637751

Tinarwo, D., Mugera, W. M., Rasimphi, T. E., Nekhubvi, V., & Machaba, H. (2014). Assessment of biogas Technology, A Case Study of a rural village in Vhembe District, Limpopo, South Africa. Paper Presented at *a Conference, Nairobi*, Kenya.

Torrey, S. (1978). Coal Ash Utilization: Fly Ash, Bottom Ash and Slag. Noyes Data.

Trenberth, K. E., & Jones, P. D. (2007). Observations: surface and atmospheric climate change. In Climate Change 2007: The Physical Science Basis. Cambridge University Press.

Trenberth, K. E., Jones, P. D., Ambenje, P., Bojariu, R. E., Parker, D., Rahimzadeh, F., … Zhai, P. (2007). Observations: surface and atmospheric climate change. In Climate Change 2007: The Physical Science Basis. Contribution of Working Group I to the Fourth Assessment Report of the Intergovernmental Panel on Climate Change. Cambridge, UK: Cambridge University Press.

Tsigaridis, K., & Kanakidou, M. (2018). The Present and Future of Secondary Organic Aerosol Direct Forcing on Climate. *Current Climate Change Reports*, 1–15.

Turner, B. D. (1970). *Workbook of Atmospheric Dispersion Estimates*. Environmental Protection Agency.

Turner, M. C., & Jerrett, M. (2016). Long-term ozone exposure and mortality in a large prospective study. *AJRCCM Issues*, *193*, 10. PMID:26680605

Turner, M. C., Jerrett, M., Pope, C. A. III, Krewski, D., Gapstur, S. M., Diver, W. R., ... Burnett, R. T. (2016). Long-term ozone exposure and mortality in a large prospective study. *American Journal of Respiratory and Critical Care Medicine*, *193*(10), 1134–1142. doi:10.1164/rccm.201508-1633OC PMID:26680605

UNCHE. (1972). *Declaration of the UN Conference on the Human Environment, Stockholm, 5–16 June 1972.* Report of the UN Conference on the Human Environment, UN Doc.A/CONF.48/14.

UNEP. (2014, June 23). Retrieved March 30, 2018 from http://drustage.unep.org/transport/astf/events/air-pollution-priority-united-nations-environment-assembly

Unger, N., Zhao, Y., & Dang, H. (2013). Mid-21st century chemical forcing of climate by the civil aviation sector. *Geophysical Research Letters*, *40*(3), 641–645.

United Nations Framework Convention on Climate Change. (2015). *Paris Agreement*. Availableat:http://unfccc.int/files/meetings/paris_nov_2015/application/pdf/paris_agreement_english_.pdf

United Nations General Assembly. (1962). *Economic Development and the Conservation of Nature*. G.A. Resolution 1831 (XVII), UN GAOR, 17th Session, Supp. no. 17, UN Doc. A/RES/1831 (XVII).

United Nations General Assembly. (2015). *Transforming Our World: The 2030 Agenda for Sustainable Development*. Draft resolution referred to the United Nations summit for the adoption of the post-2015 development agenda by the General Assembly at its sixty-ninth session. UN Doc. A/70/L.1 of 18 September (2015).

US E.P.A. (2008). Integrated Science Assessment (ISA) for Oxides of Nitrogen and Sulfur—Environmental Criteria. Final report. Washington, DC: Author.

USEPA. (2016). *Air Quality Models*. Retrieved from https://www3.epa.gov/scram001/aqmindex.htm

USEPA. (2018a). *Climate Impacts on Agriculture and Food Supply*. Accessed on October 10, https://19january2017snapshot.epa.gov/climate-impacts/climate-impacts-agriculture-and-food-supply_.html

USEPA. (2018b). *Sources of Greenhouse Gas Emissions*. Accessed on October 10, https://www.epa.gov/ghgemissions/sources-greenhouse-gas-emissions#t1fn1

Uysal, N., & Schapira, R. M. (2003). Effects of ozone on lung function and lung diseases. *Current Opinion in Pulmonary Medicine*, *9*, 144.

Vaclav Smil Energy Transitions. (2017). *Global and National Perspectives*. BP Statistical Review of World Energy.

Vallero. (2016). Air Pollution Monitoring Changes to Accompany the Transition from a Control to a Systems Focus. Sustainability, 8, 1216. doi:10.3390u8121216

Vallero, D. A. (2014). *Fundamentals of Air Pollution* (5th ed.). Waltham, MA: Elsevier Academic Press.

Vamvakas, S., Bittner, D., & Koster, U. (1993, April). Enhanced expression of the protooncogenes c-myc and c-fos in normal and malignant renal growth. *Toxicology Letters*, *67*(1-3), 161–172. doi:10.1016/0378-4274(93)90053-Z PMID:8451757

Van Aalst, M. K. (2006). The impacts of climate change on the risk of natural disasters. *Disasters*, *30*(1), 5–18. PMID:16512858

Van Grinsven, H. J. M., Holland, M., Jacobsen, B. H., Klimont, Z., Sutton, M. A., & Willems, W. J. (2013). Costs and benefits of nitrogen for Europe and implications for mitigation. *Environmental Science & Technology*, *47*(8), 3571–3579. doi:10.1021/es303804g PMID:23473305

Van Vuuren, D. P., Meinshausen, M., Plattner, G. K., Joos, F., Strassmann, K. M., Smith, S. J., ... Den Elzen, M. G. J. (2008). Temperature increase of 21st century mitigation scenarios. *Proceedings of the National Academy of Sciences of the United States of America*, *105*(40), 15258–15262. PMID:18838680

Vanos, J. K., Cakmak, S., Kalkstein, L. S., & Yagouti, A. (2015). Association of weather and air pollution interactions on daily mortality in 12 Canadian cities. *Air Quality, Atmosphere & Health*, *8*(3), 307–320. doi:10.100711869-014-0266-7 PMID:26052369

Vanos, J. K., Hebbern, C., & Cakmak, S. (2014). Risk assessment for cardiovascular and respiratory mortality due to air pollution and synoptic meteorology in 10 Canadian cities. *Environmental Pollution*, *185*, 322–332. doi:10.1016/j.envpol.2013.11.007 PMID:24355413

Van, V. (2011). Representative Concentration Pathways: An overview. *Climatic Change*, *109*(1-2), 5–31. doi:10.100710584-011-0148-z

Vassilev, S. V., & Menendez, R., Borrego, A.G., Somoano, M.D., Tarazona, M., & Rosa M. (2004). Phase-minerql and chemical composition of fly ashes as a basis for their multicomponent utilisation.-2Characterization of ceramic chemosphere and salt concentrates. *Fuel*, *83*, 1563–1583. doi:10.1016/j.fuel.2004.01.010

Vassilev, S. V., & Rosa, M. (2005). Phase-minerql and chemical composition of fly ashes as a basis for their multicomponent utilisation. 4.Characterization of heavy concentrates and improved fly ash residues. *Fuel*, *84*(7-8), 260–231. doi:10.1016/j.fuel.2004.11.021

Vautard, R., Jacob, D., Gobiet, A., & Belda, M. (2013). The simulation of European heat waves from an ensemble of regional climate models within the EURO-CORDEX project. *Climate Dynamics*, *41*(9-10), 2555–2575. doi:10.100700382-013-1714-z

Venkataraman, C., Sagar, A. D., Habib, G., Lam, N., & Smith, K. R. (2010). The Indian national initiative for advanced biomass cookstoves: The benefits of clean combustion. *Energy for Sustainable Development*, *14*(2), 63–72.

Vermylen, J., Nemmar, A., Nemery, B., & Hoylaerts, M. F. (2005). Ambient Atmospheric pollutants and acute myocardial infarction. *Journal of Thrombosis and Haemostasis*, *3*(9), 1955–1961. doi:10.1111/j.1538-7836.2005.01471.x PMID:16102102

Vikas, M., & Dwarakish, G. S. (2015). Coastal Pollution: A Review, International conference on water resources, coastal and ocean engineering (ICWRCOE 2015). *Aquatic Procedia*, *4*, 381–388.

Vingarzan, R. (2004). A review of surface ozone background levels and trends. *Atmospheric Environment*, *38*, 3431–3442.

Visscher, A. D. (2014). *Air Dispersion Modeling – Foundations and Applications*. John Wiley & Sons Inc.

Von Schneidemesser, E., Monks, P. S., Allan, J. D., Bruhwiler, L., Forster, P., Fowler, D., ... Sindelarova, K. (2015). Chemistry and the linkages between air quality and climate change. *Chemical Reviews, 115*(10), 3856–3897. PMID:25926133

Voorhees, S. A., Sakai, R., Araki, S., Sato, H., & Otsu, A. (2001, July). Cost-Benefit Analysis Methods For Assessing Air Pollution Control Programs in Urban Environments — A Review. *Environmental Health and Preventive Medicine, 6*, 63–73. Retrieved from https://slideheaven.com/cost-benefit-analysis-methods-for-assessing-air-pollution-control-programs-in-ur.html

Walther, G. R., Post, E., Convey, P., Menzel, A., Parmesank, C., Beebee, T. J. C., ... Bairlein, F. (2002). Ecological responses to recent climate change. *Nature, 416*(6879), 389–395. doi:10.1038/416389a PMID:11919621

Wang, Cheng, Lang, Li, & Tian. (2016). On-board measurements of gaseous pollutant emission characteristics under real driving conditions from light-duty diesel vehicles in Chinese cities. *Journal of Environmental Sciences, 46*, 28–37.

Wang, H. J., & Chen, H. P. (2016). Understanding the recent trend of haze pollution in eastern China: Roles of climate change. *Atmospheric Chemistry and Physics, 16*(6), 4205–4211. doi:10.5194/acp-16-4205-2016

Wang, H., Jacob, D. J., Le Sager, P., Streets, D. G., Park, R. J., Gilliland, A. B., & van Donkelaar, A. (2009). Surface ozone background in the United States: Canadian and Mexican influences. *Atmospheric Environment, 43*(6), 1310–1319. doi:10.1016/j.atmosenv.2008.11.036

Wang, Y., Shen, L., Wu, S., Mickley, L., He, J., & Hao, J. (2013). Sensitivity of surface ozone over China to 2000–2050 global changes of climate and emissions. *Atmospheric Environment, 75*, 374–382.

Wegmann, M., Fehrenbach, A., Heimann, S., Fehrenbach, H., Renz, H., Garn, H., & Herz, U. (2005). NO2-induced airway inflammation is associated with progressive airflow limitation and development of emphysema-like lesions in C57bl/6 mice. *Experimental and Toxicologic Pathology, 56*(6), 341–350. doi:10.1016/j.etp.2004.12.004 PMID:15945273

Wehner, M. (2013). Methods of Projecting Future changes in Extremes. In *Extremes in a Changing Climate: Detection, Analysis and Uncertainty*. London: Springer. doi:10.1007/978-94-007-4479-0_8

What Is Black Carbon? Center for Climate and Energy Solutions. (2010). Retrieved from https://www.c2es.org/site/assets/uploads/2010/04/what-is-black-carbon.pdf

Wheeler. (1994). How to Get Managers To Use System Dynamics. *International System Dynamics Conference*, Stirling, UK.

WHO Regional Office for Europe (DK). (2013). *Review of evidence on health aspects of Atmospheric pollutants: REVIHAAP Project*. Copenhagen: WHO Regional Office for Europe.

WHO Regional Office for Europe (DK). (2014). *WHO Expert Meeting: methods and tools for asessing the health risks of Atmospheric pollutants at local, national and international level.* Meeting report. Bonn, Germany, 12-13 May 2014. Copenhagen: WHO Regional Office for Europe.

WHO. (2018). *Air pollution and health: Summary.* Retrieved from http://www.who.int/airpollution/ambient/about/en/

Willems, P., Arnbjerg-Nielsen, K., Olsson, J., & Nguyen, V. T. V. (2012). Climate change impact assessment on urban rainfall extremes and urban drainage: Methods and shortcomings. *Atmospheric Research, 103*, 106–118.

Will-Wolf, S., Jovan, S., Neitlich, P., Peck, J. E., & Rosentreter, R. (2015). Lichen-based indices to quantify responses to climate and air pollution across northeastern U.S.A. *The Bryologist, 118*(1), 59–82. doi:10.1639/0007-2745-118.1.059

Wilson, R., & Spengler, J. (1996). *Particles in our air.* Harvard University Press.

Wittig, V. E., Ainsworth, E. A., & Long, S. P. (2007). To what extent do current and projected increases in surface ozone affects photosynthesis and stomatal conductance in trees? A metaanalytic review of the last 3 decades of experiments. *Plant, Cell & Environment, 30*(9), 1150–1162. doi:10.1111/j.1365-3040.2007.01717.x PMID:17661752

Wittneben, B. B., & Kiyar, D. (2009). Climate change basics for managers. *Management Decision, 47*(7), 1122–1132.

Wolf, J. (2011). Climate change adaptation as a social process. In J. D. Ford & L. Berrang-Ford (Eds.), *Climate change adaptation in developed nations – from theory to practice* (pp. 21–32). London: Springer. doi:10.1007/978-94-007-0567-8_2

World Conservation Union. (1980). *World Conservation Strategy: Living Resource Conservation for Sustainable Development.* Gland, Switzerland: International Union for Conservation of Nature.

World Energy Assessment. (2000). *Energy and the Challenge of Sustainability.* United Nations Development Programme.

World Health Organization (WHO). (2005). *Air Quality Guidelines Global Update 2005.* Available:http://www.euro.who.int/__data/assets/pdf_file/0005/78638/E90038.pdf?ua=1

World Health Organization. (2006). Fuel for Life Household Energy and health. Geneva: WHO.

World Meteorological Organization Greenhouse Gas Bulletin. (2017). *The state of Greenhouse Gases in the Atmosphere Based on Global Observations through 2016.* World Meteorological Organization. Retrieved on January 10, 2018, from: https://library.wmo.int/opac/doc_num.php?explnum_id=4022

Wuebbles, D. J., Fahey, D. W., Hibbard, K. A., Dokken, D. J., Stewart, B. C., & Maycock, T. K. (Eds.). USGCRP, 2017: Climate Science Special Report: Fourth National Climate Assessment (vol. 1). U.S. Global Change Research Program. doi:10.7930/J0J964J6

Wu, S., Mickley, L. J., Jacob, D. J., Rind, D., & Streets, D. G. (2008). Effects of 2000–2050 changes in climate and emissions on global tropospheric ozone and the policy-relevant background ozone in the United States. *Journal of Geophysical Research, 113*(D18), D18312. doi:10.1029/2007JD009639

Wu, S., Mickley, L. J., Jacob, D. J., Rind, D., & Streets, D. G. (2008a). Effects of 2000–2050 changes in climate and emissions on global tropospheric ozone and the policy relevant background surface ozone in the United States. *Journal of Geophysical Research, D, Atmospheres, 113*(D18).

Wu, S., Mickley, L. J., Leibensperger, E. M., Jacob, D. J., Rind, D., & Streets, D. G. (2008b). Effects of 2000–2050 global change on ozone air quality in the United States. *Journal of Geophysical Research, D, Atmospheres, 113*(D6).

Xiao, H. W., Xiao, H. Y., Long, A. M., Wang, Y. L., & Liu, C. Q. (2013). Chemical composition and source apportionment of rainwater at Guiyang, SW China. *Journal of Atmospheric Chemistry, 70*(3), 269–281. doi:10.100710874-013-9268-3

Xiao, J. (2016). Chemical composition and source identification of rainwater constituents at an urban site in Xi'an. *Environmental Earth Sciences, 75*(3), 209. doi:10.100712665-015-4997-z

Xu, M. M., Guo, Y. M., Zhang, Y. J., Westerdahl, D., Mo, Y., Liang, F., & Pan, X. (2014). Spatiotemporal analysis of particulate air pollution and ischemic heart disease mortality in Beijing, China. *Environmental Health, 13*(1), 109. doi:10.1186/1476-069X-13-109 PMID:25495440

Xu, X., Ha, S. U., & Basnet, R. (2016). A review of epidemiological research on adverse neurological effects of exposure to ambient Atmospheric pollutants. *Frontiers in Public Health, 4*, 157. doi:10.3389/fpubh.2016.00157 PMID:27547751

Yamashita, Y., Sakamoto, K., Akiyoshi, H., Takahashi, M., Nagashima, T., & Zhou, L. B. (2010). Ozone and temperature response of a chemistry climate model to the solar cycle and sea surface temperature. *Journal of Geophysical Research, D, Atmospheres, 115*(D3).

Yeom, C., & Kim, Y. (2017). Adsorption of ammonia using mesoporous alumina prepared by a templating method. *Environmental Engineering Research, 22*(4), 401–406. doi:10.4491/eer.2017.045

York, O. N. (n.d.). *The Plan for a Strong and Just City – 2016.* Disponível em: http://100resilientcities.org/wp-content/uploads/2017/07/OneNYC-ilovepdf-compressed.pdf

Young, P. J., Naik, V., Fiore, A. M., Gaudel, A., Guo, J., Lin, M. Y., ... Tilmes, S. (2018). Tropospheric Ozone Assessment Report: Assessment of global-scale model performance for global and regional ozone distributions, variability, and trends. *Elem Sci Anth, 6*(1).

Yue, X., Mickley, L. J., Logan, J. A., Hudman, R. C., Martin, M. V., & Yantosca, R. M. (2015). Impact of 2050 climate change on North American wildfire: Consequences for ozone air quality. *Atmospheric Chemistry and Physics, 15*(17), 10033–10055. doi:10.5194/acp-15-10033-2015

Yu, K., Keller, C. A., Jacob, D. J., Molod, A. M., Eastham, S. D., & Long, M. S. (2018). Errors and improvements in the use of archived meteorological data for chemical transport modeling: An analysis using GEOS-Chem v11-01 driven by GEOS-5 meteorology. *Geoscientific Model Development*, *11*(1), 305. PMID:30420911

Zannetti, P. (1990). *Air Pollution Modelling – theories, Computational Methods and Available Software*. New York: Van Nostrand Reinhold. doi:10.1007/978-1-4757-4465-1

Zeppilli, D., Sarrazin, J., & Leduc, D. (n.d.). Is the meiofauna a good indicator for climate change and anthropogenic impacts? *Mar Biodiv.*, *45*, 505.

Zhang, J., & Mu, Q. (2017). (in press). Air pollution and defensive expenditures: Evidence from particulate-filtering facemasks. *Journal of Environmental Economics and Management*.

Zhang, N., Cao, J., He, Y., & Xiao, S. (2014). Chemical composition of rainwater at Lijiang on the Southeast Tibetan Plateau: Influences from various air mass sources. *Journal of Atmospheric Chemistry*, *71*(2), 157–174. doi:10.100710874-014-9288-7

Zhang, Q. F. (2012). *Toward an Environmentally Sustainable Future: Country Environmental Analysis of the People's Republic of China*. Beijing, China: China Financial and Economic Publishing House.

Zhang, Q., Wua, L., Yang, Z., Zou, C., Liu, X., Zhang, K., & Mao, H. (2016). Characteristics of gaseous and particulate pollutants exhaust from logistics transportation vehicle on real-world conditions. *Transportation Research Part D, Transport and Environment*, *43*, 40–48. doi:10.1016/j.trd.2015.09.005

Zhao, S., Feng, T., Tie, X., Long, X., Li, G., Cao, J., ... An, Z. (2018). Impact of climate change on Siberian high and wintertime air pollution in China in past two decades. *Earth's Future*, *6*(2), 118–133. doi:10.1002/2017EF000682

Zlauddin, A., & Siddiqui, N. A. (2006). Air quality index (AQI) A tool to determine ambient air quality. *Pollution Research*, *25*(4), 885–887.

About the Contributors

G. Venkatesan is an Assistant professor in Civil Engineering Department, University College of Engineering (BIT campus), and Anna University. He graduated in Civil Engineering from Bharathiyar University in the year 1993 and obtained his Master's Degree in Water Resources Engineering and Management from National Institute of Technology, Tiruchirappalli in the year 2003. He has Recipient of Ph.D. fellowship from TEQIP, Government of India and awarded Doctoral Degree by National Institute of Technology, Tiruchirappalli for his thesis "Comparative studies on clay liner with and without microsilica, in permeation of municipal solid waste leachate" in the year 2010. He has been involved in teaching, research and industrial experience for the past 20 years. He has taught more than 20 subjects to under graduate and post graduate students. At present 3 candidates completed and 10 candidates are pursuing their Ph.D. programme under his supervision. He has published 35 papers in International and National Journals and presented 60 papers in Conferences. He published two books. He has carried out funded research project title Studies on structural element by totting tire derived aggregate sponsored by Indian Rubber Manufacturers Research Association (IRMRA), Ministry of Commerce and industry Government He have involved consultancy projects of improvements and Rehabilitation of Irrigation structures in Cauvery Basin- Preparation of Detailed project report(DPR) Government of Tamil Nadu, Air quality monitoring, water and soil analysis clients are WAPCOS Limited, Ministry of Water Resources, River Development and Ganga Rejuvenation, Government of India, Central public works, CDM smith Ltd. He organized fourteen conferences and workshops and attended seventeen workshops. He has delivered 22 invited talks conferences and seminars. He has been awarded Vishwakarma Award for the category of Academician for the year 2017 organized by Construction Industry Development Council (CIDC), Government of India. Rashtriya Gaurav Award for the year 2015 by the by India International friendship society New Delhi, best research paper on Environmental issues in the Department of Environment, Government of Tamil Nadu state award for the year of 2012. He is Indian Geotechnical Society Trichy Chapter founder Secretary. He is Council Member of Institution of Public health Engineers, India for the year of 2012 to 2018.Ground water Expert published in directory of ground water professionals, Central Ground Water Board, Ministry of Water Resources. Member in Environmental and Water Resources Systems Committee, American Society of Civil Engineers from December 2005. Apart from regular academic activities he as P.G Environmental Engineering Professor in-charge and Environmental Lab in charge, Staff in charge of Student Association of Civil Engineering, Department coordinator for students curricular activities and motivated many students to participate in conferences, symposia and workshops, Acted as chairman in Anna University central valuation, Member of the affiliation committee Anna University and served as Estate maintenance coordinator in Anna University Regional office Tiruchirappalli. Established in Environmental Microbiology Lab and part of Hydraulics Lab. He was a member of board of studies and member of the question paper scrutiny board Anna University of Technology, Tiruchirappalli. He is fellow of International society for Research and Development, Fellow in Institution of Public health Engineers and Indian Water Works Association. Member in Indian Geotechnical Society, The Indian Science Congress Association,, National Solid Waste Association of India, Institution of Engineers, Indian Society for Technical Education,, Indian Geotechnical Society Chennai Chapter and Charted Engineer in Institution of Engineers (India). His research interest is Solid waste management, Environmental Geotechniques, landfill design and Irrigation water management techniques.

* **

Katiucia Adam PhD in Water Resources and Environmental Sanitation from Instituto de Pesquisas Hidráulicas (IPH) at UFRGS; Msc in Water Resources and Environmental Sanitation IPH-UFRGS; Graduated in Sanitary Engineering on Universidade Federal do Pará (UFPA); Professor at Feevale University. Has experience working with, hydrological simulation, climate change and water resources.

Darlan Alves is a Technologist in Environmental Management (2012) and Master in Environmental Quality (2014) by Feevale University. Specialist in Environmental Education (2015) from the Federal University of Rio Grande (FURG). PhD in Environmental Quality from Feevale University. Experience as a professor at the Technological Center for Leather and Feevale University. At the University Feevale, he taught Environmental Licensing, Waste Management and Cleaner Production Technologies (P & L) disciplines. He worked in the Environmental Advisory Center of Centro Tecnológico do Couro, performing activities related to business consulting, especially in the areas of waste management. Experience as an industrial effluent treatment (ETE) technician at Curtume Bender S.A.

Ravichandran C. is Teaching Environmental Sciences for M.Sc. Students for the past 25 years; supervised 10 Ph.D. Scholars.

Md Mahfuzar Rahman Chowdhury obtained LL.M from the department of Law & Justice, University of Rajshahi, Bangladesh. He has experience of working in NGOs and law-firms. Also, he is an advocate of the Bangladesh Supreme Court, Dhaka. Mr. Chowdhury is a published author of a book chapter titled, "Bridging the Public-Private Partnership in Disaster Management in Bangladesh" by Taylor and Francis. Recently, he participated on the 18th World Congress of Criminology held at O. P. Jindal Global University, NCR, New Delhi, India in December, 2016 where he presented a paper titled "Examining the Implication of Reducing Recidivism". Now, Mr. Chowdhury is working with a reputed law-firm, "The Legal Care". His areas of interest includes, disaster management, climate change, environmental law, criminology, amongst others.

Gustavo Costa is a Biologist, PhD in Environmental Quality, Professor and Researcher. Experience in Environmental Sciences, working mainly on the following topics: atmospheric pollution, water bodies, biomonitoring active and passive, environmental chemistry and genotoxicity.

Arzu Eren Şenaras graduated from Uludağ University in 2007. She took her master and Ph.D degree in the Department of Econometrics at Uludag University.

Rajiv Ganguly completed his Bachelor of Engineering in Civil Engineering from Nagpur University, India and his Master of Civil Engineering specialising in Environmental Engineering from Jadavpur University, India. Thereafter, he was a research student at Indian Institute of Technology-Guwahati for almost two years. He has done previous work on CNG and its viability as an alternative fuel to diesel and gasoline. He completed his PhD from the Department of Civil, Structural and Environmental Engineering at Trinity College Dublin, Ireland working within the project 'Environment Transport Interface', responsible for evaluating various atmospheric dispersion models and their applicability to local environmental amendments. He also worked as a Postdoctoral Research Scientist at the School of Public Health in University of Michigan. Presently, he is working as an Associate Professor in the Department of Civil Engineering in Jaypee University of Information Technology, Waknaghat, Solan, Himachal Pradesh, India.

Şahin Inanç graduated from Trakya University in 2002. He took her master degree in the Department of Econometrics at Uludag University. And also he is a Ph.D student in the Department of Econometrics at Uludag University.

Ashok Kumar is a professor and chairman of Civil and Environmental Engineering at The University of Toledo, Toledo, Ohio. Throughout his distinguished career Dr. Kumar has received grants/contracts worth more than $ 5 million from industry and regulatory agencies, authored or co-authored more than 450 peer-reviewed papers, book chapters, conference papers and reports, edited three conference proceedings and two books, and co-authored five books. He is an honorary member of Air and Waste Management Association (A&WMA).

A. S. Maheshwari, Associate Professor completed her B.Tech. in Electrochemical Engineering & Technology and M.Tech. in Biotechnology. She obtained her Post Graduate Diploma and Doctoral degree in Bioinformatics. She is having more than sixteen years of teaching and research experience. Her research areas are Protein Engineering, Bioinformatics and Bio-batteries.

Navneet has done his Master and Ph.D. degree from Kurukshetra University, Haryana and presently working as a professor in the Department of Botany and Microbiology at Gurukula Kangri University. He has completed so many projects of UGC and UCOST.

Hamid Omidvarborna obtained his B.Sc. and M.Sc. from Iran in Chemical Engineering and Ph.D. in Environmental Engineering (Air Quality) from The University of Toledo, Ohio (US). His research focuses on combustion chemistry of various biodiesel fuels in diesel engines at Toledo. He worked at the Sultan Qaboos University (Oman) as a post-doctoral student on a variety of environmental engineering problems, including the analysis and modelling of different ambient air pollutants. Currently, he is working as a research fellow at the University of Surrey (UK) on the implementation of low-cost sensors. His research interests are in the areas of air quality, air pollution control, environmental monitoring and assessment, energy and environment. He has published over 20 journal papers and book chapters.

Daniela Osório, PhD in Ecology by the Graduate Program of the Federal University of Rio Grande do Sul (2009). Professor of the Postgraduate Program in Environmental Quality at Feevale University. Experience in Environmental Sciences, working mainly on the following topics: atmospheric pollution, rainwater, environmental chemistry and environmental management.

Sankar Ganesh Palani is an Associate Professor of Biological Sciences and Associate Dean of Practice School Division at BITS Pilani, Hyderabad Campus. He did his PhD in Environmental Science & Engineering from Pondicherry (Central) University and his post-doctoral training at Institut National des Sciences

Appliquees, Toulouse, France. Prof. Ganesh has about 20 years of sandwich experience in academia and industry across SAARC, Middle East and European countries. Prof. Ganesh has been involved in several academic and societal projects. He has published a book, 4 book chapters, and 60 research papers in reputed international and national journals and conference proceedings.

Daniela Quevedo PhD in Water Resources and Environmental Sanitation by the Institute of Hydraulic Research of UFRGS and Master in Mathematics in the Area of Probability and Statistics by UFRGS. She is an adjunct professor at Feevale University, working in undergraduate and postgraduate studies, with master's and doctoral degrees in the Postgraduate Program in Environmental Quality. It develops research in the area of monitoring directed to the analysis of the water quality of the Sinos River basin, identifying models and basin management indicators. Has experience in the area of Applied Probability and Statistics, working mainly on the following topics: Time Series, Extreme Value Theory and Georeferenced Information Systems (GIS). It guides master dissertations and doctoral theses that perform quantitative, multivariate and temporal analysis for diagnosis, risk assessment and environmental planning in watersheds.

Ruma Rano obtained her Ph.D in Chemistry from ISM Dhanbad in 2006. She is working as Assistant Professor, Department of Chemistry in NIT Silchar since 2007. He area of interest includes, characterization and utilization of Environmental solid waste materials and synthesis of nanocomposites.

Thilivhali Rasimphi was born in Ha-Kutama, Maebane Limpopo Province. He accomplished his Primary Education at the local Maebane Primary School. Thereafter, he matriculated at Kutama Secondary School, Muduluni Village. He obtained his Bachelor Degree of Environmental Management in 2012, School of Environmental Sciences.

Akash Saklani did his Bachelor and Master degree (in Botany) form HNB Garhwal University, Uttarakhand, India and pursuing his Ph.D.(in Botany) from Gurukul Kangri University under the supervision of Professor Navneet. His research work is on the Estimation of Phytodiversity of Rajaji Tiger reserve under Shivalik Hill. He has published so many research paper on Phytodiversity of Rajaji Tiger reserve and adjacent to Shivalik Hill.

Hayrettin Sezen is professor at the department of Econometrics, Operations Research MSB at Uludag University in Bursa.

Kaushik Shandilya is an environmental executive working in the State of West Virginia. His research interests include exploring different matrices of environmental engineering and sciences. He is presently focusing on the reclamation of natural environment using verified sediment and erosion control technology. He is a specialist in Article 3 and Article 11 of USEPA. His work on algae, energy efficiency, air and water quality, biodiesel and renewable energy are of wide application. His degrees are a MS (Devi Ahilya Vishwavidyalaya) in Chemistry, ME (Malaviya National Institute of Technology, Jaipur) and PhD (The University of Toledo) in environmental engineering. He is author/co-author of a number of peer-reviewed articles and book chapters related to air and water quality. Dr. Shandilya has taught graduate/undergraduate courses at The University of Toledo, Baker College, Owens Community College and Govindram Seksaria Institute of Technology and Sciences.

Rajagopalan Varadarajan is an Assistant professor in the Department of civil, over 16 years experience in the field of teaching and research. Graduated Civil Engg. at UG level, Environmental Management in PG and has a Ph.D. Specialized in the area of Environmental Engg and has done various research and consultancy works.

Sabrina A. Vieira is a Environmental Engineer (2014) by UPF and Master in Environmental Quality (2017) by Feevale University. Currently is a doctoral student of the postgraduate program in Environmental Quality - Feevale.

Index

E

F

G

H

I

L

M

N

O

P

R

S

T

U

V

W

Encyclopedia of
Information Science and Technology
Fourth Edition

www.ingramcontent.com/pod-product-compliance
Lightning Source LLC
LaVergne TN
LVHW081344300926
849161LV00008B/2064